Cryogenic Heat Management

Cryogenic Heat Management

Technology and Applications for Science and Industry

Jonathan A. Demko, James E. Fesmire and Quan-Sheng Shu

CRC Press
Taylor & Francis Group
Boca Raton   London   New York

CRC Press is an imprint of the
Taylor & Francis Group, an **informa** business

First edition published 2022
by CRC Press
6000 Broken Sound Parkway NW, Suite 300, Boca Raton, FL 33487–2742

and by CRC Press
4 Park Square, Milton Park, Abingdon, Oxon, OX14 4RN

CRC Press is an imprint of Taylor & Francis Group, LLC

© 2022 Jonathan A. Demko, James E. Fesmire and Quan-Sheng Shu

Library of Congress Cataloging-in-Publication Data
Names: Demko, Jonathan, author. | Fesmire, James E., author. | Shu,
    Quan-Sheng, author.
Title: Cryogenic heat management / Jonathan Demko, James E. Fesmire and Quan-Sheng
    Shu.
Description: First edition. | Boca Raton : CRC Press, 2022. | Includes bibliographical
    references and index.
Identifiers: LCCN 2021051993 (print) | LCCN 2021051994 (ebook) |
    ISBN 9780367542351 (hardback) | ISBN 9780367565251 (paperback) |
    ISBN 9781003098188 (ebook)
Subjects: LCSH: Heat—Transmission.
Classification: LCC QC320 .D45 2022 (print) | LCC QC320 (ebook) |
    DDC 621.5/9—dc23/eng/20211223
LC record available at https://lccn.loc.gov/2021051993
LC ebook record available at https://lccn.loc.gov/2021051994

ISBN: 978-0-367-54235-1 (hbk)
ISBN: 978-0-367-56525-1 (pbk)
ISBN: 978-1-003-09818-8 (ebk)

DOI: 10.1201/9781003098188

Typeset in Times
by Apex CoVantage, LLC

Brief Contents

Contents

Preface

In cryogenics, like all other fields in the known universe, heat flows from a hotter region to a colder region if there are two different temperatures and some physical proximity or connection between the two. How to preserve and efficiently use the "cold", in other words directing or redirecting, organizing or orchestrating, and promoting or inhibiting this flow of heat, is the job of cryogenic heat management. Heat management is central to designing and operating cryogenic systems that will function and work safely and efficiently. The range of systems includes, although not exclusively, systems such as superconducting magnets and cavities for MRI, high-energy physics, fusion (tokamak), and free electron laser systems; space launch and exploration; energy and transportation use of liquid hydrogen; transportation and storage of liquefied gases; and food preservation, as well as potential future applications in cryo-life-sciences and chemical industries.

Cryogenic Heat Management addresses these crucial tasks in its primary topics but always with a synergetic approach with both the total system design and end-use application in mind. Thermal management information is scattered across many different previous publications. One goal of this book is to meet the reader's needs by gathering almost all aspects of cryogenic heat management in one book, updated with the latest design, applications, and realizations in sufficient technical detail.

The book provides simplified but useful and practical equations that can be applied in estimating the performance and design of energy-efficient systems in low temperatures or cryogenics. Materials, design methodologies, mechanical structures, thermal analyses, computer simulations, qualification tests, and component integrations are also systematically introduced.

A comprehensive introduction to the necessary theory and models needed to find solutions for common difficulties in the fields of thermodynamics, heat transfer, thermal insulation, and fluid mechanics is presented. Most of the book addresses specific technical systems, and each single chapter (from 2 to 12) will be appealing to both experts and recent graduates working in that specific field. To help the reader understand these topics, practical approaches and advanced design materials for insulation, shields/anchors, cryogen vessels/pipes, calorimeters, cryogenic heat switches, cryostats, current leads, and RF couplers are introduced with about 300 Figures. Finally, several extra-large cryogenic machines based on SC magnets & SRF cavities, and space projects are selected to demonstrate these technologies in the applications.

Cryogenic Heat Management is written for engineers, scientists, professors, and technical staff whose work is related to cryogenic subjects. In addition, senior college students, graduate students, and technicians will also find the book a valuable reference in dealing with low-temperature systems and cryogenics.

Quan-Sheng Shu
James E. Fesmire
Jonathan A. Demko

Authors

Jonathan A. Demko began his career in industry on the thermal management of the X-30 National Aerospace Plane (NASP). He transitioned to the Super Collider Laboratory Cryogenics Department. At Oak Ridge National Laboratory, he worked on cryogenics for high temperature superconducting (HTS) power equipment. He is currently a Professor of Mechanical Engineering with LeTourneau University in Texas.

James E. Fesmire renowned expert in cryogenic systems design and thermal insulation, is President of Energy Evolution LLC, Chief Architect and CTO of GenH2 Corp., and founder of the Cryogenics Test Laboratory at Kennedy Space Center (NASA-retired). Distinctions include the NASA Distinguished Service Medal, an R&D 100 Award, 20 US Patents, and inductee of the NASA Inventors Hall of Fame.

Quan-Sheng Shu is a leading expert in cryogenics and has authored four monographs on cryogenics and superconductivity. He has led scientific teams at Fermilab, SSC Lab, Cornell University, DESY, and Zhejiang University. His achievements include: the world's first prototype of HTS Maglev Cryo-pipe; SRF cavities reaching the highest accelerating fields; created the Enhanced Black Cavity Theory for MLI with crack/slot; and cryogenic devices for SC magnets.

Abbreviations / Acronyms / Nomenclature

ABC	aerogel blanket composite insulation
ADR	adiabatic demagnetization refrigerator
AEP	American Electric Power Company
AeroFoamTM	Aerogel-Foam Composite
AeroPlasticTM	Aerogel-Plastic Composite
AeroFiberTM	Aerogel-Fiber Laminate Composite
AIAA	American Institute of Aeronautics and Astronautics
APSU	Advanced Photon Source undulator
ASME	American Society of Mechanical Engineers
ASTM	ASTM International
ATLAS	A Toroidal LHC ApparatuS
BCS	Bardeen–Cooper–Schrieffer theory
BOG	boiloff gas
BOMC	boil-off meter calorimeters
BOR	boiloff rate (% of tank volume per day)
CADR	continuous adiabatic demagnetization refrigerators
CBT	cold boundary temperature
CCC	cryocooler cooled cryostat
CCWCF	cryostats cooled with Continuing Flow Cryogen
ca-H$_2$	cryo-adsorbed hydrogen (solid-state molecular)
cc-H$_2$	cryo-compressed hydrogen (supercritical)
CEBAF	Continuous Electron Beam Accelerator Facility
CEC	Cryogenic Engineering Conference
CERN	European Organization for Nuclear Research
CESR	Cornell Electron-positron Storage Ring
CFC	cryogen free cryostat
CFSM	cryogen-free SC magnet
CHP	cryogenic heat pipe
CMS	Compact Muon Solenoid
CS-100	Cryostat-100 Test Instrument (cylindrical—absolute) [ASTM C1774]
CS-200	Cryostat-200 Test Instrument (cylindrical—comparative)
CS-400	Cryostat-400 Test Instrument (flat plate—comparative)
CS-500	Cryostat-500 Test Instrument (flat plate—absolute)
CS-600	Cryostat-600 Test Instrument (flat plate—absolute—w/ structural)
CS-900	Cryostat-900 Test Instrument (cylindrical—absolute—LH$_2$)
CPT	cold pipeline tester
CMU	cryogenic moisture uptake
CSA	Cryogenic Society of America
CSP	Cryogenic Spill Protection
CTL	Cryogenics Test Laboratory
CVP	cold vacuum pressure
CWWB	cryostat with warm bore
DAM	double-aluminized mylar
DESY	Deutsches Elektronen-Synchrotron (German Electron Synchrotron)

DHS	diode heat switch
DN	diameter nominal
DR	dilution refrigerator
DTE-HS	differential thermal expansion heat switch
DW	double-walled
EBCM-MLI	Enhanced Black Cavity Model for MLI with Cracks/Slots
EEC	energy efficient cryogenics
EAST	Experimental Advanced Superconducting Tokamak
FEL	free electron laser
FIC	factional insertion component
GN$_2$	gaseous nitrogen
GGHS	gas gap heat switch
GHe	gaseous helium
GH$_2$	gaseous hydrogen
GHP	guarded hot plate (ASTM C177)
GHS	gas heat switch
HERA	Hadron Electron Ring Accelerator
HSR	heat switch ratio
HFM	heat flux meter (ASTM C745)
HOM	high order mode
HOMC	high-order mode coupler
HV	high vacuum
HV	high vacuum (<1 millitorr or <0.1 millitorr for MLI systems)
HVP	hot vacuum pressure (bake-out)
HTS	high-temperature superconducting
ICC	insulative-conductive composite
ICEC	International Cryogenic Engineering Conference
IIR	International Institute of Refrigeration
ILC	International Linear Collider
IQF	insulation quality factor for total tank system
ISO	International Standards Organization
IRaS	Integrated Refrigeration and Storage
ITER	International Thermonuclear Experimental Reactor
JT	Joule Thomson
LB-MLI	load-bearing MLI
LEL	lower explosive limit
LEP	Large Electron–Positron Collider
LCI	Layered Composite Insulation
LCX	Layered Composite Extreme
LETC	local equivalent thermal conductivity
LHC	Large Hadron Collider
LINAC	Linear Accelerator
LIPA	Long Island Power Authority
LNG	liquefied natural gas (111 K @ NBP)
LOFA	loss of flow accident
LO2	liquid oxygen (90 K @ NBP)
LN2	liquid nitrogen (77 K @ NBP)
LH$_2$	liquid nitrogen (20 K @ NBP)
LHe	liquid nitrogen (4 K @ NBP)
lpm	liters per minute
LTS	low-temperature superconducting

Macroflash	Cup Cryostat thermal conductivity test instrument (ASTM C1774)
Maglev	magnetic levitation
MLI	multi-layer insulation
MLI	multilayer insulation
MRHS	magneto-resistive heat switch
MRI	magnetic Resonance Imaging
MV	moderate vacuum (~10 millitorr)
NBP	normal boiling point (at 1 atm)
NER	normal evaporation rate
NMR	Nuclear magnetic resonance
NPS	nominal pipe size
NV	no vacuum (ambient pressure)
OD	outer diameter
PCM	Patch -Covering Method
PHP	pulsating heat pipe
PM	permanent magnet
PTCL-X	prototype cryoline – group X
PUF	polyurethane foam insulation
P&ID	piping and instrumentation diagram
REPM	rare earth permanent magnet
RFIC	radio frequency input coupler
RGA	residual gas analysis
SMSHS	Shape memory alloy heat switch
ML-SBMHS	Maglev smart bimetal heat switch
SC undulator	(SCU)
SCC/S	standard cubic centimeters per second
SCHS	superconducting heat switch
SDMC	specially designed multipurpose calorimeter
SHS	Solid heat switch
SI	super insulation (same as MLI)
SMA	shape memory alloy
SNS	Spallation Neutron Source
SRF	superconducting radio frequency
SSCL	superconducting super collider laboratory
STIC	structural-thermal insulating composite
SOFI	spray-on foam insulation (polyisocyanurate)
STM	structural-thermal materials
STP	standard temperature and pressure (0 °C and 1 atm)
SV	soft vacuum
TCMC	thermal conductive meter calorimeters
TESLA	TeV-Energy Superconducting Linear Accelerator
TIS	thermal insulation system
t-MLI	traditional MLI
TTT	Transient Thermal Tester
TISCALC	design tool calculator for cryogenic insulation systems
UEL	upper explosive limit
VDA	vacuum deposited aluminum
VIP	vacuum insulation panel
VJ	vacuum-jacketed
VJ	vacuum jacket
VTI	Variable temperature inserting

VSR	voltage sensitive relay	
VSWR	Voltage Standing Wave Ratio	
WB	warm bore	
WBT	warm boundary temperature	
WG	waveguide	
WVP	warm vacuum pressure	
XFEL	X-ray free electron laser	
ZBO	zero boil-off	

SYMBOLS AND UNITS OF MEASUREMENT

α	accommodation coefficient, radiation absorption	
A	Area	m^2
A_e	effective area of heat transmission	m^2
A_i	inner surface area (inner tank)	m^2
A_o	outer surface area (outer jacket)	m^2
B	Thermal radiation absorption factor	—
C	Volumetric heat capacity	J/m^3-K
CVP	cold vacuum pressure	millitorr
CBT	cold boundary temperature	K
d_o	outside diameter of test specimen	m
d_i	outside diameter of test specimen	m
d_e	effective diameter of test specimen	m
d_e	mean diameter	m
δ	Skin depth	m
E	Emitted thermal radiation	W
ε	effective emissivity	—
$Eacc$	accelerating electric field	MV/m
F	View factor	—
F_{ST}	Structural-Thermal Figure of merit	K-m-s/g
Gr	Grashoff number	—
γ	coefficient of performance	—
h	convective heat transfer coefficient	W/m^2-K
h	enthalpy	J/g
h_c	thermal contact conductance	W/K
h_{fg}	heat of vaporization	J/g
H	magnetic field	Tesla
H_c	superconductor critical magnetic field	tesla
I	second moment of the cross section	m^4
I	electrical current	Amp
I_c	superconductor critical current	Amp
J	polar moment of inertia	m^4
k	thermal conductivity	W/m-K
K	thermal conductivity integral	W/K
Kn	Knudsen number	
k_e	effective thermal conductivity	mW/m-K
ke	kinetic energy	J/g
k_s	total system thermal conductivity	mW/m-K
L_e	effective length of test specimen	m
λ	Mean free path	m

$\dot{m}$	mass flow rate	g/s
M	bending moment	N-m
Ne/Nc	normal electrons/superfluid coop-pair electron	
μ	vacuum pressure	millitorr
μ	micron	millitorr
μ	permeability	H/m
n	number of layers (or layer-pairs)	—
Nu	Nusselt number	—
η	thermal efficiency	
η	impedance	Ω
ρ	bulk density of material	kg/m^3
ρ	electrical resistivity	ohm-m
P	fluid pressure	bar
P	perimeter	m
pe	potential energy	J/g
Pe	Peclet number	—
Pr	Prandtl number	—
P	residual gas pressure (vacuum level)	torr
q	heat flux	W/m^2
Q	heat flow rate (or heat leak or heat load)	W or J/s
Q	quality factor of oscillator	—
Q_{total}	total heat load into tank	W
Q_i	heat leak through insulation	W
Q_s	heat leak through support structures	W
Q_p	heat leak through piping and penetrations	W
Q_{IQF}	heat leak due to system IQF	W
R	Thermal resistance	K/W
R	RF electrical resistivity	Ω
RRR	residual resistivity ratio	—
Ra	Rayleigh number	—
Re	Reynolds number	—
R_h	Heat switch ratio	—
R_s	surface resistance	Ω
σ	entropy production	J/g·K
σ	Stefan-Boltzmann constant (5.67x10-8)	W-m/K^4
σ	Surface tension	N-m
σ	material compressive strength	MPa
s	entropy	J/g·K
T	temperature	K
τ	shear stress	Pa
T_h	warm boundary temperature (hot side)	K
T_c	cold boundary temperature (cold side)	K
T_c	superconductor critical temperature	K
ΔT	temperature difference	K
U	Internal energy	j/g
V	vertical shear force	N
V	volume	m^3
$\dot{V}$	volumetric flow rate	m^3/h
w	work	j/g
ω	circular frequency	rad/s
WVP	warm vacuum pressure	torr or millitorr

WBT	warm boundary temperature	K
x	thickness of test specimen $[(d_o-d_i)/2]$	mm
Δx	thickness	m
z	layer density	layers/mm
Ω	solid angle	m^2/m^2

Introduction

James E. Fesmire, Quan-Sheng Shu, and Jonathan A. Demko

To function in the ambient environment, a cryogenic system requires that a suitable "island of cold" be created. The first definition of the word "insulate" (from the Latin *insulatus*, made like an island) is "to set apart, detach from the rest, isolate". Most cryogenic systems, therefore, could generally be called high-performance thermal insulation systems. Unlike the island of cold shown in Figure 0.1, the cryogenic system "island of cold" is situated on the relatively warm Earth and expected to operate in that "high temperature" environment.

The engineering steps to designing a cryogenic system to operate in the ambient environment involve bringing together a combination of pressure vessels and piping, vacuum-jacketing techniques, materials, welding, seals, thermal insulation systems, instrumentation feedthroughs, process components, structural support systems, functional insertions, heat switches, cryocooler interfaces, and so forth. While the job of the cryogenic system is either to provide a low-temperature environment or else convey or store a cryogen, the cryogenic engineer must meet the challenge of creating a boundary between the cold working elements (the cold mass) and the hot working elements (the ambient world). With temperature being relative, the cold mass is the "cold side" (T_c), and the ambient is the "hot side" (T_h). And as there is always a temperature difference ($T_h - T_c = \Delta T$), energy will flow, in the form of heat transfer, from the hot side to the cold side. The left-hand and right-hand views of Figure 0.2 represent the same situation. One can only seek to minimize or manage this flow of heat. The rate of heat transfer (Q) for a given design is directly proportional to the temperature difference (ΔT).

Heat management in cryogenic systems involves many interrelated topics depending on the specific functional requirements. The type of cryogen (for example LHe, LH_2, LN_2, LO_2, or LNG), cryogenic purpose, pressure vessel ratings, use of vacuum-jacketing and corresponding vacuum levels, size and weight, control system, and many other factors start the list of items to consider. The

FIGURE 0.1 Island of cold.

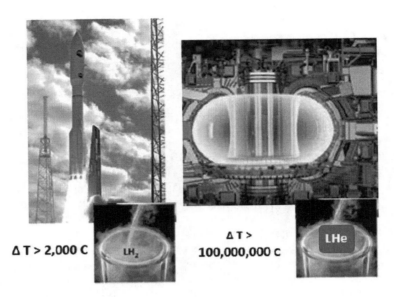

$\Delta T > 2,000\ C$ LH$_2$ $\Delta T >$ 100,000,000 C   LHe

FIGURE 0.2 Hot or cold, it is the temperature difference that makes the heat flow, and cryogenic heat management takes care of it.

system as an integrated whole, with the heat flow reduced to a manageable level and controlled to a suitable degree, is the goal.

To underscore the importance of a total system approach to design, examples of engineering integration are also a common theme. Although the subject matter does build from one chapter to the next, the aim is for the reader to quickly access any specific topic of interest. Data in the appendices provide a quick reference for key information. While some of the data are unique, much of the information is meant to be complementary and provide that initial guidance that is not always easy to find.

The subject of cryogenic heat management can be approached along the line of four broad categories: (1) heat transfer and thermal insulation systems; (2) structural supports, thermal anchors, shields, piping, tanks, and vacuum; (3) test cryostats, heat switches, current leads, and RF couplers; and (4) special lab cryostats, space cryostats, and large-scale applications. These categories and subjects, in order of the chapters of this book, are outlined in the following to serve as a guide to accessing the information of interest.

0.1 HEAT TRANSFER, THERMAL INSULATION, AND MLI

The nature of low-temperature or cryogenic systems makes them highly susceptible to incoming energy in the form of heat due to the large temperature difference between the system and the warm environment. Applications of the principles of energy conservation, that is, the first law of thermodynamics, and the allowable direction of change for a process and losses that are a result of the second law of thermodynamics are fundamental to the heat management of cryogenic systems. Maintenance of low temperatures is achieved by supplying refrigeration, either through a mechanical refrigerator or a low-temperature fluid. The modes of heat transfer, including conduction, convection, and thermal radiation, are addressed, with an emphasis on large temperature differences. In most situations, a combination of all heat transfer modes is present, and applications to maintain the cold space are presented by functional components in separate chapters.

Thermal insulation systems are the first line of defense in managing heat leak into cryogenic systems. The total system design must be approached with the goal of thermal isolation in mind. Heat transmission depends first on the vacuum level of the system. The right design approach to complex thermal insulation systems depends on the shape, size, and operational duty cycle. New and advanced materials in recent decades have likewise enabled advances in thermal insulation systems for storage and transfer of cryogens. These advances include a variety of aerogels, aerogel blanket composites, hybrid aerogel-multilayer insulation (MLI) composites, layered composite insulation (LCI), layered composite extreme (LCX), prefabricated MLI, discrete spacer MLI, new structural composites, bulk-fill glass bubbles, aerogel-foam composites, vacuum panels, and more. Traditional materials include several types of MLI, bulk-fill perlite powder, fiberglass, cellular glass, polyisocyanurate foams, and others. In addition to the thermophysical and chemical properties of a given material, the practical matters of mechanical structure, installation, and environmental conditions can have an overwhelming effect on the final system performance.

Multilayer insulation systems are also known as superinsulation. These systems are characterized by a number of radiation (or reflective) shields separated by low thermal conductivity spacers within an evacuated environment. MLI systems are used when lower heat leakage rates are required than those obtained with other evacuated insulations, for example, perlite powder, glass bubbles, foams, and aerogels. The question of what constitutes the best MLI system is often asked in one form or another. Reducing the heat load into the system is the goal. An unacceptable heat load can cost too much money over time, cost too much time in operations, or render a system inoperable. Material selection and design are crucial, but the two main factors in the result are almost always level of vacuum (including the level of vacuum between the layers) and layer density. In turn, the layer density depends on the materials used, the design, and the installation. Experimental baseline data then become paramount for design and analysis of new cryogenic equipment that will rely on an MLI system of some kind.

0.2 STRUCTURAL SUPPORTS, THERMAL ANCHORS, SHIELDS, PIPING, TANKS, AND VACUUM

Supports in cold masses and storage volumes of cryogenic fluids are designed to meet the mechanical loading requirements to position the cold space at a desired location and to provide thermal isolation from the warm environment through the design of thermally isolating support structures. Thermal insulation is important, but supports contribute a significant portion of the total heat load. The optimal selection of materials used in construction of a support depends on the thermal conductivity and the mechanical strength of the materials. Supports come in many forms, including rigid, such as tension rods or support posts, or flexible, such as chains, cables, or straps. Practical examples are provided for applications such as supports for MRI and accelerator magnets, pipe supports when multiple cryogenic fluid lines are enclosed in a single vacuum enclosure, supports for storage vessels, and advanced support designs that use high-temperature superconductors for cold mass levitation.

Two special design features used to efficiently manage the heat transfer in cryogenic devices are thermal anchors and thermal shields. Thermal anchors provide a low-loss connection to the source of the cold to maintain the cold mass at its operating temperature or to remove heat load at an intermediate temperature that comes from the warm boundary. Thermal anchors are connected to a source of cooling such as a refrigerator; cryocooler; or supply of a cryogenic fluid such as helium, hydrogen, or nitrogen. In addition to making a good thermal connection to a heat sink, thermal anchor technology is applied to minimize errors in temperature sensing through anchoring of instrumentation leads and thermometers.

Thermal shields are placed between a high temperature boundary and a cold surface to limit the flow of heat by thermal radiation. There are two main types, floating shields, which passively reach thermal equilibrium and reflect heat away from the cold space, and actively cooled shields, which must be anchored to a low-temperature thermal sink to remove the intercepted heat load reaching the shield to maintain it at the operating temperature. Analysis methods for estimating the performance of thermal shields are outlined. The determination of the number of passively cooled shields required to achieve a specified reduction is described. For actively cooled shields, a high thermal conductance anchor is required to efficiently transfer the heat captured by the shield to the low-temperature heat sink. Practical examples include thermal anchors for superconducting magnet support posts, thermometer mounts, and the cold ends of current leads.

During the transfer and storage of cryogens, changes of the temperature, vapor quality, and pressure of the cryogens should be negligibly small. Therefore, minimizing heat flow from the ambient environment to the inner pipe or vessel containing cryogens is crucial. The latest developments and successful examples of design methodologies and unique integrated structures of cryogenic piping systems and vessels are systematically introduced and demonstrated as the following: (1) basic cryogenic piping for transfer of LNG, LO_2, LN_2, LH_2, or LHe with various thermal insulation systems; (2) efficient connections (bayonets) for cryogenic piping in laboratories and industrial fields; (3) complex cryogenic piping systems with multi-cryogen channels in a common vacuum enclosure for extra-large and/or extra-complicated cryogenic applications; (4) basic storage vessels and tanks, which are insulated by multilayer insulation, aerogels, glass bubbles, powders, or foams; (5) design and construction of large-scale cryogenic tanks for LH_2, LO_2, LHe, and LNG (storage and shipping); (6) zero boil-off of storage vessels and extra-large cryogenic tanks; and (7) field diagnosis, repair, and modification of large-scale tanks.

Vacuum techniques are a mainstay of cryogenic heat management and cryogenic engineering. Vacuum technology is the central enabling feature of cryogenic engineering on Earth. As we understand that the temperature difference is the driver for heat flow, we know that vacuum is the necessary second line of defense against reducing the flow of heat. Proper design is always the first line of defense in reducing heat conduction to a manageable level, but it is the vacuum enclosure or vacuum jacket that commands the overall design, look, and operational characteristics of the cryogenic system. Vacuum can be an important component for the thermal isolation of cryostats, tanks, and piping to enable basic functionality of the system and control of the cryogenic fluid. Vacuum is essential for conducting many scientific measurements. Vacuum comes into play for a range of experimental physical and chemical processes. The word "vacuum" means the absence of matter, but the reality is always an approximation. Thus, the level of vacuum must be clearly defined and understood for each specific application.

0.3   TEST CRYOSTATS, HEAT SWITCHES, CURRENT LEADS, AND RF COUPLERS

Various thermal insulation materials and systems are widely utilized in cryogenic systems. However, reliable calorimeter systems and methods are crucial to test the performance of thermal insulations under conditions like the real cryogenic system and for scaled-up applications. First, the design methodologies and structures of several boil-off calorimeters (BOCs) with LHe, LH_2, and LN_2 as cryogens are introduced. These calorimeters can test cylindrical and flat-plate specimens with boundary temperatures of 77 K and 300 K, 77 K and 4 K, and many special intermediate temperatures in this range. Thermal conductance calorimeters with multi-boundary temperatures for various specimen shapes are addressed. Specially designed multipurpose calorimeters include one for MLI with penetrations, cracks, or holes and another for a 1,000-liter spherical-calorimetric tank for LH_2 boil-off testing. Successful demonstrations of cryogenic heat management with the assistance of calorimeters such as the Structural Heat Intercept, Insulation, and Vibration Evaluation Rig (SHIIVER) for large space applications are summarized.

Cryogenic heat switches (CHSs) are novel instruments with controlled variable heat conduction, working in a certain temperature region about from 50 mK up to 400 K. Heat switches can alternatively provide high thermal connection or ideal thermal isolation to the cold mass. These cryogenic heat switches are widely applied in a variety of unique cryogenic devices and critical space applications. Based on their working principles, cryogenic heat switches can be characterized into solid heat switches (SHSs) and gas heat switches (GHSs). Solid heat switches relying on the superconducting-normal phase change, magnetic levitation suspension, shape memory alloys, differential thermal expansion, and piezoelectric effect have been successfully used. Switches based on gas properties, helium or hydrogen gap-gap heat switches, cryogenic diode heat switches, and heat pipe heat switches are available. Comparisons of advantages and limitations of the different cryogenic heat switches give an outlook for future thermal management solutions.

Many cryogenic systems require that electric power be supplied to the cold space, which must be done with minimal heat loads to the cold space. A current lead provides an efficient means of transmitting electrical power between ambient temperature electrical connections to low temperature connections. Current leads can generally be divided into two categories: (1) conduction cooled and (2) vapor cooled. For cold end temperatures of the lead above 70 K or liquid nitrogen temperature, current leads are made only from conventional materials such as copper, brass, or aluminum. For conduction-cooled leads, the conventional portion can be designed using results from an analytical method. The design length to diameter ratio for a given current rating that minimizes the heat load is presented for copper leads. Many different types of heat transfer surfaces are used in vapor-cooled current leads, including various fin configurations, braided copper, or the flow between plates. When the low-end temperature is significantly colder, as found in many superconducting magnet applications, the low-temperature end of the lead very often incorporates the use of high-temperature superconductors (HTSs) to minimize the heat load to the low temperature. Descriptions of many applications of current leads of various types, such as research magnet systems, MRI magnets, high-energy physics (HEP) applications, fusion magnet systems, superconducting power applications, and specialty current lead designs, are provided. Important considerations in the design of current leads include heat loads when no current is applied, heat loads at full current, design for over-currents that are several times the design operating current for short durations, and loss of cooling flow situations.

Superconducting radio-frequency (SRF) cavities can deliver a quality value, Q, higher than copper cavity by ~10^5 and accelerating fields, E_{acc}, higher by an order of magnitude in continuous wave mode. The RF input coupler (IC) transports huge radio frequency power to SRF cavities, and high-order mode couplers (HOMCs) extract and dissipate unwanted RF energy present in the cavity to protect particle beams from instability. These couplers are directly connected between SRF cavities at 2 K–4 K and the ambient environment. However, the couplers also bring a huge heat leak (RF-dynamic and static) to the cold mass, which is usually much greater than that through the entire MLI system and solid supports combined. The chapter focuses on the comprehensive trade-offs and optimizations in the designs, particularly the thermal aspects. Methodology in balance between the technical functions demanded and minimization of heat load is introduced. Selected examples from around the world are also summarized.

0.4 SPECIAL LAB CRYOSTATS, SPACE CRYOSTATS, AND LARGE-SCALE APPLICATIONS

The cryostat is a device designed to hold the sample and apparatus at cryogenic temperature while providing all the interfaces (cryogen feeding, powering, diagnostics instruments, safety devices, etc.) for reliable testing or performing special tasks. The design, configuration, and operation of the cryostats are strong demonstrations of various technologies in cryogenic heat management. From a broad perspective, cryostats have numerous applications within science, engineering, and biomedicine. Therefore, cryostats have a great number of designs, structures, and functions for various unique applications. Special applications within the domains of laboratories and space exploration include both commercially available cryostats and many one-of-a-kind designs. Through design considerations and methodologies, representative examples of advanced cryostats are presented with their design schematics, operational techniques, and performance data.

Due to the unique and irreplaceable advantages of superconducting devices, more and more large machines have been successfully developed, operated, and constructed for high-energy physics, nuclear physics, fusion reactors, materials sciences, free electron lasers, and other applications. Sophisticated superconducting (SC) magnets and SRF cavities in various systems operate at temperatures of 1.8 K and lower and are enclosed inside high-vacuum environments. Some machines have masses of thousands of tons and lengths of tens of kilometers.

Hydrogen is the energy carrier of the future, and liquid hydrogen (LH_2) is the basis for the logistics of moving large amounts of renewable energy around the world at all scales and sizes of application. To store and move around large amounts of energy, coupled with the burgeoning electrification of the world, LH_2 is at the heart of it all as the means for providing a practical, high-density method of energy storage, conveyance, and usage. LH_2 remains the signature fuel for space launch and exploration, both government and commercial, and has been successfully used since 1963 in the evolution of upper-stage Centaur rocket engines, followed by Saturn V upper stages for the Apollo missions and the Space Shuttle. Liquefied natural gas continues to be a major business throughout the world, with container ships with capacities of over 200,000 m^3 in service.

1 Heat Transfer at Low Temperatures

Jonathan A. Demko, James E. Fesmire, and Quan-Sheng Shu

1.1 INTRODUCTION

Thermal management is a term used to describe how energy interactions between a system and its surroundings or within a system are moderated to maintain the system at a certain thermodynamic state. For low-temperature or cryogenic systems, the term is applied to describe how the operational temperature of the system is maintained when it may be surrounded by temperatures hundreds of Kelvins hotter. This could refer to how a vessel used for storage or transport of a cryogen is kept cold or to maintaining the operating temperature of a superconducting device. For all of these situations, principles of energy conservation, or the first law of thermodynamics, cannot be violated.

1.2 REVIEW OF THERMODYNAMICS

Managing the exchange of thermal energy, or the heat transfer, involves finding the temperature distribution in the system. This starts with the application of the first law of thermodynamics, which is reviewed here.

The first law of thermodynamics concerns energy conservation. Energy has several forms. The most common forms of energy may be taken as thermal, kinetic, and potential energy. Heat transfer and work are energy transfers that occur across the boundary of a system. Work can be defined as energy that is transferred across the boundary of a system as a result of a force applied at the system boundary. This can take several forms, including shaft work, electrical work, and boundary work, which is related to a change in the system volume. Heat transfer is the transfer of energy across the boundary of a system as a result of a temperature difference, which is the driving potential, at the system boundary. In low-temperature systems, the temperature difference is significant. Controlling the conductance of this driving potential can be thought of as a goal of thermal management.

Considering the simple situation of a unit mass system, as shown in Figure 1.1, the first law of thermodynamics can be stated as the net heat transfer to the system, δq, in units of kJ/kg, minus the work done by the system, δw, in units of kJ/kg, balanced by the change of energy of the system, de, in units of kJ/kg. The equation for a closed system constituting a unit mass is provided in Equation 1.1.

$$\delta q - \delta w = de_{sys} = du + dke + dpe \cong du \tag{1.1}$$

In Equation 1.1, dke is a differential quantity of kinetic energy, and dpe is a differential quantity of potential energy terms that can be assumed to be negligible. The internal energy, u, or its differential, du, represents the thermal energy stored in a system. The differential heat transfer, δq, and work, δw, are inexact differentials, hence the use of the δ operator. These two terms in the energy balance depend on the process path, unlike the other energy properties, which only depend on the beginning and end states. The differential form of the first law of thermodynamics can be applied over a system with a mass of m kilograms, resulting in Equation 1.2, with all energy quantities in units of kilojoules.

DOI: 10.1201/9781003098188-1

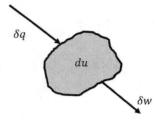

FIGURE 1.1   Energy balance of a unit mass undergoing heat transfer and work interactions.

$$dE_{sys} = \delta Q - \delta W \tag{1.2}$$

By taking a time derivative of Equation 1.2, the instantaneous time rate form of the energy balance is given as:

$$\frac{dE_{sys}}{dt} = \dot{Q} - \dot{W} = \frac{dU}{dt} + \frac{dKE}{dt} + \frac{dPE}{dt} \tag{1.3}$$

The sign convention in this expression is that heat transfer entering the system is positive, and work leaving the system is positive, as illustrated by the direction of the energy terms in Figure 1.1.

For a control volume where there is a mass flow, the work term can be separated into multiple terms, the simplest consisting of shaft work, δw_{shaft}, and flow work, δw_{flow}. Shaft work consists of the work supplied by a pump or extracted from an engine such as a turbine. Flow work is the work necessary to push a unit mass into or out of the control volume for a particular flow stream against the pressure force. Flow work is given by:

$$\delta w_{flow} = pdv \tag{1.4}$$

where p is the pressure, and v is the specific volume. Separating the flow work from other mechanical work terms results in the definition of the thermodynamic property called the enthalpy, h, which is defined as:

$$h = u + pv \tag{1.5}$$

The enthalpy is conveniently used in the energy balance for a control volume. The time-dependent first law of thermodynamics for a control volume is given by Equation 1.6. This form is appropriate for cooldown and warm-up simulations. Under steady state, the rate of change of energy stored in the control volume, $\dfrac{dE_{cv}}{dt}$, is zero.

$$\dot{Q} - \dot{W} = \sum \dot{m}\left(h + ke + pe\right)_{out} - \sum \dot{m}\left(h + ke + pe\right)_{in} + \frac{dE_{cv}}{dt} \tag{1.6}$$

The concept of entropy and the second law of thermodynamics provide a direction for processes to go and a means to determine the efficiency of a process. Some relevant statements of the second law of thermodynamics are:

    I. Kelvin-Planck—It is impossible to construct a heat engine which, while operating in a cycle, produces no effects except to do work and exchange heat with a single reservoir.

    II. Clausius—It is impossible for a self-acting machine unaided by an external agency to move heat from one body to another at a higher temperature.

The property entropy can be defined for a reversible process by:

$$ds = \left.\frac{\delta Q}{T}\right)_{rev} \tag{1.7}$$

For an irreversible process, the inequality is given by:

$$ds > \left.\frac{\delta Q}{T}\right)_{irrev} \tag{1.8}$$

The entropy balance for a closed system states that the change in the amount of entropy contained within some system during a time interval equals the net amount of entropy transferred in across the system boundary plus the entropy production. Mathematically, this can be expressed as:

$$S_2 - S_1 = \int_1^2 \left(\frac{\delta Q}{T}\right)_{boundary} + \sigma \tag{1.9}$$

The entropy crossing the boundary is given by the integral, and the last term, σ, is called the entropy production. The entropy production contributes to the irreversibility, or lost work, of a system. A differential form for the entropy balance is provided by:

$$dS = \left(\frac{\delta Q}{T}\right)_{boundary} + \delta\sigma \tag{1.10}$$

A simple application of the concept is described in Figure 1.2, which consists of a solid heated at one end to high temperature and cooled at the other end to a low temperature. If there is no heat lost along the sides of the solid, then, according to the first law, the heat entering at the hot end, Q, has the same magnitude as the heat leaving at the cold end.

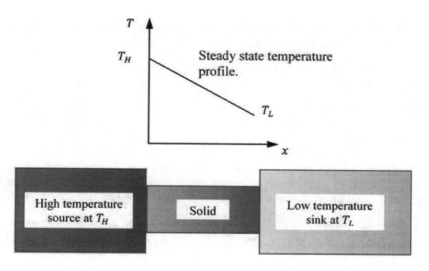

FIGURE 1.2   Application of entropy generation to a one-dimensional heat conduction case.

For the solid bar, the entropy change during a steady-state heat conduction process will be zero, since the state of the bar is not changing. At the warm end, the entropy entering the solid is:

$$S_{in} = \frac{Q}{T_H} \tag{1.11}$$

For the cold end, the entropy leaves the system at a rate of:

$$S_{out} = -\frac{Q}{T_L} \tag{1.12}$$

The entropy generation can be simply calculated from Equations 1.13 and 1.14.

$$S_2 - S_1 = 0 = \frac{Q}{T_H} - \frac{Q}{T_L} + \sigma \tag{1.13}$$

$$\sigma = -\frac{Q}{T_H} + \frac{Q}{T_L} = Q\left(\frac{1}{T_L} - \frac{1}{T_H}\right) \geq 0 \tag{1.14}$$

This result shows two important applications of the second law. First is the simple observation that heat must flow from the high to low temperature if the second law is not to be violated. The second, more subtle, observation is that there always will be entropy produced if there is a difference in temperature involved in the transfer of heat. It might be said a goal of thermal management is to minimize the entropy produced, hence the irreversibility of the system. The temperature extremes are fixed, so the only parameter left is the heat transfer.

1.3   THERMODYNAMIC CYCLES

A system can undergo a cyclic process that returns a system to the same starting state. This is the thermodynamic principle used to describe heat engines and refrigerators. For a thermodynamic cycle, the heat and work differentials can be integrated around the cycle according to Equation 1.15.

$$\delta q - \delta w = de = 0 \tag{1.15}$$

The first law of thermodynamics applied to any cycle basically states that the net heat transfer for the cycle equals the net work. This can be expressed as Equation 1.16.

$$q_{net} = w_{net} \tag{1.16}$$

Of importance to low-temperature systems are refrigeration cycles. Principles that arise from entropy and the second law of thermodynamics are needed to develop a complete view of the thermal management concept.

A heat engine is a device that takes energy from a high temperature source and produces work while rejecting heat to a low-temperature sink according to the Kelvin-Planck statement of the second law. A block diagram of a heat engine is provided in Figure 1.3(a). This consists of a high-temperature heat reservoir at T_H, which transfers an amount of heat, Q_{in}, to a heat engine that rejects heat, Q_{out}, to a low-temperature reservoir at T_L. The net work produced by this cycle is the difference between the heat supplied and the heat rejected. The thermal efficiency is the ratio of the net work produced to the heat supplied given in Equation 1.17.

$$\eta = \frac{W_{cycle}}{Q_{in}} = \frac{Q_{in} - Q_{out}}{Q_{in}} = 1 - \frac{Q_{out}}{Q_{in}} \tag{1.17}$$

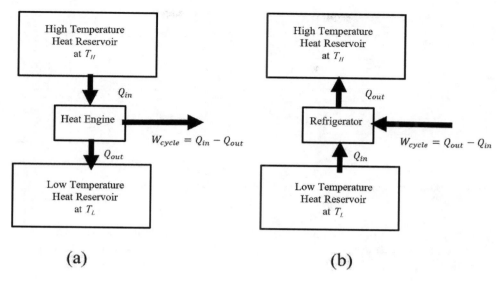

FIGURE 1.3 Block diagram of (a): a heat engine and (b): a refrigerator.

For refrigeration cycles, heat is transferred in reverse. Heat is removed from the low temperature reservoir, and, with the input of work, heat is rejected at a high temperature. This is illustrated in Figure 1.3(b). The measure of performance is given by the coefficient of performance, γ, and is the ratio of the heat removed from the low-temperature reservoir, or the refrigeration, to the work input. The coefficient of performance can be greater than 1 for systems that operate near or above room temperature. For low-temperature refrigerators, this is not the case.

$$\gamma = \frac{Q_{in}}{W_{cycle}} = \frac{Q_{in}}{Q_{out} - Q_{in}}$$

The inverse of the coefficient of performance is sometimes called the specific power consumption, or:

$$SPC = \frac{1}{\gamma} \equiv \frac{W_{input}}{Q_{refrigeration}} \tag{1.18}$$

This expresses the ratio of the work required per unit of refrigeration supplied and so is sometimes a useful way to express refrigerator performance.

1.4 THE CARNOT CYCLE

The theoretical maximum thermodynamic performance is given by the Carnot cycle. A Carnot cycle engine is composed of four reversible processes, as shown in Figure 1.4. Process 1–2 is an isentropic compression. This is followed by an isothermal heat addition from 2–3. Work is output during process 3–4, and some of the work may be put back into the cycle from 1–2. The cycle is restored to its initial state by an isothermal heat rejection from 4–1.

The inequality of Clausius is a corollary or consequence of the second law of thermodynamics. It is valid for all possible cycles, reversible or irreversible, heat engines or refrigerators. The Clausius inequality is given as:

$$\oint \frac{\delta Q}{T} \leq 0 \tag{1.19}$$

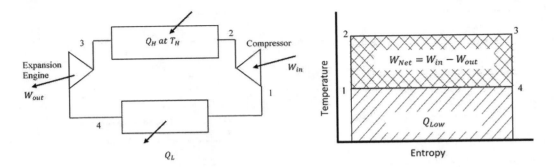

FIGURE 1.4 Block diagram and temperature entropy diagram of a Carnot cycle engine.

The integral is equal to zero for all reversible cycles, such as the Carnot cycle, and is less than zero for irreversible (practical) cycles. The application of the Clausius inequality to the Carnot cycle results in the following relation:

$$\frac{Q_H}{T_H} = \frac{Q_L}{T_L}$$ (1.20)

From the definition of entropy:

$$\Delta S = \int \frac{\delta q}{T} = \frac{Q}{T}$$ (1.21)

Thus, the heat transfer can be expressed in terms of the entropy changes as:

$$Q_H = T_H \left(S_3 - S_2 \right) = T_H \Delta S_H$$ (1.22)

$$Q_L = -T_L \left(S_4 - S_1 \right) = -T_L \Delta S_L$$ (1.23)

Applying the Clausius inequality for a reversible cycle to the Carnot cycle results in:

$$\Delta S_H = \Delta S_L = \Delta S$$ (1.24)

The thermal efficiency of a Carnot engine can be determined from the heat transfers and is only a function of the high and low temperatures:

$$\eta_{Carnot} = \frac{W_{cycle}}{Q_{in}} = \frac{Q_H - Q_L}{Q_H} = \frac{T_H \Delta S_H - T_L \Delta S_L}{T_H \Delta S_H} = \frac{T_H - T_L}{T_H}$$ (1.25)

For refrigeration cycles, heat is removed from the low temperature sink and discharged to a high temperature reservoir by supplying work to the cycle in accordance with the Clausius statement of the second law. This was shown in Figure 1.3(b). A Carnot refrigerator and liquefier are shown schematically in Figures 1.5 and 1.6.

Refrigeration and liquefaction cycles differ in some respects with regard to their operation. In a refrigeration cycle, the working fluid is circulated through a load that is being cooled. The flow on the supply and return leg of the cycle is balanced. In a liquefaction cycle, the liquid produced is removed, leaving less mass flow in the low-pressure return leg of the cycle. Warm make-up gas is supplied at the beginning of the compression process. The principles of design and operation of large-scale systems are discussed in Ganni [1].

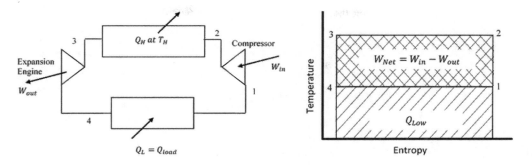

FIGURE 1.5   Schematic diagram of a reversed Carnot refrigeration cycle and T-s diagram.

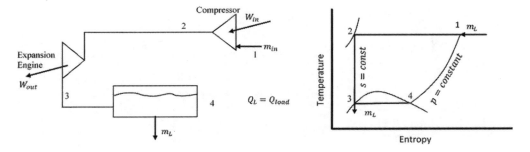

FIGURE 1.6   Carnot liquefier and liquefaction process on the temperature entropy plane.

For the ideal or Carnot refrigerator, the same approach can be used to express the Carnot coefficient of performance that was used in developing thermal efficiency. For the Carnot refrigerator, the performance can be determined knowing only the temperature extremes as:

$$\gamma_{Carnot} = \frac{T_L}{T_H - T_L} \tag{1.26}$$

and

$$SPC_{Carnot} = \frac{1}{\gamma_{Carnot}} = \frac{T_H - T_L}{T_L} \tag{1.27}$$

Often the performance of an actual refrigeration system is given in terms of the percent Carnot. This is simply determined from the ratio:

$$\%Carnot = \frac{\gamma_{actual}}{\gamma_{Carnot}} \times 100\% \tag{1.28}$$

Applying the definition for specific power consumption for a real system, the ratio of work input required per unit of refrigeration for an actual system can be determined by:

$$SPC_{actual} = \frac{100}{\%Carnot \times \gamma_{Carnot}} \tag{1.29}$$

The increase of the ideal specific power consumption as refrigeration temperature decreases is shown in Figure 1.7. An early summary of the performance of actual cryogenic refrigeration systems was made by Strobridge [2]. Figure 1.8 shows a range of operating efficiencies for different-capacity

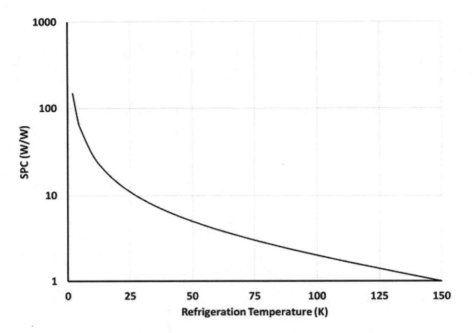

FIGURE 1.7   Ideal work required per unit of refrigeration from $T_H = 300$ K as a function of refrigeration temperature.

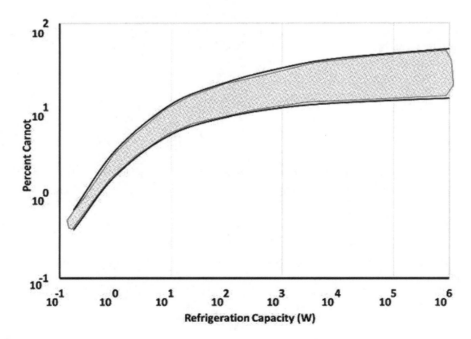

FIGURE 1.8   Approximate range of the percent Carnot for existing cryogenic refrigerators and liquefiers as a function of refrigeration capacity based on Strobridge [2].

TABLE 1.1

Normal Boiling Point and Ideal Liquefaction Work for Several Gases from 300 K and 101.325 kPa

Gas	Normal Boiling Point (K)	Ideal Work Input (kJ/kg)
He-4	4.2	6,820
H_2	20.3	12,020
Ne	27.1	1,335
N_2	77.4	768
Air	78.8	739
Ar	87.3	479
O_2	90.2	636
CH_4	111.7	1,091

systems. Strobridge collected data for refrigeration systems, which he separated into three operating temperature ranges: (1) 1.8 K to 9 K, (2) 10 K to 30 K, and (3) 30 K to 90 K. The performance of all three temperature ranges is covered in the figure. Small-scale cryogenic refrigerators such as cryocoolers tend to have a lower percent Carnot. These performance levels should be used as a starting place, and the actual performance depends on the final system selected.

The thermal management of cryogenic systems seeks not only to minimize refrigeration but also to minimize boil-off of stored cryogens. A brief discussion of the energy requirements of the liquefaction process is warranted. The Carnot liquefier is presented in Figure 1.6 on the temperature entropy plane. The liquefaction for a unit mass of a gas at room temperature can be calculated from an energy balance along the process path 1–2–3. The ideal liquefaction can also be determined by the exergy change from state 1 to state 3 given by Equation 1.28.

$$w_{ideal,liq} = \left(h_1 - h_3\right) - T_1\left(s_1 - s_3\right) \tag{1.30}$$

Table 1.1 lists the normal boiling point at 1 atmosphere for several commonly used liquefied gases and the ideal liquefaction work required. Helium and hydrogen require the most energy to liquefy and are also frequently used in many low-temperature systems.

As mentioned previously, there are different capacity ranges for cryogenic refrigerators and liquefiers. An arbitrary grouping of different refrigeration technologies can be made by system capacity, as done in Table 1.2. The table also includes some of the types of machines used in each of these refrigeration levels.

Some of the important issues with selecting a cryogenic refrigeration system for an application are the cycle thermal efficiency, cost, and reliability. The thermal efficiency affects the operating cost of the unit. It may not be a main driver, depending on the importance of the application. The thermal

TABLE 1.2

Classification of Cryogenic Refrigerator/Liquefier Systems for Temperatures from 4 K to 110 K

Small Scale (0.1 to 100 W)	Medium Scale 0.1 to 1 kW	Large Scale Turbomachinery Based > 1 kW
Gifford-McMahon	Reciprocating engines	Turbo-Brayton
Pulse tube	Stirling cycle	Claude cycle

efficiency for a refrigerator contributes to the operating cost through the cost of the power supplied. In some applications, this may not be the main driver.

The initial cost of a refrigerator is important but does not prevent use of the technology. Reliability can be a main driver for applications such as superconducting power devices. Since all machinery requires periodic maintenance, a reliable back-up system is frequently applied in such cases.

1.5 CONDUCTION HEAT TRANSFER

Heat transfer is defined as the energy transfer at the boundary of a system (control volume) resulting from a temperature difference at the boundary. It is related to the thermodynamics of a system. A very rudimentary look at the topic is provided here, but readers are encouraged to investigate the material more comprehensively in literature dedicated to heat transfer (see Berman [3] and Kreith [4]). Barron [5] specifically deals with cryogenic heat transfer.

There are three modes of heat transfer: conduction, convection, and thermal radiation. Conduction is a process by which heat flows from a region of higher temperature to a region of lower temperature within a medium (solid, liquid, or gaseous) or between mediums in direct physical contact. In conduction heat flow, the energy is transmitted by direct molecular communication without appreciable displacement of the molecules. Convection is a process of energy transport by the combined action of heat conduction, energy storage, and mixing motion. Radiation is a process by which heat flows from a high-temperature body to a body at a lower temperature when the bodies are separated in space, even when a vacuum exists between them. The term "radiation" is generally applied to all kinds of electromagnetic-wave phenomena, but in heat transfer, only those phenomena that are a result of temperature and can transport energy through a transparent medium or through space are of interest.

Conduction heat transfer is modeled with Fourier's law of heat conduction. At a surface (on the surface or an imaginary one inside the material), the general vector form of Fourier's law is given by Equation 1.31.

$$q'' = \frac{Q}{A} = -k\nabla T \tag{1.31}$$

where the mathematical ∇ operator has the following forms depending on geometry:

Cartesian

$$\vec{\nabla} = \hat{e}_x \frac{\partial}{\partial x} + \hat{e}_y \frac{\partial}{\partial y} + \hat{e}_z \frac{\partial}{\partial z}$$

Cylindrical-polar

$$\vec{\nabla} = \hat{e}_r \frac{\partial}{\partial r} + \hat{e}_\theta \frac{1}{r} \frac{\partial}{\partial \theta} + \hat{e}_z \frac{\partial}{\partial z}$$

The time-dependent differential energy balance or heat conduction equation for a solid in three dimensions can be found according to Equation 1.32.

$$\nabla (k\nabla T) + \dot{q}'' = \rho c \frac{\partial T}{\partial t} \tag{1.32}$$

General solutions to practical problems are difficult to obtain analytically, so most often these types of problems are done with numerical procedures. For the purposes of this discussion, one dimensional formulation is sufficient. The one-dimensional conduction (plane wall, x-direction only) is given by:

$$\dot{Q} = -kA \frac{dT}{dx} \cong -kA \frac{T_2 - T_1}{X_2 - X_1} \tag{1.33}$$

The right side of the equation is valid when the thermal conductivity is constant. This is rarely the case for the temperature variation between low temperatures and room temperature. In many situations, it is convenient to discuss the heat flux, which is simply the heat transfer divided by the area over which it occurs.

$$\dot{q}'' = \frac{Q}{A} \; Heat \; Flux \tag{1.34}$$

One-dimensional heat conduction is frequently encountered in common geometries such as the plane wall, hollow cylinder, and sphere. Table 1.3 provides a summary of the conduction equations with constant properties. The conduction heat transfer and temperature distribution for a plane wall is illustrated in Figure 1.9.

Low-temperature applications occur over large ranges of temperature. As a result, property values, such as the thermal conductivity, vary considerably for most materials. The low temperature variation of thermal conductivity with temperature for some selected materials are shown in Figure 1.10.

To account for the variation of thermal conductivity in low-temperature applications, a thermal conductivity integral [5], sometimes referred to as the Kirchoff transformation (Özis̗ik [6]), can be applied. The integral is found by rearranging Fourier's law into a form with spatial variables on one side of the equation and temperature-dependent terms on the other side, as shown in Equation 1.35.

$$Q \frac{dx}{A} = -k(T) dT \tag{1.35}$$

TABLE 1.3

One-Dimensional Heat Conduction Relationships for Constant Thermal Conductivity

Geometry	Heat Conduction Equation		Thermal Resistance
Plane wall	$Q_x = -kA \dfrac{dT}{dx}$	$Q_x = kA \dfrac{T_2 - T_1}{x_2 - x_1}$	$R_T = \dfrac{\Delta x}{kA}$
Cylinder	$Q_r = -kA \dfrac{dT}{dr}$	$Q_r = \dfrac{2\pi L k (T_2 - T_1)}{\ln(r_2 / r_1)}$	$R_T = \dfrac{\ln(r_2 / r_1)}{2\pi L k}$
Sphere	$Q_r = -4\pi r^2 k \dfrac{dT}{dr}$	$Q_r = \dfrac{4\pi k (T_2 - T_1)}{(1/r_1) - (1/r_2)}$	$R_T = \dfrac{(1/r_1) - (1/r_2)}{4\pi k}$

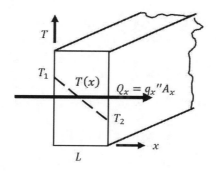

FIGURE 1.9 Relationship between one-dimensional heat conduction and temperature distribution in a plane wall.

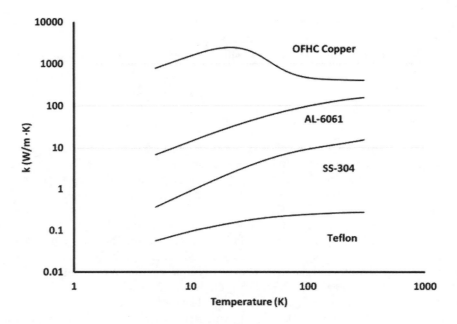

FIGURE 1.10 Temperature variation of the thermal conductivity of selected materials.

The left-hand term is only in terms of the geometry, and the right-hand term involves only temperature. Integrating both sides of the equation gives Equation 1.36.

$$Q\int_{x_1}^{x_2} \frac{dx}{A} = -\int_{T_H}^{T_C} k(T)\,dT = \int_{T_C}^{T_H} k(T)\,dT \tag{1.36}$$

This enables a shape factor and thermal conductivity integral, defined by Equations 1.37 and 1.38:

$$\frac{1}{S} = \int_{x_1}^{x_2} \frac{dx}{A} \tag{1.37}$$

$$K = \int_{T_{ref}}^{T} k(T)\,dT \tag{1.38}$$

Using these transformations permits the one-dimensional conduction heat transfer with variable properties to be calculated using a simple equation:

$$Q = S(K_H - K_C) \tag{1.39}$$

Shape factors are provided in Table 1.4 for some common geometries.

1.6 CONVECTION HEAT TRANSFER

Convection heat transfer is a process of energy transport by the combined action of heat conduction, energy storage, and mixing motion. An illustration of the velocity and temperature distribution in a heated plate is shown in Figure 1.11. It is most important as the mechanism of energy transfer

TABLE 1.4

Conduction Shape Factors for Common Shapes Used in the Determination of the Thermal Resistance or Conductance for Conduction Heat Transfer

Geometry	Shape Factor
Plane wall	$S = \dfrac{A}{\Delta x}$
Hollow cylinder	$S = \dfrac{2\pi L}{\ln\left(D_o / D_i\right)}$
Hollow sphere	$S = \dfrac{2\pi D_o D_i}{D_o - D_i}$

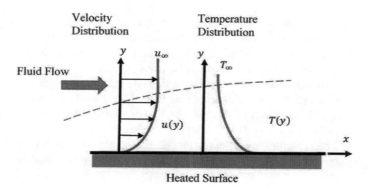

FIGURE 1.11 Convection heat transfer.

between a solid surface and a liquid or a gas. The transfer of energy by convection from a surface whose temperature is higher than that of the surrounding fluid takes place in several steps.

1. Heat will flow by conduction from the surface to adjacent particles of fluid.
2. The energy transfer will increase the temperature and the internal energy of these fluid particles.
3. Fluid particles will move to a region of lower temperature in the fluid, where they mix with and transfer part of their energy to other fluid particles.

The flow is of the fluid and the energy. Energy is stored in fluid particles and carried as a result of their motion. Convection heat transfer can occur by a forced circulation of the fluid or through fluid motion that occurs as a result of density differences caused by the temperature distribution in the fluid. The terms forced and natural or free convection are given to these two situations.

Convection heat transfer is modeled using Newton's law of cooling, which states the convection heat transfer is proportional to a convective heat transfer coefficient, h; the surface area, A; and the temperature difference $(T_\infty - T_{Surface})$. In high-speed flows approaching or exceeding the speed of sound, compressibility and kinetic energy recovery effects can add additional recovery factors to the expression. This is beyond the scope of this work, so for the purposes of this work, the convection equation is written as:

$$\dot{Q} = hA\left(T_{\infty} - T_{Surface}\right) \tag{1.40}$$

Or, in terms of a heat flux:

$$\dot{q}'' = \frac{Q}{A} = h\left(T_{\infty} - T_{Surface}\right) \tag{1.41}$$

The convective heat transfer coefficient is in general a function of the flow velocity and the thermodynamic and transport properties of the fluid.

$$h = h\left(\rho, V, P, \mu, k, C_p, \sigma, h_{fg}\right) \tag{1.42}$$

Table 1.5 provides some orders of magnitude of the convective heat transfer coefficient for common situations. The determination of the convective heat transfer coefficient is most often performed using correlations from experimental data, except in very simple cases such as laminar flow in a regular-shaped passage of flow over a flat plate, where the fluid mechanics and thermal transport can be simply analyzed. The correlations of heat transfer data are commonly expressed in terms of dimensionless numbers, as given in Table 1.6. Heat transfer by convection depends on whether the flow is laminar or turbulent. Turbulent flow has significantly higher mixing of the fluid and results in significantly higher values of the heat transfer coefficient, h.

Convective heat transfer correlations are available for many situations, including flat surfaces, flow over tubes, or flow through cooling channels. These correlations depend on whether the flow is laminar or turbulent. A few simple situations are presented as examples, but other more comprehensive references should be consulted when analyzing more complex geometries. For laminar flow for a flat plate, the local Nusselt number and heat transfer coefficient can be found from Equation 1.43.

$$Nu_x = \frac{h_x x}{k} = 0.332 Re_x^{1/2} Pr^{1/3} \tag{1.43}$$

The expression can be integrated to provide an average for the total distance, x, along the plate, resulting in Equation 1.44.

$$\overline{Nu_x} = \frac{\overline{h_x} x}{k} = 0.664 Re_x^{1/2} Pr^{1/3} \tag{1.44}$$

TABLE 1.5

Order of Magnitude of Convective Heat Transfer Coefficients, $\overline{h}_c$

Situation	Btu / hr · ft² · °F	W/m² · K
Air, free convection	1–5	6–30
Superheated steam or air, forced convection	5–50	30–300
Oil, forced convection	10–300	60–1,800
Water, forced convection	50–2,000	300–6,000
Water, boiling	500, 10,000	3,000–60,000
Steam, condensing	1,000–20,000	6,000–120,000

TABLE 1.6

Dimensionless Numbers Commonly Used in Convective Heat Transfer Correlations

Dimensionless Number	Definition
Reynolds number	$Re = \dfrac{\rho \bar{V} D}{\mu}$
Prandtl number	$Pr = \dfrac{\mu c_p}{k}$
Grashoff number	$Gr = \dfrac{g \beta \rho^2 \left(T_s - T_\infty\right) L^3}{\mu^2}$
Rayleigh number	$Ra = GrPr$
Peclet number	$Pe = RePr$
Nusselt number	$Nu = \dfrac{hD}{k}$

The expression has been shown to be valid for $Pr \geq 0.6$. Transition to turbulence over a flat plate depends on factors such as the freestream turbulence of the flow and the surface roughness. For a laminar flow at the start of a flat plate, the transition to turbulence has been correlated to a critical Reynolds number based on the distance from the leading edge of the plate, which is in the range:

$$10^5 \leq Re_{x,crit} = \frac{\rho V_\infty x_{crit}}{\mu} \leq 3 \times 10^6 \tag{1.45}$$

A representative value is commonly applied for many analyses, as suggested by Berman:

$$Re_{x,crit} = 5 \times 10^5 \tag{1.46}$$

For the turbulent boundary layer over a flat plate:

$$Nu_x = 0.032 Re_x^{0.8} Pr^{0.43} \tag{1.47}$$

which can be integrated to get an average heat transfer coefficient (ignoring any laminar region at the first part of the plate):

$$Nu_L = 0.0370 Re_L^{0.8} Pr^{.43} \tag{1.48}$$

For forced flow inside of tubes, the fully developed laminar flow case can be shown to have a constant Nusselt number that depends on whether the wall is at a constant temperature or constant heat flux. For these circular tubes, these values are:

$$Nu_T = 3.66 \quad Nu_H = 4.36 \tag{1.49}$$

The values of the Nusselt number for other geometries can be found in Kays [7].

1.7 THERMAL RADIATION HEAT TRANSFER

Radiation is a process by which heat flows from a high-temperature body to a body at a lower temperature when the bodies are separated in space, even when a vacuum exists between them. The term "radiation" is generally applied to all kinds of electromagnetic-wave phenomena, but in heat transfer, only those phenomena that are a result of temperature and can transport energy through a transparent medium or through space are of interest.

Black body thermal radiation is thermal energy emitted by the surface that originates from the thermal energy of matter bound by the surface. The upper limit on the emissive power from the ideal radiator, called a black body radiator for a surface with an area, A, and a surface temperature, T_S, is determined by Equation 1.50.

$$Q = E_B = \sigma A T_S^4 \tag{1.50}$$

where the Stefan-Boltzmann constant is given by $\sigma = 5.669 \ (10^{-8})$ W/m²·K = 0.1714 (10^{-8}) Btu/hr·ft²·°R. Many factors affect radiation heat transfer, including geometric and material properties. Heat flux emitted by a real surface is less than that of a black body at the same temperature and is given by Equation 1.51.

$$E = \epsilon E_B = \epsilon \sigma A T_S^4 \tag{1.51}$$

where ϵ is the emissivity of the surface, which is the fraction of the thermal radiation emitted from an actual surface compared to a black surface. The emissivity range is $0 \le \epsilon \le 1$. Thermal radiation can be incident on a surface from the surroundings (the sun, other surfaces, etc.). An example of the thermal radiation energy balance is given in Figure 1.12 for a surface.

Irradiation, G, is the rate at which thermal radiation is incident on a unit area of a surface. All or part of this may be absorbed by the surface. Absorptivity, α, of a surface is the fraction of the irradiation absorbed, $0 \le \alpha \le 1$:

$$G_{abs} = \alpha G \tag{1.52}$$

If $\alpha < 1$, a portion of the radiation is reflected (ρ is the reflectivity, the fraction of the incident radiation reflected). If it is semi-transparent, some may also be transmitted with a transmittance of $\tau < 1$.

In general, the sum of the absorption, reflection, and transmission is given in Equation 1.53.

$$\alpha + \rho + \tau = 1 \tag{1.53}$$

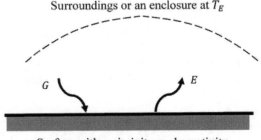

FIGURE 1.12   Thermal radiation heat transfer surface energy balance.

In many engineering applications, the surfaces are opaque ($\tau = 0$), so the irradiation may be approximated as from a black surface.

$$G = \sigma T_E^{\,4} \tag{1.54}$$

It can be shown (see Kirchoff's Law in Berman) that for many enclosures when the surfaces have no transmission, $\alpha = \epsilon$, so:

$$G_{abs} = \alpha G = \epsilon \sigma T_E^{\,4} \tag{1.55}$$

This is assuming the surroundings are treated as black. The net radiation from a surface per unit area of the surface can be expressed as:

$$q''_{rad} = \frac{q}{A} = \epsilon E_B(T_S) - \alpha G = \epsilon \sigma T_S^4 - \epsilon \sigma T_E^{\,4} \tag{1.56}$$

$$q''_{rad} = \epsilon \sigma \left(T_S^4 - T_E^{\,4} \right) \tag{1.57}$$

Engineers like to use a radiation heat transfer coefficient, h_r, in modeling, which would take the form of Equation 1.58. As shown in Equations 1.59–1.61:

$$q_{rad} = h_r A (T_S - T_E) \tag{1.58}$$

$$h_r = \frac{\epsilon \sigma \left(T_S^4 - T_E^{\,4} \right)}{(T_S - T_E)} = \frac{\epsilon \sigma \left(T_S^2 + T_E^{\,2} \right) \left(T_S^2 - T_E^{\,2} \right)}{(T_S - T_E)} = \frac{\epsilon \sigma \left(T_S^2 + T_E^{\,2} \right) (T_S + T_E)(T_S - T_E)}{(T_S - T_E)} \tag{1.59}$$

$$h_r = \epsilon \sigma \left(T_S^2 + T_E^{\,2} \right) (T_S + T_E) \tag{1.60}$$

This is a way to linearize thermal radiation so that it can be treated like heat convection.

When convection and radiation are present:

$$q = q_{conv} + q_{rad} = hA(T_S - T_\infty) + \epsilon \sigma A \left(T_S^4 - T_{Sur}^4 \right) \tag{1.61}$$

Then radiation heat transfer between a surface 1 that is completely enclosed by a surface 2, like an idealized double-walled vessel, is a common situation. For the case where both surfaces are black, the net heat transfer can be calculated by:

$$Q = \sigma A_1 \left(T_1^4 - T_2^4 \right) \tag{1.62}$$

Real bodies do not meet the specifications of an ideal radiator but emit thermal radiation at a reduced rate of black bodies. These are called gray bodies. The net rate of heat transfer from a gray body at a temperature T_1 to a black surrounding body at T_2 is:

$$Q = \sigma A_1 \epsilon_1 \left(T_1^4 - T_2^4 \right) \tag{1.63}$$

where ϵ_1 is the emittance of the gray surface and is equal to the ratio of the emission from a gray surface to the emission of a black body at the same temperature.

If neither of the two bodies is a perfect radiator and the two bodies possess a geometrical relationship to each other, the net heat transfer is given by:

$$Q = \sigma A_1 \mathcal{F}_{1-2} \left(T_1^4 - T_2^4 \right) \tag{1.64}$$

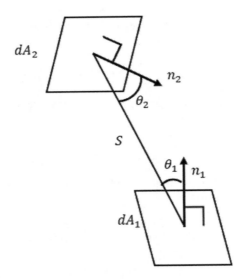

FIGURE 1.13 Geometry of view factors for a differential geometry.

where $\mathcal{F}_{1-2}$ is the view factor from surface 1 to surface 2 calculated to account for the geometry and gray body emittance of the actual bodies. The view factor or shape factor is the fraction of the total diffuse radiation leaving one surface that is intercepted by another surface. This is illustrated in Figure 1.13.

For two differential areas, the view factor is given by:

$$dF_{dA_1-dA_2} = \frac{\cos\theta_1 \cos\theta_2 dA_2}{\pi S^2} \tag{1.65}$$

A reciprocity relation can be found by writing this for both surfaces 1 and 2.

$$dA_1 dF_{dA_1-dA_2} = dA_2 dF_{dA_2-dA_1} \tag{1.66}$$

Integrating over the surface areas provides a simple but useful relationship.

$$A_1 F_{12} = A_2 F_{21} \tag{1.67}$$

There are many compilations of view factors available. Some examples are provided in the following tables that may be useful. In addition, the concept of view factor algebra provides a means to work with thermal radiation in enclosures. The reciprocity relation, already discussed, is the first part of view factor algebra. For an enclosure of many surfaces, the view factors from a surface to all others surrounding it must sum to 1, or:

$$F_{11} + F_{12} + F_{13} + \cdots + F_{1n} = 1$$
$$F_{21} + F_{22} + F_{23} + \cdots + F_{2n} = 1$$

$$\vdots$$

$$F_{n1} + F_{n2} + F_{n3} + \cdots + F_{nn} = 1 \tag{1.68),(1.69),(1.70)}$$

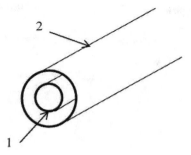

FIGURE 1.14   Infinitely long concentric cylinder geometry.

The reciprocity relationship can be written between any two surfaces:

$$A_i F_{ij} = A_j F_{ji} \tag{1.71}$$

As an example of the application of this, consider an infinitely long cylinder inside another infinitely long cylinder, similar to a vacuum-insulated double-walled pipe (Figure 1.14).

The inner cylinder sees only the outer cylinder, so the view factor from surface 1 to 2 is unity, and the outer surface of the inner pipe cannot radiate to itself, so:

$$F_{12} = 1 \tag{1.72}$$

$$F_{11} = 0 \tag{1.73}$$

But the outer cylinder sees both itself and the smaller-diameter inner cylinder, so, applying the reciprocity relation:

$$A_1 F_{12} = \pi D_1 \left(1\right) = A_2 F_{21} = \pi D_2 F_{21} \tag{1.74}$$

$$F_{21} = \frac{D_1}{D_2} \tag{1.75}$$

Since the sum of the view factors from a surface must equal unity:

$$F_{21} + F_{22} = 1 \tag{1.76}$$

$$F_{22} = 1 - \frac{D_1}{D_2} \tag{1.77}$$

As a result, all the view factors have been found.

When multiple surfaces are stacked to form a series of thermal radiation shields, the thermal radiation analysis has been compiled into some simplified forms for flat, cylindrical, and spherical radiation shields. These are given in Table 1.7.

Within an enclosure, typical for low-temperature devices, there are two gray surfaces, and one sees only the second, but the second sees itself and the first surface. A generic representation is given in Figure 1.10, Figure 1.15, where $E = \epsilon E_b = \epsilon \sigma T^4$, or the black body radiation for a surface. It is seen

TABLE 1.7

View Factors for Some Common Two- and Three-Dimensional Situations

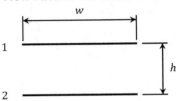

$$F_{1-2} = F_{2-1} = \sqrt{1 + \left(\frac{h}{w}\right)^2} - \left(\frac{h}{w}\right)$$

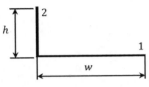

$$F_{1-2} = \frac{1}{2}\left[1 + \left(\frac{h}{w}\right) - \sqrt{1 + \left(\frac{h}{w}\right)^2}\right]$$

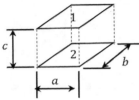

Let $X = a/c$ and $Y = b/c$. Then:

$$F_{1-2} = \frac{2}{\pi XY}\left\{\begin{matrix}\ln\left[\frac{(1+X^2)(1+Y^2)}{1+X^2+Y^2}\right]^{1/2} - X\tan^{-1}X - Y\tan^{-1}Y + \\ X\sqrt{1+Y^2}\tan^{-1}\frac{X}{\sqrt{1+Y^2}} + Y\sqrt{1+X^2}\tan^{-1}\frac{Y}{\sqrt{1+X^2}}\end{matrix}\right\}$$

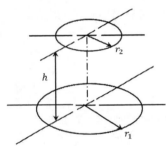

Let $R_1 = r_1/h$, $R_2 = r_2/h$ and $X = 1 + (1 + R_2^2)/R_1^2$. Then:

$$F_{1-2} = \frac{1}{2}\left[X - \sqrt{X^2 - 4(R_2/R_1)^2}\right]$$

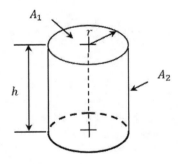

$$H = \frac{h}{2r}$$

$$F_{1-2} = 2H\left[(1+H^2)^{\frac{1}{2}} - H\right]$$

that there are an infinite number of reflections and emissions possible. Even though this appears to be an impossible task, it can be shown that the heat transfer between the surfaces simplifies to:

$$q_1 = \frac{A_1\sigma\left(T_1^4 - T_2^4\right)}{\dfrac{1}{\epsilon_1} + \dfrac{A_1}{A_2}\left(\dfrac{1}{\epsilon_2} - 1\right)} = -q_2 \tag{1.78}$$

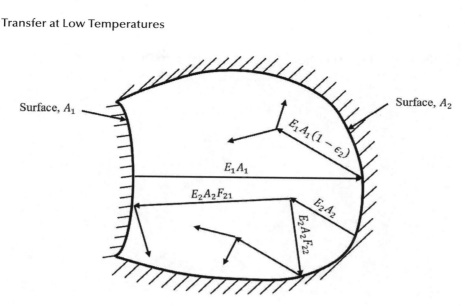

FIGURE 1.15 Radiation interchanges for an enclosure consisting of two surfaces from Gebhart [8]. Only first reflections are shown.

When multiple surfaces (more than two) are involved, a procedure developed by Gebhart can be applied. First, an absorption factor is defined, B_{ij}, that is the fraction of emission from A_i that is received at A_j. The emission from a surface is:

$$E_j = \epsilon E_b = \epsilon \sigma T^4 \tag{1.79}$$

The net heat transfer to a surface from all surrounding surfaces is:

$$q_j = E_j A_j - B_{1j} E_1 A_1 - B_{2j} E_2 A_2 - \cdots - B_{ij} E_i A_i - \cdots - B_{nj} E_n A_n \tag{1.80}$$

or:

$$q_j = E_j A_j - \sum_{i=1}^{n} B_{ij} E_i A_i \tag{1.81}$$

where j may be taken as any one of the n surfaces of the enclosure. Note that in general, B_{ij} is not zero; some of $E_j A_j$ may be reabsorbed at A_j. For the determination of the absorption factors

$$B_{1j} = F_{1j}\epsilon_j + F_{11}\rho_1 B_{1j} + F_{12}\rho_2 B_{2j} + F_{13}\rho_3 B_{3j} + \cdots + \cdots + F_{1n}\rho_n B_{n_j} \tag{1.82}$$

For each of the other surfaces, a similar relation may be written:

$$B_{2j} = F_{2j}\epsilon_j + F_{21}\rho_1 B_{1j} + F_{22}\rho_2 B_{2j} + F_{23}\rho_3 B_{3j} + \cdots + \cdots + F_{2n}\rho_n B_{n_j}$$

$$B_{3j} = F_{3j}\epsilon_j + F_{31}\rho_1 B_{1j} + F_{32}\rho_2 B_{2j} + F_{33}\rho_3 B_{3j} + \cdots + \cdots + F_{3n}\rho_n B_{n_j}$$

$$\cdots\cdots\cdots\cdots\cdots\cdots\cdots\cdots\cdots\cdots\cdots\cdots\cdots\cdots\cdots\cdots\cdots\cdots$$

$$B_{nj} = F_{nj}\epsilon_j + F_{n1}\rho_1 B_{1j} + F_{n2}\rho_2 B_{2j} + F_{n3}\rho_3 B_{3j} + \cdots + \cdots + F_{nn}\rho_n B_{n_j}$$

A total of n linear equations is obtained. These can be transposed and rearranged by introducing:

$$F_{rs}\rho_{rs} = \alpha_{rs}$$

$$\vdots$$

$$\rho_{rs} = \left(1 - \epsilon_s\right)$$

$$\left(\alpha_{11} - 1\right)B_{1j} + \alpha_{12}B_{2j} + \alpha_{13}B_{3j} + \cdots + \alpha_{1n}B_{nj} + F_{1j}\epsilon_j = 0$$

$$\alpha_{21}B_{1j} + \left(\alpha_{22} - 1\right)B_{2j} + \alpha_{23}B_{3j} + \cdots + \alpha_{2n}B_{nj} + F_{2j}\epsilon_j = 0$$

$$\alpha_{31}B_{1j} + \alpha_{32}B_{2j} + \left(\alpha_{33} - 1\right)B_{3j} + \cdots + \alpha_{3n}B_{nj} + F_{3j}\epsilon_j = 0$$

$$\cdots\cdots\cdots\cdots\cdots\cdots\cdots\cdots\cdots\cdots\cdots\cdots\cdots\cdots\cdots\cdots\cdots\cdots$$

$$\alpha_{n1}B_{1j} + \alpha_{n2}B_{2j} + \alpha_{n3}B_{3j} + \cdots + \left(\alpha_{nn} - 1\right)B_{nj} + F_{nj}\epsilon_j = 0$$

There are variations on the problem. If a surface is assumed to be in radiant balance, set the emissivity = 0. If the temperature of the surface is wanted, then use the real emissivity and adjust the surface temperature until the heat transfer is 0. Frequently many situations can be closely modeled with simple geometries. Some useful common geometries are presented in Table 1.8. Approximations

---

TABLE 1.8

Some Useful Common Geometries for Radiation Heat Transfer Calculations

Geometry		Energy Rate, Q_1
Infinite parallel plates 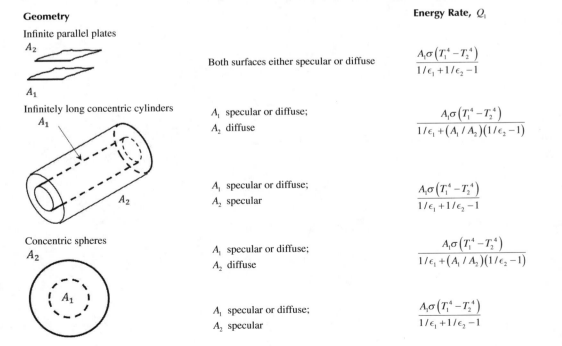	Both surfaces either specular or diffuse	$\dfrac{A_1\sigma\left(T_1^4 - T_2^4\right)}{1/\epsilon_1 + 1/\epsilon_2 - 1}$
Infinitely long concentric cylinders	A_1 specular or diffuse; A_2 diffuse	$\dfrac{A_1\sigma\left(T_1^4 - T_2^4\right)}{1/\epsilon_1 + \left(A_1/A_2\right)\left(1/\epsilon_2 - 1\right)}$
	A_1 specular or diffuse; A_2 specular	$\dfrac{A_1\sigma\left(T_1^4 - T_2^4\right)}{1/\epsilon_1 + 1/\epsilon_2 - 1}$
Concentric spheres	A_1 specular or diffuse; A_2 diffuse	$\dfrac{A_1\sigma\left(T_1^4 - T_2^4\right)}{1/\epsilon_1 + \left(A_1/A_2\right)\left(1/\epsilon_2 - 1\right)}$
	A_1 specular or diffuse; A_2 specular	$\dfrac{A_1\sigma\left(T_1^4 - T_2^4\right)}{1/\epsilon_1 + 1/\epsilon_2 - 1}$

Source: From Siegel and Howell [9]

for thermal radiation shields for planar, cylindrical, and spherical geometries are shown in Figures 1.16 and 1.17.

In many cases, the surface emissivities are the same for the shield layers. Let all the shield layers have emissivity ϵ_s; then the heat transfer becomes:

$$\frac{Q}{A} = q = \frac{\sigma\left(T_1^4 - T_2^4\right)}{1/\epsilon_1 + 1/\epsilon_2 - 1 + N\left(2/\epsilon_s - 1\right)}$$

$$Q = \frac{A_1\sigma\left(T_1^4 - T_2^4\right)}{1/\epsilon_1 + \left(A_1/A_2\right)\left(1/\epsilon_2 - 1\right) + \sum_{n=1}^{N}\left(A_1/A_{sn}\right)\left(1/\epsilon_{n1} + 1/\epsilon_{n2} - 1\right)} \tag{1.83}$$

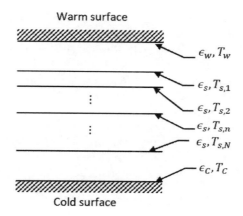

FIGURE 1.16   Radiation shields between flat surfaces.

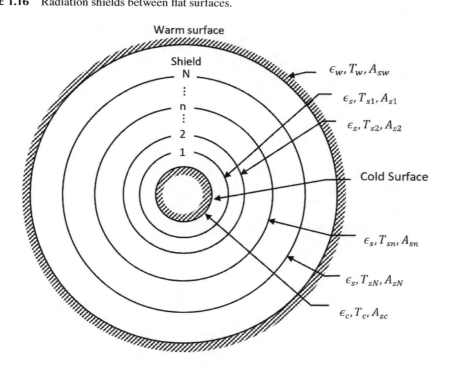

FIGURE 1.17   Radiation shields between concentric cylinders or spheres.

1.7   GAS CONDUCTION

A special case that frequently occurs in low-temperature systems is when there is a low-pressure gas, such as when high vacuum is used as a thermal insulation. The level or degrees of vacuum have been suggested in Guthrie [10] and are summarized in Table 1.9.

At low pressures, the density of the gas decreases, or the distance between gas molecules increases. The mean free path, λ, is defined as the average distance traveled by molecules between collisions; see Tien and Lienhard [11]. Under a rough vacuum, λ is very small compared to the distance between the boundaries between which the gas is transporting heat, d. This is commonly expressed in terms of the dimensionless Knudsen number, Kn, defined as:

$$Kn = \frac{\lambda}{d} \tag{1.84}$$

Table 1.10 lists the flow regimes that occur for various ranges of Kn. For values of $Kn \ll 1$, continuum flow is present, and conventional heat transfer analysis applies. When high vacuum is present, the mean free path is much larger than the distance between the boundaries between which the gas is transporting heat, d, and $Kn \gg 1$, which is called the free molecular flow regime. The regimes between the continuum and free molecular flow regime are the slip flow and transitional regimes, which occur in fine or soft vacuum. No formula for heat conduction is known in these regimes, and if operation is required, one must rely on experimental data.

Corruccini [13] provides an approach for calculating a heat transfer coefficient between two surfaces at different temperatures if there is a high vacuum in the gap separating the surfaces, that is, when $Kn \gg 1$. This approach has been applied to model gas conduction in multilayer insulation systems by Augustynowicz et al. [14]. According to Corruccini [13], the heat transfer by gas conduction

TABLE 1.9
Degrees of Vacuum

Classification	Pressure (torrs)
Rough vacuum (or no vacuum)	760 to 1
Fine vacuum (sometimes called soft vacuum)	1 to 10^{-3}
High vacuum	10^{-3} to 10^{-6}
Very high vacuum	10^{-6} to 10^{-9}
Ultra-high vacuum	10^{-9} and less

Source: According to Guthrie [10]

TABLE 1.10
Flow Regimes Based on the Knudsen Number

Knudsen Number	Regime
$Kn < 0.01$	Continuum
$0.01 < Kn < 0.1$	Slip flow
$0.1 < Kn < 3$	Transition
$3 < Kn$	Free molecular flow

Source: From Zucrow and Hoffman [12]

to the inner surface, 1, from the outer surface, 2, for coaxial cylinders, concentric spheres, or parallel plane walls can be calculated according to the relation:

$$\dot{Q}_1 = \alpha \left(\frac{\gamma+1}{\gamma-1} \right) \sqrt{\frac{R}{8\pi}} \frac{p}{\sqrt{MT}} (T_2 - T_1) \tag{1.85}$$

where the overall accommodation coefficient, α, depends on the accommodation coefficients for the inner surface, α_1, and the outer surface, α_2, and the area of the inner surface, A_1, and the outer surface, A_2.

$$\alpha = \frac{\alpha_1 \alpha_2}{\alpha_2 + \alpha_1 (1 - \alpha_2) A_1 / A_2} \tag{1.86}$$

The specific heat ratio for the gas is

$$\gamma = \frac{c_p}{c_v} \tag{1.87}$$

where R is the universal gas constant, T_1 is the inner surface temperature, and T_2 is the outer surface temperature. The molecular weight of the gas is M. The value of the pressure, p, corresponds to the temperature, T, at which a value would be measured. Typically, the vacuum level is measured by a gauge mounted on an external surface and would be at room temperature.

1.8 BOILING AND CONDENSATION

Boiling of a liquefied gas or condensation to liquefy a gas can be used when high heat flux situations are encountered in cryogenic heat management. For boiling to take place, the surface being cooled must be above the saturation temperature of the liquid, and the temperature difference, ΔT, is given by:

$$\Delta T = T_w - T_l \tag{1.88}$$

where T_w is the surface temperature, and T_l is the liquid saturation temperature. Boiling occurs in different regimes, as shown in Figure 1.18. At small temperature differences, heat transfer is only by convection. As ΔT increases, nucleate boiling occurs, where vapor bubbles form at the surface. At some point, ΔT is large enough that a critical heat flux (CHF) or boiling crisis occurs. Additional increases in ΔT cause an unstable transition from nucleate boiling to film boiling, and the heat flux decreases with increasing ΔT. A minimum heat flux occurs at a point known as the Leidenfrost point, where stable film boiling takes place. In film boiling, the surface temperature is at a temperature such that no liquid reaches the surface and there is only vapor. As ΔT increases in this region, stable film boiling takes place.

For nucleate pool boiling, the Kutateladze correlation is recommended for liquid nitrogen by McAshan et al. [15] and is given by:

$$\frac{h_{boil}}{k_l} \left(\frac{\sigma}{g\rho_l} \right)^{1/2} = 3.25 \times 10^{-4} \left\{ \frac{(Q/A)c_{p,l}\rho_l}{h_{fg}\rho_v k_l} \left(\frac{\sigma}{g\rho_l} \right)^{1/2} \right\}^{0.6} \left\{ g \left(\frac{\rho_l}{\mu_l} \right)^2 \left(\frac{\sigma}{g\rho_l} \right)^{3/2} \right\}^{0.125} \left\{ \frac{p}{(\sigma g\rho_l)} \right\}^{0.7} \tag{1.89}$$

where h_{boil} is the nucleate pool boiling coefficient in (W/cm²·K), k_l is the thermal conductivity of the liquid in (W/cm·K), $c_{p,l}$ is the constant pressure specific heat for the liquid (J/g·K), ρ_l is the

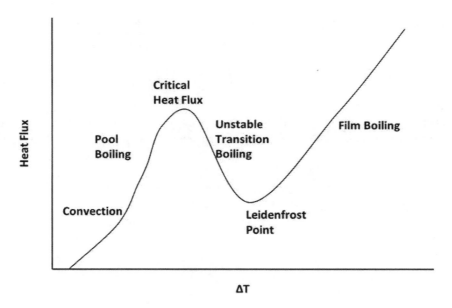

FIGURE 1.18   Boiling regimes for different values of ΔT indicating relative heat flux for these regions.

liquid density (g/cm³), ρ_v is the vapor density (g/cm³), μ_l is the liquid viscosity (g/cm·sec), h_{fg} is the latent heat of vaporization (J/g), σ is the surface tension between the liquid and its own vapor evaluated at the temperature of the boiling liquid (dynes/cm), g is the acceleration of gravity (cm/sec²), (Q/A) is the rate of heat transfer per unit area (W/cm²), and p is the pressure of the boiling system (dynes/cm²). A detailed discussion of other liquefied gases can be found in Brentari and Smith [16].

Condensation heat transfer occurs when a vapor is exposed to a surface at a temperature below the saturation temperature at the system pressure. Condensation occurs on smooth clean surfaces by the formation of a film of liquid. Surface treatments can be applied to limit wetting of the surface to produce dropwise condensation. For the thermal management of cryogenic systems, filmwise condensation is most often desired. Condensation can also take place in the situation where a failure occurs in the thermal insulating vacuum (Lehman and Zahn [17]).

The average Nusselt number for laminar film condensation on a vertical plate can be determined from the following correlation from Berman et al. [3].

$$\overline{Nu_L} = \frac{\overline{h_L} L}{k_l} = 0.943 \left[\frac{\rho_l g (\rho_l - \rho_v) h'_{fg} L^3}{\mu_l k_l (T_l - T_w)} \right] \tag{1.90}$$

where h'_{fg} is a modified latent heat given by the following expression:

$$h'_{fg} = h_{fg} + 0.68 c_{p,l} (T_l - T_w) \tag{1.91}$$

The heat transfer from the plate, Q, can be calculated from

$$Q = \overline{h_L} (L \times w)(T_l - T_w) \tag{1.92}$$

where w is the width of the plate, L is the length of the plate, $\overline{h_L}$ is the average heat transfer coefficient for condensation over the length of the plate, and μ_l is the viscosity of the liquid.

In this equation, all liquid properties should be evaluated at the film temperature given by:

$$T_f = \frac{T_l + T_w}{2} \tag{1.93}$$

The vapor density, ρ_v, and the latent heat of vaporization, h_{fg}, should be evaluated at the saturation temperature, T_l.

1.9 APPLICATION OF HEAT TRANSFER TO HEAT MANAGEMENT

Understanding the modes of heat transfer is a fundamental concept in cryogenic heat management. In order to maintain a component at a prescribed temperature, two aspects must be properly configured. First, the component needs to have a very efficient link to the low temperature sink that permits the cold component to reach the desired temperature. Second, the production of refrigeration at low temperatures has been shown to be costly in terms of energy and, in turn, of financial costs. Thermal management also involves the blocking of high-temperature sources of energy from entering the cold space. This can be achieved through the concept of thermal resistances or conductances and the stack-up of temperature differences, or ΔTs. As was shown earlier, minimizing the temperature differences is equivalent to minimizing the entropy generation.

An analogy with the features of an electrical circuit is the best way to approach this topic. This network analogy can be explained using the parameters provided in Table 1.11.

The electrical resistance is determined from Ohm's law, as in Equation 1.94. The thermal resistance by analogy is given in Equation 1.95 and the conductance in Equation 1.96.

$$R = \frac{V}{I} \tag{1.94}$$

$$R_T = \frac{\Delta T}{Q} \tag{1.95}$$

$$C_T = \frac{1}{R_T} = \frac{Q}{\Delta T} \tag{1.96}$$

For conduction heat transfer, the thermal resistance is described in Table 1.3 for some common shapes. In Figure 1.19, it is shown that the thermal resistances can be arranged in series, as in Figure 1.19(a), or in parallel, as in Figure 1.19(b), as shown in Augustynowicz for heat transfer through multilayer insulation.

TABLE 1.11

Comparison of Quantities Used in the Electrical-Thermal Analogy

Generic	Electrical Quantity	Thermal Quantity
Balanced quantity	Charge, q	Energy, e
Driving potential	Voltage, V	Temperature, T
Flowed quantity	Current, I	Heat transfer, Q
Resistance	Resistance	Thermal resistance
Stored property	Capacitance, C	Thermal heat capacity, mC_l

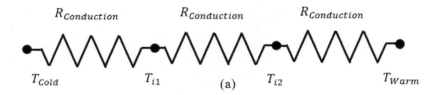

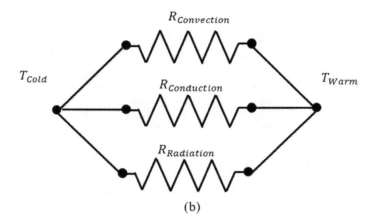

FIGURE 1.19 Thermal resistances in (a): series as for conduction through several materials and (b): parallel as heat transfer through multilayer superinsulation.

Thermal resistances can be defined for the three modes of heat transfer. For conduction in three common geometries, such as a one-dimensional plane wall:

$$R_{cond} = \frac{t}{kA} \tag{1.97}$$

where t is the thickness, k is the thermal conductivity, and A is the surface area. For one dimension in a hollow cylinder:

$$R_{cond} = \frac{\ln\left(r_o / r_i\right)}{2\pi kL} \tag{1.98}$$

where L is the length of the cylinder, r_o is the outer radius of the cylinder, and r_i is the inner radius of the hollow cylinder. A third common situation is the spherical shell; here the thermal resistance is given as:

$$R_{cond} = \frac{1}{4\pi k}\left(\frac{1}{r_i} - \frac{1}{r_o}\right) \tag{1.99}$$

As with conduction heat transfer, convection can be expressed as a thermal resistance in terms of the convective heat transfer coefficient as:

$$R_{conv} = \frac{1}{hA} \tag{1.100}$$

where h is the convective heat transfer coefficient, and A is the surface area. This form also applies to the contact resistance where the contact coefficient, h_c, is used instead of the convection coefficient.

For the application of thermal resistance to radiation heat transfer and assuming the simple geometries discussed earlier, the radiation exchange between two surfaces can be expressed as:

$$Q_{rad,1-2} = \sigma A_1 F \left(T_1^4 - T_2^4 \right) \tag{1.101}$$

The exchange factor, F, includes the emissivity and view factor.

This can be arranged into a form that represents the thermal resistance by reducing the expression to the following form:

$$Q_{rad,1-2} = \sigma A F \left(T_1^4 - T_2^4 \right) = \sigma A F \left(T_1^2 + T_2^2 \right)\left(T_1 + T_2 \right)\left(T_1 - T_2 \right) \tag{1.102}$$

For example, if the radiation exchange between two long concentric cylinders were needed, from Table 1.8, one would get Equation 1.103.

$$Q_{Rad} = \frac{A_1 \sigma \left(T_1^4 - T_2^4 \right)}{1/\epsilon_1 + \left(A_1 / A_2 \right)\left(1/\epsilon_2 - 1 \right)} = \frac{A_1 \sigma \left(T_1^2 + T_2^2 \right)\left(T_1 + T_2 \right)\left(T_1 - T_2 \right)}{1/\epsilon_1 + \left(A_1 / A_2 \right)\left(1/\epsilon_2 - 1 \right)} \tag{1.103}$$

In general, for thermal radiation calculations, the thermal resistance is temperature dependent, as shown in Equation 1.104.

$$F = \frac{\left(T_1^2 + T_2^2 \right)\left(T_1 + T_2 \right)}{1/\epsilon_1 + \left(A_1 / A_2 \right)\left(1/\epsilon_2 - 1 \right)} \tag{1.104}$$

The application of thermal resistances in cryogenic heat management depends on the application. For limiting the heat transfer, as through the walls of a storage vessel or along a structural support, the objective is to maximize the thermal resistance. An equally challenging task, also part of cryogenic heat management, is to minimize the thermal resistance when connecting to a cooling system, such as a cryocooler, or between a surface and a temperature measurement. The thermal resistance is minimized when the function is thermal anchoring.

REFERENCES

1. Ganni, V., "Design of Optimal Helium Refrigeration and Liquefaction Systems: Simplified Concepts and Practical Viewpoints," *Cryogenic Society of America Short Course Notes*, Presented at the Cryogenic Engineering Conference, Chattanooga, TN, July 2007.
2. Strobridge, T. R., "Refrigeration for Superconducting and Cryogenic Systems," *IEEE Transactions on Nuclear Science*, Volume NS-16, No. 3, June 1969, pp. 1104–1108.
3. Berman, T. L., Lavine, A. S., Incroprera, F. P., et al., *Fundamentals of Heat and Mass Transfer*, Seventh Edition, Wiley Press, New York, 2011.
4. Kreith, F., *Principles of Heat Transfer*, Third Edition, Intext Press, Inc, New York, 1973.
5. Barron, R. F., *Cryogenic Heat Transfer*, Taylor and Francis, Philadelphia, PA, 1999.
6. Özişik, M. N., *Heat Conduction*, John Wiley & Sons, New York, 1980.
7. Kays, W. M., *Convective Heat and Mass Transfer*, McGraw-Hill Inc, New York, 1966.
8. Gebhart, B., "Surface Temperature Calculations in Radiant Surroundings of Arbitrary Complexity—for Gray, Diffuse Radiation," *Int. J. Heat Mass Transfer*, Vol. 3, pp. 341–346, 1961.
9. Seigel, R. and Howell, J. R., *Thermal Radiation Heat Transfer*, Second Edition, McGraw-Hill, New York, 1981.

10. Guthrie, A., *Vacuum Technology,* John Wiley and Sons Inc., New York, 1963.
11. Tein, C. L. and Lienhard, J. H., *Statistical Thermodynamics*, McGraw-Hill, New York, 1979.
12. Zucrow, M. J. and Hoffman, J. D., *Gas Dynamics Volume I*, John Wiley and Sons, New York, 1976.
13. Corruccini, R. J., "Gaseous Heat Conduction at Low Pressure and Temperatures," *Vacuum*, Vol. VII &VIII, pp. 19–29, 1959.
14. Augustynowicz, S. D., Demko, J. A., and Datskov, V. I., "Analysis of Multi-Layer Insulation between 80 K and 300 K," *Advances in Cryogenic Engineering*, Vol. 39, Plenum Press, New York, 1994.
15. McAshan, M., Thirumaleshwar, M., Abramovich, S. and Ganni, V., "Nitrogen System for the SSC," *Superconducting Super Collider Technical Report SSCL-592*, October 1992.
16. Brentari, E. G. and Smith R. V., "Nucleate and Film Pool Boil Design Correlations for O_2, N_2, H_2, and He," *International Advances in Cryogenic Engineering*, Plenum Press, New York, 1965.
17. Lehmann, W. and Zahn, G., "Safety Aspects for LHE Cryostats and LHE Transport Containers," *Proceedings of the 7th Annual ICEC*, IPC Science and Technology Press, London, pp. 569–579, 1978.

2 Thermal Insulation Materials and Systems

James E. Fesmire, Quan-Sheng Shu, and Jonathan A. Demko

2.1 INTRODUCTION TO THERMAL INSULATION

Heat from the ambient environment is always being transmitted into the cold regions of the cryogenic system (or "cold mass"). To reduce this, heat flow is crucial for any cryogenic system. Thermal insulation is not an important part of a cryogenic system, because it is not a part at all. The goal is to isolate or insulate the cold mass from the world of heat we call the ambient environment. The approach of this thermal isolation is driven by two factors: (1) minimizing the conduction of heat from the ambient environment and (2) minimizing the incoming heat through the surface area of the system. The latter factor is governed, when possible, by providing a vacuum barrier, or vacuum jacket, around the cold mass.

2.1.1 THREE KEY QUESTIONS

The first and foremost question for determining the best approach to thermal insulation is the target range of cold vacuum pressure (CVP). Cryogenic thermal insulation systems can be divided into three main categories according to four CVP ranges: high vacuum (HV), moderate vacuum (MV), soft vacuum (SV), and no vacuum (NV). The latter category of NV, or ambient pressure, is also referred to as "mechanical insulation" among general industry applications.

The second question is to define the environment. The composition of any residual gas or water vapor content, as applicable, is an important factor in defining the environment. The temperatures involved are the key factors in defining the operational environment. That is, what are the boundary temperatures and temperature excursions for cooldown, normal operation, and warm up? The normal operating temperatures include the warm boundary temperature (WBT) and the cold boundary temperature (CBT). The apparent thermal conductivity (λ) for a given material will indeed vary with the mean temperature (T_m) in many cases. However, it is important to keep in mind that heat does not flow according to a temperature but only according to a difference in temperature (ΔT). Therefore, the boundary temperatures, warm and cold, are the crucial information. Thermal performance data for materials should be reported either with either T_m or boundary temperatures (WBT and CBT) according to the test methodology employed. Be aware that these two reporting methods are not interchangeable and do not suffice for one another.

The third key question to address is the system design approach and the installation process. Here the integrated system aspect becomes apparent, as the limiting factors in thermal performance become the way the materials are physically located or how fittings, valves, flanges, expansion joints, supports, instrumentation, sensors, feedthroughs, and other components can be accommodated.

2.1.2 FULL RANGE VACUUM PRESSURE

This chapter addresses thermal insulation systems for the full range of environments from non-vacuum to soft vacuum to high vacuum and for temperatures from about 4 K to 400 K. Many different materials and combinations of materials are available for use depending on the design and

DOI: 10.1201/9781003098188-2

operational requirements. Included in these choices is the approach of the multilayer insulation (MLI) system. MLI is mainly targeted for radiation shielding in a high vacuum environment. This chapter includes a look at all proven cryogenic insulation materials and systems, including MLI systems, for the entire range of cold vacuum pressure environments. This chapter also examines MLI systems in high vacuum environments in greater detail due to their prominent position as, potentially, the "ultimate" insulation.

Most systems that are designed to be operated at below-ambient temperatures require thermal insulation for both control and energy-efficiency concerns. For cryogenic systems, the need is amplified, and the level of performance must be much higher to properly limit heat transmission from the ambient environment. Relative to foam-type insulation materials in air, aerogels can reduce the heat flow by about one-half (see Fesmire and Coffman et al. [1]), and moderately evacuated ("soft vacuum") layered composites can reduce the heat by ten times (see Fesmire [2]). The latest evacuated bulk-fill systems (glass bubbles) can reduce the heat flow by almost 100 times (see Scholtens et al. [3]), and highly evacuated multilayer insulation systems can reduce the heat transmission by more than 1,000 times compared to foams (see Fesmire and Johnson [4]). With each higher level of thermal performance, the design approach, engineering, and manufacturing execution become progressively more challenging. A fitting combination of materials, testing, and engineering is key to meet the objectives of system control, safety, and performance. Most importantly, as discussed by Augustynowicz et al. [5], the thermal performance must justify the cost.

The total heat leakage rate (Q_{total}), or "heat leak," into any cryogenic system is composed of three main parts in a cold triangle of integration: (1) heat leak through the insulation, (2) heat leak through the support structures, and (3) heat leak attributed to piping penetrations and feedthroughs. But real systems almost always have additional heat leak due to practical limitations of fabrication and installation, as well as the negative effects of supports and piping. This additional real-world heat leak for cryogenic systems is accounted for using the insulation quality factor (IQF). The "cold triangle" approach, with the use of the IQF, can be readily applied to piping systems, piping connections, cryostats, storage tanks, and so on to determine the best insulation system for a particular case.

2.2 TYPES OF THERMAL INSULATION SYSTEMS

With the architectural concept of approach followed by design followed by materials, listings of types, applications, and materials are provided in this section. To begin with, a generalized listings of thermal insulation system types and insulation application types are given as follows:

Thermal insulation system (TIS) types:

- Vacuum jacketed, independent
- Vacuum jacketed, integral
- Vacuum panel
- Cryopumped vacuum panels/jacket
- Non-vacuum

Insulation application types:

- Evacuated (vacuum-jacketed)
- Evacuated panels
- Bulk-fill
- Spray-applied
- Panel-applied
- Modular boxes
- Combination systems
- Hybrid systems

Material listings are divided into two categories: insulation materials and structural-thermal materials. Structural-thermal materials (STMs) are predominately load-carrying materials but with low thermal conductivity relative to true structural materials such as stainless steel (SST) and other metals and alloys. In this case, the distinction between insulation and ST materials is established according to a thermal conductivity of approximately 30 mW/m-K: that is, materials below this value are considered insulation materials, and materials above this value are considered ST materials. The rationale is that only insulation materials will be considered for the broad area insulation service, while ST materials are still of critical importance for the complete TIS design. Listings of a variety of different insulation materials and STM are given as follows:

Insulation materials:

- Glass bubble bulk-fill such as K1
- Perlite powder bulk-fill
- Cellular glass foam blocks such as Foamglas
- Aerogel particle bulk-fill such as Lumira
- Aerogel blanket composite (ABC) such as Cryogel, Spaceloft, and Pyrogel
- Polyisocyanurate spray-on foam insulation (SOFI)
- Polyisocyanurate (polyiso) and polyurethane foam (PUF) panels
- Phenolic and other rigid foam panels
- Vacuum insulation panels (VIPs)
- Multilayer insulation (MLI)
- Layered composite insulation (LCI)
- Layered composite extreme (LCX)
- AeroFoam molded shapes
- Combination materials and stack-ups
- Hybrid composite materials

Structural-thermal materials:

- Glass fiber-reinforced plastic (GFRP) composites: G10 and Durostone
- Polytetrafluoroethylene (PTFE) such as Teflon and other fluoropolymers
- Polychlorotrifluoroethylene such as Kel-F or Neoflon
- Ultra-high molecular weight polyethylene (UHMWPE)
- Polymethyl methacrylate (PMMA) acrylic such as Plexiglas
- Polycarbonate such as Lexan
- Polyetherimide (PEI) such as Ultem
- High-density reinforced polyurethane foam
- Polymethacrylimide (PMI) foam
- Polyimide aerogel composites such as AeroFoam
- Composite aerogel panels such as AeroFiber and AeroPlastic
- Balsa wood
- Plywood

The approach to high-efficiency, lowest practical heat leak tank designs is underscored by the counterintuitive theme that "insulating is not about insulation": insulating is about gaining some level of control over the state of the liquid. Insulating is not about minimizing heat leak: insulating is about maximizing control and efficiency. Put another way, insulating is about efficiency in energy and operation (which also has a strong bearing on safety). Of course, there is a threshold of heat leak above which there is no control at all, and the designed heat leak must be well below this level. This threshold is usually expressed in terms of heat flux (q) in W/m^2 for the prescribed boundary temperatures and environment.

TABLE 2.1

Evaluation of Insulation Materials/Systems

Material/System	*k_e [mW/m-K] for different †CVP			Bulk Density	Form	Load Bearing
	HV	SV	NV	kg/m³		
Glass bubble bulk-fill	0.7	1.7	26	65	fluid/granular	yes
Perlite powder bulk-fill	1	3.8	35	132	granular	no
Cellular glass foam blocks	n/a	n/a	32	118	rigid	yes
Aerogel particle bulk-fill	1.7	4.3	14	80	granular	yes
Aerogel blanket composite	1.5	2.9	11	133	blanket	yes
Fiberglass blanket or glass wool	1.9	7.3	26	16	blanket	no
Polyiso spray-on foam insulation	7.8	18	21	42	rigid (spray)	no
Polyiso/polyurethane foam panels and reinforced foams			24		rigid panel	yes
Phenolic and other rigid foam panels			20		rigid panel	no
Vacuum insulation panels		3.6			rigid panel	yes
Multilayer insulation	<0.1	3	18	45	blanket	no
Layered composite insulation	0.1	0.6	16	50	blanket	yes
Layered composite extreme	0.1	1	14	80	blanket or panel	yes
AeroFoam blocks			30		rigid panel	yes

*Boundary temperatures 293 K/78 K; residual gas nitrogen.

†HV is <1 millitorr; SV is 100 millitorr; NV is 760,000 millitorr [note: 1 millitorr = 0.1333 Pa].

While materials are of course a key piece of the puzzle, there are many more pieces to achieve the total system, or thermal insulation system. Crucial facets of the design approach are that the TIS must be integral with mechanical structural design as well as with the processes involved in tank operation.

A preliminary evaluation of different insulation materials/systems is presented in Table 2.1. The effective thermal conductivity (k_e) and bulk density values are typical and based on experimental test data for real systems. However, it must be understood that wide variations exist for specific changes in bulk density, composition, manufacturing form, aging, environmental exposure, and layer density (to name a few). See Section 3.3.2 for more details.

The major factor in thermal performance is the level of vacuum of the entire thermal insulation system under steady-state operating conditions, or the cold vacuum pressure, as defined in ASTM C1774—Standard Guide for Thermal Performance Testing of Cryogenic Insulation Systems. The optimum CVP depends on the materials involved and how they are arranged and installed. The practicality of achieving a desired CVP for a given TIS depends on manufacturing, fabrication, installation, verification, maintenance, monitoring, and a host of other factors, chief among them being the cost (both time and money).

For simplicity and first-order comparison among the very wide range of insulation systems given here, the effective thermal conductivity data are divided, as relevant for a given material/system, into three categories of CVP: high vacuum [<1 millitorr], soft vacuum [~100 millitorr], and no vacuum [760 torr]. The big difference between the SV range and the HV range is the cross-over into the free-molecular gaseous heat conduction that occurs at about 50 millitorr depending on the physical size and geometry of the system. Evacuation to ~100 millitorr is accomplished by relatively simple vacuum pumping equipment and processes, while achieving <1 millitorr is much more technically difficult, especially for large-scale systems.

2.3 CALCULATIONS, TESTING, AND MATERIALS

Engineering analysis and thermal testing go hand in hand for determining the best insulation system for a given design situation. Conducting analysis and calculations according to standard methods is essential for fair comparison of different materials and accurate applications of results, as detailed by Fesmire [6]. Testing involves the measure of the total heat flow rate through an insulation system under relevant conditions. Relevant conditions typically include a large temperature difference (ΔT) and a controlled, steady-state test environment or vacuum level. The specific vacuum level is the key driver for thermal performance of any system.

2.3.1 CALCULATIONS OF HEAT TRANSMISSION

Calculations for cryogenic insulation systems follow the guidance given in ASTM C1774, Standard Guide for Thermal Performance Testing of Cryogenic Insulation Systems. The heat flow rate (Q), also known as "heat leak," is calculated as given in Eq. 2.1, and the effective thermal conductivity (k_e) is further calculated in Eq. 2.2.

$$Q = \dot{m}h_{fg} = k_e A_e \frac{\Delta T}{\Delta x} \qquad \left[\frac{J}{s} \text{ or } W\right] \tag{2.1}$$

$$k_e = \frac{Q}{A_e}\frac{\Delta x}{\Delta T} \qquad \left[\frac{mW}{m-K}\right] \tag{2.2}$$

where $\dot{m}$ is the mass flow rate in g/s, h_{fg} is the heat of vaporization in J/g, A_e is the effective area for heat transmission in m², ΔT is the temperature difference in K, and Δx is the thickness in m. For flat plate test apparatuses, A_e is constant through the thickness of the thermal insulation system. For cylindrical apparatuses or spherical geometries, A_e is the log-mean area between the inner and outer diameters of the insulation test specimen. Q is obtained directly by boil-off calorimetry, while other methods are indirect, measuring a combination of electrical power and temperatures.

Reducing the total heat transfer is the primary objective. The challenge in most cryogenic thermal insulation systems is that in many cases, there several different modes of heat transfer, each having a significant effect that must be addressed. The four modes of heat transfer are: solid conduction, convection, gaseous conduction, and radiation. Convection is not a part of vacuum-based systems but does have to be considered for evacuation or loss of vacuum processes. After solid conduction is reduced to its practical minimum, radiation becomes the main mode of heat transfer in vacuum-based systems.

This real-world heat flow rate includes all modes of heat transmission (radiation, solid conduction, gaseous conduction, and convection). While solid conduction is usually fixed by the system design, the test environment determines the relative amounts of the other components of the total flow of heat. The test environment is defined by the cold vacuum pressure (that is, the vacuum level under steady-state operating conditions of the system) and the composition of the residual gas. From the total heat flow rate, the heat flux (q) is calculated; which is the heat flow rate, under steady-state conditions, through a unit area, in a direction perpendicular to the plane of the thermal insulation system, as given in Eq. 2.3.

$$q = \frac{Q}{A_e} \qquad \left[\frac{W}{m^2}\right] \tag{2.3}$$

2.3.2 OVERVIEW OF TESTING OF CRYOGENIC INSULATION SYSTEMS

Standard laboratory cryostats in both cylindrical (Cryostat-100) and flat-plate (Cryostat-500) geometries are used for testing of various insulation materials and systems. Both are steady-state liquid nitrogen boil-off calorimeters designed for absolute thermal performance measurements. Test conditions include a cold boundary temperature of 78 K and a warm boundary temperature of 293 K for a temperature difference (ΔT) of approximately 215 K. The vacuum level in millitorr (μ), or CVP, is set via a residual flow of nitrogen (or other gas) with continuous pumping and an electronic control-loop to maintain the desired pressure.

Cryostat-100 has a vertical cylindrical cold mass assembly that is 1 m tall by 167 mm in diameter (see Fesmire [7]), while the Cryostat-500 has a horizontal flat plate cold mass assembly of 204 mm diameter (see Fesmire [8]). The Macroflash (Cup Cryostat) is also a flat-plate instrument for small specimens of 76 mm diameter as described by Fesmire, Bateman, and Thomas [9]. For high-vacuum tests (<0.1 μ), each test specimen is heated and evacuated according to standard laboratory procedures. The test methodology follows the guidance of ASTM C1774, Annex A1 and Annex A3, respectively.

From the direct experimental test results for heat flux, different heat transmission values can be calculated. The point is to gauge the transmission of heat from a defined "hot surface" to a defined "cold surface" according Fourier's theory of heat. Although there does not exist the convenience of intrinsic or constant thermal properties of materials in the context of these complex insulation systems, calculation of "thermal conductivity" values is useful for the following reasons: comparison of different systems tested under similar conditions, comparison of the same systems but for varying thickness, to analyze temperature dependence and extend to different boundary temperatures, and to enable analysis and modeling for specific engineering design cases.

Two key definitions from ASTM C1774 are used. The effective thermal conductivity (k_e) is defined as the thermal conductivity value through the total thickness of the insulation test specimen (one material, homogeneous or non-homogeneous, or a combination of materials) between specified boundary temperatures and in a specified environment. The system thermal conductivity (k_s) is defined as the thermal conductivity value through the total thickness of the insulation test specimen and all ancillary elements such as packaging, supports, seams, joints, getter packages, enclosures, and so on—in other words, k_s is always determined using the total distance between the hot and cold surfaces. The k_e or k_s, as applicable, and q are calculated for the standard ΔT of 215 K, unless otherwise noted. Each test is for a given CVP environment of nitrogen gas as indicated. Physical characteristics include the as-tested measures of thickness and density.

2.3.3 Overview of Insulation Material Data

Data in terms of k_e for a variety of insulation materials and systems, tested under representative and similar conditions, are presented as a variation on CVP in Figure 2.1. Included are two MLI systems, one with 80 layers of aluminum foil and micro-fiberglass paper and another with 20 layers of double-aluminized Mylar (DAM) and polyester netting (see Fesmire and Johnson [4]). Two different layered composite insulation systems are given: LCI with a fumed silica and DAM (see Fesmire et al. [10]) and LCI with aerogel paper (0.7 mm) and DAM (see Fesmire et al. [11]). Data for an aerogel blanket material (Cryogel by Aspen Aerogels) are presented by Fesmire et al. [12]. Two bulk-fill materials are given: aerogel particle (white, 1 mm diameter) material by Cabot Corp. and glass bubble (Type K1) material by 3M (see Scholtens et al. [3]).

For general comparison, the data for "vacuum only" and a polyisocyanurate foam (BX-265) by NCFI are also given by Fesmire and Coffman et al. [1]. The boundary temperatures are 293 K (WBT) and 78 K (CBT) in all cases. The physical parameters of thickness, number of layers, and bulk density are also provided in the legend. Table 2.2 lists select data for the CVP points of high vacuum, moderate vacuum, and no vacuum.

2.3.4 Structural-Thermal Material Data

Structural-thermal materials also play an important role in the design and execution of thermal insulation systems. Such materials are needed for stand-offs, supports, straps, bands, transition joints,

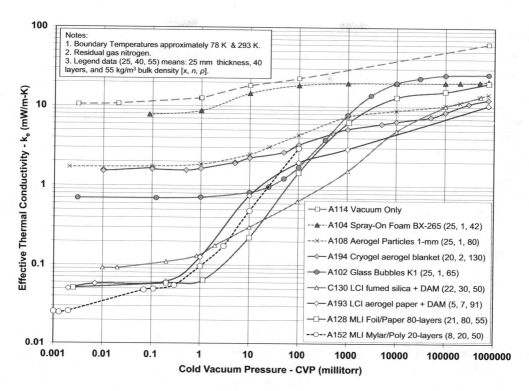

Notes:
1. Boundary Temperatures approximately 78 K & 293 K.
2. Residual gas nitrogen.
3. Legend data (25, 40, 55) means: 25 mm thickness, 40 layers, and 55 kg/m³ bulk density [x, *n*, *ρ*].

—□— A114 Vacuum Only
--▲-- A104 Spray-On Foam BX-265 (25, 1, 42)
--✕-- A108 Aerogel Particles 1-mm (25, 1, 80)
—◇— A194 Cryogel aerogel blanket (20, 2, 130)
—●— A102 Glass Bubbles K1 (25, 1, 65)
—△— C130 LCI fumed silica + DAM (22, 30, 50)
—◇— A193 LCI aerogel paper + DAM (5, 7, 91)
—□— A128 MLI Foil/Paper 80-layers (21, 80, 55)
--○-- A152 MLI Mylar/Poly 20-layers (8, 20, 50)

FIGURE 2.1 Cryostat test data for thermal insulation systems tested at 293 K/78 K.

TABLE 2.2
Select Thermal Conductivity Data for Cryogenic Thermal Insulation Systems

Thermal Insulation System	Ref. No.	†Density kg/m³	CVP μ	*k_e mW/m-K
Glass bubble type K1	A102	65	<0.1	0.70
			100	1.7
			760,000	26
Aerogel particles (1 mm diameter)	A108	80	<0.1	1.7
			100	4.3
			760,000	14
LCI with fumed silica and Mylar	C130	50	<0.1	0.09
			100	0.64
			530,000	13.4
MLI 80 layers foil and paper	A128	55	<0.1	0.051
			100	1.5
			760,000	20
MLI 20 layers Mylar and poly net	A152	50	<0.1	0.026
			100	3.0
			760,000	18 (est.)

†As tested.
*For boundary temperatures of 293 K and 78 K.

TABLE 2.3

Thermophysical Data for Structural-Thermal Materials Used in Cryogenic Systems

Material	$^\dagger\sigma$ MPa	ρ kg/m³	$^*k_e$ mW/m-K	F_{ST} K-m-s/g
G-10 (transverse direction)	448	1,939	467	495
Ultem 2300 glass-filled PEI	221	1500	212	695
Teflon PTFE	24.1	2,120	253	45
Rohacell WF-300 PMI foam (2 psi)	17.8	324	42.1	1,305
Balsa wood (transverse direction)	7.0	166	45.9	919
AeroZero polyimide aerogel	1.6	150	28.1	380
Foamglas cellular glass foam	0.8	118	32.3	210
Divinycell H45 PVC foam (2 psi)	0.6	50	23.8	504
Spray foam polyiso BX-265 (2 psi)	0.4	37	22.6	483

†At ambient temperature.

*Boundary temperatures 293 K/78 K; compressive load 5 psi or as noted.

fasteners, feedthroughs, and other mechanical elements. Thermophysical data for a range of different structural-thermal materials used in cryogenic systems data are given in Table 2.3. Included are polyimide aerogel AeroZero, Ultem, Foamglas, Divinycell, and Rohacell. Also included are data for G10 composite, Teflon, balsa wood, and polyiso spray foam for general reference (see NIST [13] and Flynn [14]). These k_e data were produced using a Macroflash (Cup Cryostat), per ASTM C1774 Annex A4, for boundary temperatures of 78 K (CBT) and 293 K (WBT) and under a compressive load of 34 kPa. The structural-thermal figure-of-merit (F_{ST}) is calculated via Eq. 2.4.

$$F_{ST} = \frac{\sigma}{\rho k_e} \times 10^6 \qquad \left[\frac{K \cdot m \cdot s}{g}\right] \qquad (2.4)$$

where ρ is the bulk density in kg/m³, k_e is the effective thermal conductivity in mW/m-K (with the prescribed CBT and WBT), and σ is the compressive strength in MPa (at ambient temperature) as described by Fesmire, Bateman, and Thomas [9].

Newer materials developed in the last decade include AeroFoam, AeroPlastic, and AeroFiber composites as summarized by Williams and Fesmire [15]. These structural-thermal composites composed of hybrid combinations of aerogels, polyimides, and other engineering polymers further demonstrate the possibilities for tailoring systems to achieve specific combinations of light weight, low thermal conductivity, and high strength, along with excellent fire properties as shown by Smith et al. [16].

2.4 ENGINEERED SYSTEM ANALYSIS APPROACH

To show the interplay among different materials and design approaches for a thermal insulation, Fesmire and Swanger [17] five examples of cryogenic systems, three tanks and two piping systems, are shown in Figure 2.2. In each case, the optimum approach involves a different combination of materials, and for simplicity and comparison, uniform hot and cold surfaces are assumed. The CVP, ranging from high vacuum (<0.1μ), to normal vacuum (~1μ), to moderate vacuum (~10μ), to soft vacuum (~100μ), to no vacuum (760,000μ), are also uniquely representative of actual working systems in each case. Demko et al. [18] present a design tool for heat leak analysis of various geometries including insulation data as a function of CVP.

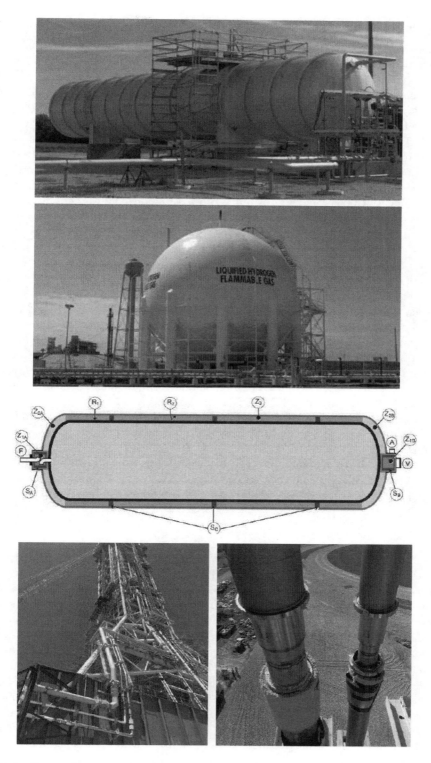

FIGURE 2.2 Cryogenic system examples: 1) cylindrical tank (u. left), 2) large spherical tank (u. right), 3) small composite tank (middle), 4) transfer piping (l. left), 5) piping field joint (l. right). Source: *NASA*

2.4.1 Comparative Analysis of Example Systems

The first example is a 125-m^3 horizontal cylindrical SST tank with a carbon steel outer jacket and a 200-mm annular space. This factory-built, medium-size tank includes an 80-layer MLI system for high vacuum, four G10 pad supports, several piping penetrations, and a 0.6-m-diameter manway for internal access and multiple instrumentation feedthroughs. Boil-off testing using LN$_2$ and liquid hydrogen (LH$_2$) is described by Swanger [19].

The second example is a 3200-m^3 spherical 304 SST tank (30-mm thickness) with a carbon steel outer jacket (17-mm thickness) and a 1200-mm annular space. This tank includes 40 vertical SST support rods connecting the inner vessel to the outer jacket, four piping penetrations on the bottom, and a 0.5-m-diameter manway for internal access. The specified boil-off rate is 0.075% per day (LH$_2$) with the original perlite power insulation [36]. In this example, the insulation system is glass bubbles, which provides an estimated 46% reduction in boil-off compared to perlite at a moderate vacuum level of 10 millitorr based on field test data by Sass [20]. An even larger LH$_2$ storage sphere (4,700 m^3), using the glass bubble insulation system, is being constructed at the Kennedy Space Center to supply future space launch programs as described by Fesmire et al. [21].

The third tank example is a 100-liter carbon composite tank in a 4:1 cylindrical geometry with an SST outer jacket and an annular space of 23 mm thickness. The LCI system is designed for soft vacuum of 100μ for low cost and to avoid vacuum maintenance. Three polyimide aerogel rings support and isolate the inner vessel, and a single 12.7-mm-diameter piping penetration is also included as depicted for this generalized storage tank concept as presented by Adams et al. [22].

In addition to the three tanks, two piping systems are also analyzed. Example four is a DN 25 × 80-mm all-SST vacuum-jacketed (VJ) pipe segment of 18.3-m length. The ends are thermally guarded with LN$_2$ so that an absolute boil-off test is achieved. The insulation is 20 layers of MLI (aluminized Mylar and paper), with a CVP in the normal vacuum range of approximately 1μ (see Fesmire, Augustynowicz, and Nagy [23]).

Finally, example five is a DN 250 × 300-mm, double-walled field joint connection (1 m overall length) between two VJ pipe segments. The ends, 425-mm-length cones, are treated as the supports, while a straight middle portion is 150 mm in length. The insulation system is a complete bulk-fill by aerogel particles (1 mm nominal diameter) of the sealed annular space, which is designed for non-vacuum. The space is optionally backfilled with CO$_2$ to achieve enhanced performance at a cryo-pumped soft vacuum (~100μ) under cold operating conditions as shown by Johnson et al. [24].

2.4.2 The Insulation Quality Factor in System Design

The insulation materials chosen and the data given represent preferred solutions in the relative sense among a vast landscape of different applications. There are many more details and material choices to be made for an optimum solution in the final design, but the basic approach to thermal insulation system design, in an integral manner, is outlined by Fesmire and Swanger [17]. The following examples give first-order approximations and show the interplay among materials, design, installation, manufacturing, and operation.

There is another potentially crucial part of the Q_{total} for a cryogenic system: the additional heat leak due to the insulation quality factor. The IQF is first-order means of capturing the combined degradation of the insulation's thermal performance due to practical limitations of its installation and the negative effects of supports and piping. A high-fidelity IQF can only be determined experimentally, but estimates can be obtained via a combination of historical information and analytical methods. As warranted by the design objectives and fidelity required, an individual IQF can be applied to different zones of insulation area, to each support structure, and to each penetration.

For the most basic case, a single IQF can be applied for the entire cryogenic assembly as a separate heat leak element of Q_{total}, as given in Eq. 2.5.

$$Q_{total} = Q_i + Q_s + Q_p + Q_{IQF} \quad [\text{W}] \tag{2.5}$$

where Q_i is the heat leak for the insulation, Q_s is the heat leak attributed to the supports, Q_p is the heat leak due to the piping penetrations, and the additional Q_{IQF} is the heat leak associated with the IQF. Breakdowns of the heat leaks for the five example systems are summarized in Table 2.4. The k_s is also calculated for each case.

The best thermal insulation system under idealized laboratory conditions is obviously not always the best one to use in the real world. In the example of the field joint, most of the heat leak is fixed by solid conduction (75% of Q_{total}), and the annular space is mostly inaccessible for wrapping with MLI. Added to the geometric constraints are the difficulties regarding field fabrication, for which achieving a long-term high-vacuum in the annular space is not a reasonable expectation. With the option of adding a CO_2 backfill to achieve an effect of cryopumping to a soft vacuum, the Q_{total} drops to 30.6 W and the k_s to 8.1 mW/m-K, which would be comparable to a high-vacuum MLI system in this case.

In another example, evacuated aerogel compression packs or panels can be manufactured which allow for wrapping around a pipe and inserting within an outer pipe or structural containment as shown by Koravos [25]. After installation, the vacuum pack can be allowed to pressurize resulting in a mechanically strong system. The additional compressive loading on the aerogel actually improves the thermal insulation performance as the interstitial space for the air is reduced.

These cases represent idealized solutions for each situation. The Q_{IQF} ranges from zero to 34% of the total. In many real cases, the Q_{IQF} can be as high as or higher than the Q_i. For example, Johnson et al. [26] show that a small single penetration (12 mm diameter) through an MLI blanket results in an Q_{IQF} that is from one to five times the Q_i. Even in bulk fill systems, the two instances of zero IQF in the two cases given, there can be a problem. Two identical-design 3,200-m^3 VJ LH$_2$ storage tanks were built at Kennedy Space Center in 1965, and one ended up with more than triple the heat leak of the other (3.8 vs 1.1 m^3 per day boil-off losses). Krenn [27] shows that this problem was eventually determined to be due to a large void within the perlite annular space filling. Therefore, the Q_{IQF} is not necessarily just a deviation from the ideal or a degradation of the insulation material that can be mitigated but a substantial heat leak potential that must be addressed up front in the design process.

The elements are highly interdependent, showing that the analysis of the total heat leak is an iterative process. The right design approach depends on the shape, size, operational duty cycle, and of course economic objectives.

2.4.3 METHODOLOGY AND KEY TO SUCCESS

A practical methodology for designing cryogenic thermal insulation systems is based on a "cold triangle" approach of insulation, supports, and piping, as well as the insulation quality factor. The total heat leak of the end product is what matters, but to minimize this heat leak in the most cost-effective way, the calculations, testing, and materials data must be covered and the thermal performance understood as a summation of its parts. The approach also provides a basis for evaluating performance benefits of new materials and analyzing the cost effectiveness in overall system design.

Energy efficiency, system control, and operational safety are interrelated aspects of deciding the best thermal insulation system for a given application. Emerging cryofuel enterprises, including liquefied natural gas (LNG) and LH$_2$, are a particular challenge due to the transient operational processes to be addressed and the competitive economic targets to be met. New ASTM standards such as C1774 and C740 are part of providing thermal performance data and material specifications for the cryogenic industry. Benchmark thermal performance data can be used to calibrate comparative instruments and support detailed studies of insulation system designs for specific applications. Future technical consensus standards are envisioned for both test methods and material practices to support the cryogenic industry and further the proliferation of new industrial opportunities in the areas of transportation and energy.

TABLE 2.4

Total Heat Flow Rate Analysis for Different Examples of Cryogenic Systems

Case	Cryogenic System Description	Thermal Insulation System	CVP μ	Dimensions t mm	Dimensions *A_e m²	Q_i W	Q_i %	Components of Q_{total} Q_s W	Q_s %	Q_p W	Q_p %	Q_{IQF} W	Q_{IQF} %	Q_{total} W	q W/m²	†k_s mW/m-K
1	Medium tank, VJ, factory built (125 m³ capacity)	MLI (80 layers) for high vacuum	<0.1	250	229	101	33.7	50	17	130	43.3	19	6.3	300	1.3	1.5
2	Large tank, VJ, site-built (3,200 m³ capacity)	Bulk-fill glass bubbles for moderate vacuum	10	1,200	1,194	178	74.1	31	12.9	31	12.9	0	0.0	240	0.2	1.1
3	Small composite tank, VJ, conceptual (0.1 m³ capacity)	LCI for soft vacuum	100	23	1.6	10	68.9	3	21	1.5	10.4	?	?	14.4	9.0	1.0
4	Piping system, DN 25 × 80-mm, with two bayonet joints, factory built (18.3 m length)	VJ, MLI (20 layers) for normal vacuum	1	26	3.2	2.5	12.0	3.2	15	8	38.3	7.2	34.4	20.9	6.5	0.8
5	Piping field joint, DN 250 × 300-mm, double-walled, field fabricated (1 m length)	Bulk-fill aerogel for non-vacuum (optional CO_2 cryopumped)	760,000	50	0.88	26.5	54.1	13	27	9.5	19	0	0.0	49.0	55.7	12.9

*Effective surface area between inner and outer shells (log-mean area).

†$\Delta T = 293$ K − 78 K = 215 K.

2.5 AEROGELS AND AEROGEL-BASED SYSTEMS

Thermal insulation systems for cryogenic applications span a wide range of requirements and call for demanding levels of performance. Aerogels and their composites continue to be developed to take advantage of highly tailorable processing techniques that enable specialized end products and unique thermo-physical properties attributed to their nano-porous internal networks. Aerogel materials have become commercially available in three categories: flexible composite blankets, bulk-fill particles, and layered composite systems. Examples of field applications include LNG (111 K), liquid oxygen (90 K), liquid nitrogen (77 K), cryo-compressed hydrogen (~40 K), and liquid hydrogen (20 K).

2.5.1 AEROGEL MATERIALS

Both bulk-fill and composite blanket type silica-based aerogel systems have been extensively investigated for cryogenic thermal performance across the range of environments from non-vacuum to high vacuum and in the temperature range of from approximately 4 K to 400 K. The research and development of specific applications for bulk fill aerogels on liquid hydrogen tanks are summarized in the literature [28–29]. These results show that the aerogel insulation systems, while breathable and nano-porous, are not characterized by the heat and mass transport mechanisms associated with cellular type materials (open or closed cell foams) but rather by the high degree of steady-state thermalization across the temperature gradient combined with the cryoadsorption capability provided the surface energy of the material. Without the direct interaction between the cold side and the warm side, there is minimal "cryopumping" that can occur at the macroscopic scale, and the breathable aerogel materials can be successfully used as cryogenic insulation even in the case of open atmospheric exposures with a cold boundary of 4 K or 20 K.

Six different aerogel materials were tested for cryogenic-vacuum thermal performance as described by Fesmire [11]. Aerogel blanket materials are composite materials that include a fiber reinforcement material in the production process (see Coffman et al. [30]). Two systems are layered composite insulation systems of radiation shield layers in combination with aerogel composite blanket materials. System G1–191 is a layered composite of pairs of ultra-low-density (ULD) aerogel blanket and double-aluminized Mylar film. System A193 is a layered composite of pairs of aerogel composite paper (0.7 mm thick) and double-aluminized Mylar film. The aerogel materials/systems tested, along with their basic physical properties, are summarized in Table 2.5.

Additional insulation materials/systems, as summarized by Fesmire [6], were tested for comparison. The descriptions of these additional test specimens are given in Table 2.6. These materials include "vacuum only" and a closed-cell spray-on foam on one end of the spectrum of thermal performance, followed by a glass bubble system and a baseline of multilayer insulation on the other.

Limitations in using MLI systems are summarized as follows: (1) high vacuum is required for operation (and in the first place, it is not possible to vacuum-jacket all hardware), (2) not all hardware can be suitably wrapped or properly covered, and (3) localized compression will ruin the thermal performance. MLI cannot withstand mechanical loading. Compared to the no-load condition for six different MLI systems tested (average heat flux of 0.6 W/m²), a mere 0.7 kPa (0.1 psi) causes a ~15× increase in heat flux, a small 7-kPa load causes a ~40× increase, and a modest 70-kPa load causes a >100× increase in heat flux as reported by Flynn [14].

Compared to MLI, bulk-fill, or foam material types, aerogel-based systems may be advantageous depending on the vacuum/pressure environment. Other key factors include physical design of system, installation build process, and operational requirements. For non-vacuum applications, an alternative to closed-cell foam is the layered composite extreme system, which takes advantage of the unique nano-porous, compressible, and hydrophobic properties of aerogel blanket material as reported by Fesmire [2]. The breathable LCX system, an MLI for the ambient-air environment, has been proven on operational LH$_2$ systems (20 K). Aerogel blanket materials such as the Pyrogel provide high temperature capability to 923 K (1,200°F) where fire protection might be needed for cryofuel systems.

TABLE 2.5

Physical Properties of Aerogel-Based Test Specimens

Cryostat	Test Series	Test Specimen	No. of Layers	Total Thickness* (mm)	Density* (kg/m³)
C100	A108	[1]Bulk-fill aerogel beads	1	25	80
C100	A111	[2]Pyrogel aerogel blanket (black)	6	18	125
C100	A194	[2]Cyrogel aerogel blanket	2	20	130
C500	G2–109	[2]Spaceloft Subsea (grey)	4	20	152
C500	G1–190	[2]ULD^ aerogel blanket white	8	23	55
C500	G2–113	[2]ULD^ melamine flexible aerogel grey	8	21	65
C500	G1–191	ULD^ Aerogel MLI layered composite	8	23	52
C100	A193	Aerogel MLI layered composite (0.7-mm aerogel paper)	7	5	91

*As tested.
^Ultra-low density.
[1]Cabot Corp. (manufacturer).
[2]Aspen Aerogels, Inc. (manufacturer).

TABLE 2.6

Physical Properties of Additional Insulation Test Specimens for Comparison

Cryostat	Test Series	Test Specimen	No. of Layers	Total Thickness* (mm)	Density* (kg/m³)
C100	A114	Vacuum only (black surfaces)	1	25	n/a
C500	G1–157	SOFI BX-265 (polyiso foam)	1	25	42
C100	A102	Glass bubbles K1	1	25	65
C100	various	Kaganer line (MLI baseline); average of 26 different MLI test specimens	From 10 to 80	~22 typical	~50 typical

*As tested.

2.5.2 Experimental Method and Apparatus for Aerogel Testing

Two cryostat test instruments were used for absolute thermal performance measurements, Cryostat-100 and Cryostat-500 [7, 8]. The cold and warm boundary temperatures for all tests are 78 K and 293 K, respectively. The effective thermal conductivity (k_e in mW/m-K) and heat flux (q in W/m²) are calculated for the standard ΔT of 215 K. The test methodology follows the guidance of ASTM C1774, Annex A1 and Annex A3, respectively [31]. For high-vacuum tests, each test specimen was heated and evacuated according to standard laboratory procedures.

Three additional temperature sensors (Type E, 30 gage thermocouples) are placed within the specimen at specific intervals through the thickness from the cold side to the warm side. Because the heat flow rate through the test specimen is constant for all layers, intermediate thermal conductivity values (λ) can be calculated and reported with the mean temperature (T_m) for each layer. In this way, a single test yields nine λ points for determining the temperature dependence of heat transmission through a test specimen operating between the large ΔT values.

2.5.3 Cryogenic-Vacuum Test Results for Aerogels

The cryostat test results for all test series, that is, for aerogel-based systems compared to other cryogenic insulation systems, are presented in Figure 2.3. Included for comparison are the additional materials/systems as discussed. For all tests, the boundary temperatures are approximately 293 K and 78 K and the residual gas is nitrogen.

In higher vacuum, the lowest k_e systems are the MLI baseline and layered composites of aerogel/MLI. The data for glass bubbles, dominated by gas conduction heat transfer, show the transition from free-molecular to continuum heat transfer at ~50 millitorr. In the soft vacuum region up to ambient pressure, all aerogel materials are superior. The two layered composites (A193 and G1–191) are examples of combining the advantages of two different material systems (aerogel and MLI) for applications requiring tolerance for vacuum degradation and/or mechanical loading.

The results for the six aerogel materials are given in Figure 2.4 for detailed examination. In high vacuum, the ULD aerogel and Spaceloft Subsea Gray gave the lowest k_e. Through the moderate vacuum (from 50 to 1,000 millitorr) and soft vacuum (from 1,000 to 50,000 millitorr) ranges is where some significant differences appear according to the characteristics of the aerogel on the nanoscopic level, the fiber matrix (if any) and physical arrangement on the microscopic level, and the interstitial spaces (if any) on the macroscopic level. Even with this simplistic analytical view, the differences can at least be somewhat resolved by considering the four modes of heat transfer (solid conduction, radiation, gas conduction, and convection) are all in play for this region of CVP.

Additional test data points were taken in all cases to verify the unique shapes of the data for each aerogel material. For example, the aerogel bead specimen has more free interstitial space for gas conduction and convection to occur, while the ULD aerogel specimen has the finest pore size and hence lowest gaseous conduction. For ambient pressure conditions, the two ULD aerogels show the highest k_e, while the twice-heavier Cryogel is superior. Detailed data sets for the Cryogel material, including different gas environments, are given by Fesmire et al. [17].

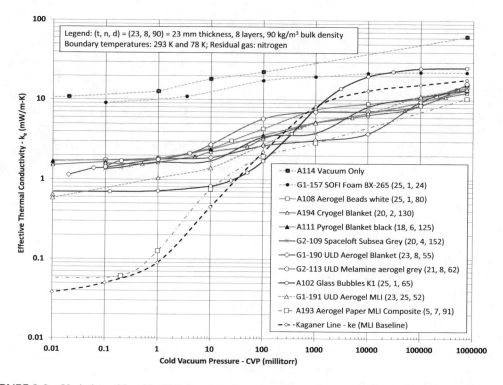

FIGURE 2.3 Variation of k_e with CVP for aerogels compared to other cryogenic insulation systems.

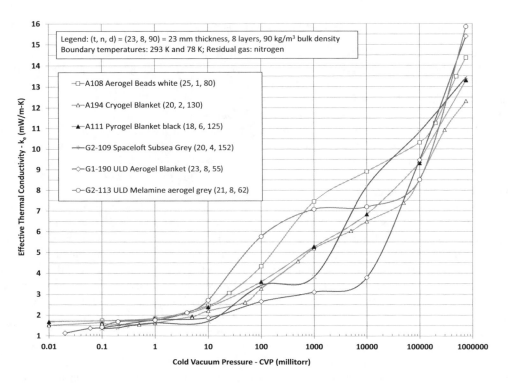

FIGURE 2.4 Variation of k_e with CVP for aerogel materials.

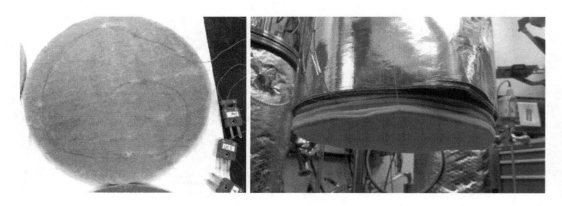

FIGURE 2.5 Preparation and temperature sensor instrumentation for Aerogel disk stack-up test specimen.

Plots of the layer temperature distributions were also produced for all layered test specimens. The key distinction is the progression of inflections in the curves, where the steepest inflection always occurs at the first layer (coldest layer) and under higher vacuum conditions. An example of the preparation and temperature sensor instrumentation for an aerogel disk stack-up type test specimen for the Cryostat-500 is shown in Figure 2.5.

Compared to MLI, bulk-fill, or foam material types, aerogel-based systems may be advantageous depending on the vacuum/pressure environment. Other key factors include: physical design of system; installation build process; and operational requirements. For non-vacuum applications, an alternative to closed-cell foam is the layered composite extreme (LCX) system which takes advantage of the unique nano-porous, compressible, and hydrophobic properties of aerogel blanket material. The breathable LCX system, an MLI for the ambient-air environment, has been proven on operational LH2 systems (20 K). Example applications of LCX systems on both LH2 and LO2 piping systems for a space launch system are given in Figure 2.6 and Figure 2.7. Aerogel blanket materials such as the Pyrogel® provide high temperature capability to 923 K (1200 °F) where fire protection might be needed for cryofuel systems. Another aerogel-based layered composite insulation system for LH2 tanks has also been investigated. The goal of this LCX variant is to solve the problem of "external insulation" on cryogenic upper stages of launch vehicles, such as shown in Figure 2.8, for the keeping of LH2. This complex insulation system must function in all three different environments: ground (moisture, liquid air formation), flight (aerodynamic forces), and space (on-orbit, high-vacuum insulation). The lightweight, robust LCX system addresses this triple problem in a synergetic approach. Cryogenic-vacuum testing shows approximately 50 times better performance (lower heat flux) in vacuum compared to state-of-the-art foam.

FIGURE 2.6 Space Launch System (SLS) cryogenic umbilical systems, LH$_2$ piping and components.

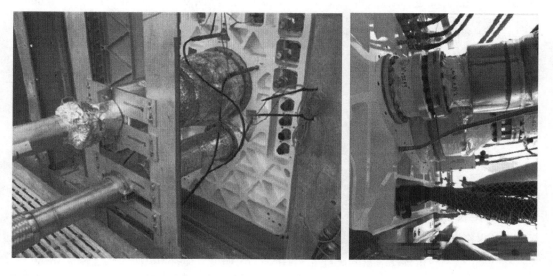

FIGURE 2.7 Custom aerogel bulk-fill system (ground side) and LCX solution (flight side) successfully tested with multiple LH$_2$ operations.

FIGURE 2.8   Custom LCX solution on LO$_2$ umbilical for Space Launch System propellant loading system and potential for future space launch and service vehicles.

TABLE 2.7
Increase in Heat Flux for Increasing WBT

WBT (K)	ΔT	% increase, ΔT	% increase, q	Factor b_w
293	215	baseline	baseline	1.00
305	227	6	14	1.14
325	247	15	32	1.32
350	272	27	46	1.46

Source: For MLI System with Constant CBT = 78 K [4]

2.5.4   THERMAL ANALYSIS OF AEROGELS (ESTIMATING FOR DIFFERENT BOUNDARY TEMPERATURES)

The intermediate or interlayer temperature sensors used in cryostat testing provide additional data to determine the temperature dependence of thermal conductivity within the two prescribed boundary temperatures. The use of three intermediate temperature sensors creates four layers, numbered from one to four, from the cold side. Equations 2.6 through 2.9 give the basic nomenclature:

$$Q = k_e \times A_e \times \Delta T / \Delta x \qquad \text{Fourier equation} \qquad (2.6)$$

$$q = Q / A_e \qquad \text{constant} \qquad (2.7)$$

$$q = q_1 = q_2 = q_3 = q_4 = \lambda_4 \times \Delta T_4 / \Delta x_4 \quad \text{and so forth} \qquad (2.8)$$

$$T_m = \left(T_{colder} + T_{warmer}\right)/2 \text{ or } T_{m4} = \left(T_{c4} + T_{w4}\right)/2 \quad \text{and so forth} \qquad (2.9)$$

The q is constant for a steady-state heat flow rate, and therefore the q through each layer is constant. Because the ΔT and Δx for each layer are known, the λ for each layer can be calculated and reported for the corresponding T_m.

Often an estimate of insulation system thermal performance is needed for specific boundary temperatures. Test data for MLI under high vacuum (<10^{-5} torr) provide a preliminary way of estimating the increases in heat transmission for WBT up to 350 K. Given in Table 2.7 are the increases in heat flux, on a percentage basis, for an average of 12 different MLI systems from a baseline WBT of 293 K. Likewise, Table 2.8 shows the reductions in heat flux for an MLI system with decreasing cold boundary temperatures from a 78 K baseline.

TABLE 2.8

Decrease in Heat Flux for Decreasing CBT

CBT (K)	ΔT	% decrease, ΔT	% decrease, q	Factor b_c
76	224	baseline	baseline	1.00
40	260	16	14	0.86
20	280	25	21	0.79
4	296	32	33	0.67

Source: For MLI System w/Constant WBT = 300 K [5]

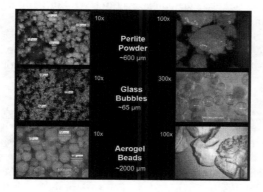

FIGURE 2.9 Microscope comparison of the three bulk fill insulations. Note: Glass bubbles are shown at 300×.

From baseline heat flux (q_{base}) test data at the standard boundary temperatures of 293 K and 78 K, a first-order estimation of the thermal performance for a specific layered system design is calculated using a WBT factor (b_w) and a CBT factor (b_c) as follows:

$$q_{design} = b_c * b_w * q_{base} \qquad (2.10)$$

For example, the heat flux estimate for a system operating at boundary temperatures of 325 K/20 K is approximately the same thermal performance as the baseline of 293 K/78 K (q_{design}=1.32 × 0.79 × q_{base} = 1.04 × q_{base}). The theoretical heat flux is proportional to the ΔT (and T^4 for the radiation portion), but the more important and influential factor is the materials' heat transmission characteristics that occur at the progressively lower temperatures combined with the likely improvement of the level of vacuum.

2.6 BULK-FILL INSULATION MATERIALS

Three different bulk-fill insulation materials—glass bubbles, perlite powder, and aerogel beads—were tested in the Cryostat-100 test apparatus. Microscopic photos of the three materials are shown in Figure 2.9. The glass bubble material is manufactured by 3M under the name 3M Scotchlite Type K1 Glass Bubbles. The perlite powder is an evacuated cryogenic-grade perlite processed by Silbrico Corp. under the trade name Ryolex grade no. 39. The aerogel beads are commercially available from

TABLE 2.9

Installed Densities for the Bulk-Fill Insulation Test Materials in Cryostat-100

Ref. No.	Material	Thickness mm	Mass g	Tap Volume Cm³	Tap Density g/cc
A102	Glass bubbles	25.4	885.6	11,073	0.080
A103	Perlite powder	25.4	1875	11,268	0.166
A108	Aerogel beads	25.4	967	11,268	0.086

Cabot Corporation under the trade name Nanogel. All three materials are a white powder. However, noticeable differences are visible at a microscopic level, as shown in Table 2.8. The morphology, microstructure, and submicroscopic features of a given material determine how heat energy will be transmitted through the material at a particular vacuum level. Further details have been reported in the literature by Kropschot [32] and Fesmire [33–34].

Each material was carefully measured and poured cup by cup into the black sleeve around the cold mass. Thickness and density details are shown in Table 2.9. Each material was heated above 340 K and evacuated to a WVP of approximately 0.13 Pa (1 millitorr) or less prior to the start of testing.

2.6.1 BULK-FILL MATERIAL TEST DATA

Extensive testing of three material systems are summarized here. Additional tests were run without any insulation material inside the black sleeve for "vacuum-only" reference data. The corresponding steady-state run time on Cryostat-100 was over 500 hours. Variation of k_e with CVP is presented in Figure 2.10. The repeatability was within 2%. The variation of heat flux with CVP is given in Figure 2.11. The k_e rises sharply in the soft-vacuum range as gas conduction begins to dominate the heat transfer. The glass bubbles have better thermal performance than perlite at all vacuum levels. The glass bubbles also perform better than aerogel beads from high vacuum to soft vacuum. Above 133 Pa (1,000 millitorr), the aerogel beads perform significantly better than glass bubbles and perlite. The vacuum-only test illustrates how using an insulation material reduces heat flux through the system. These data provide a basis for the heat transfer analysis of materials.

An example test, aerogel beads at 0.13 Pa (1 millitorr), is presented in Figure 2.12. The final k value and heat flux are calculated from the averaged test chamber boil-off flow rate when the liquid nitrogen level is between 88% and 92% full (or between about 17 and 21 hours in the example case). The standard requirement is to achieve a fine thermal equilibrium that is coincident with a liquid level of about 90%. As Figure 2.12 shows, one or two refills are sometimes required beyond the initial cooldown phase.

2.6.2 ANALYSIS AND DISCUSSION OF BULK-FILL MATERIALS

Complementary test methods are necessary for a complete understanding of the heat transfer properties of a given material system and to establish credibility of the heat measurement results. Table 2.10 bears this out by comparing the Cryostat-100 results for glass bubbles, perlite powder, and aerogel beads at high vacuum to the results of Cunnington and Tien [35], the NASA CESAT research testing program [36], the Florida State University team led by Barrios [37], and the original researchers such as Fulk [38]. Extrapolating the data to the Cryostat-100 mean temperature gives a good correspondence of the results.

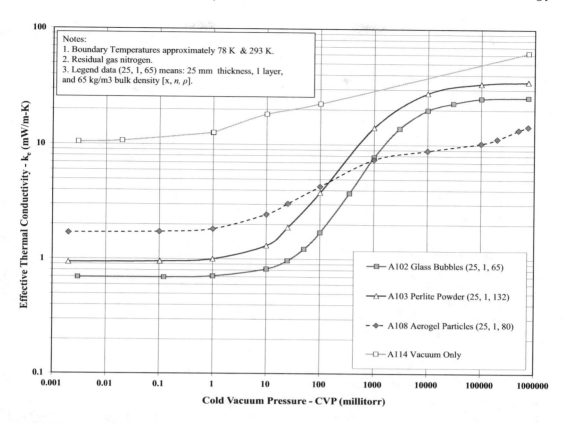

FIGURE 2.10 Variation of apparent thermal conductivity (absolute *k*-value) with cold vacuum pressure (1 millitorr = 0.13 Pa) for bulk-fill cryogenic insulation materials. The boundary temperatures are approximately 78 K and 293 K. The residual gas is nitrogen.

The vacuum test series and the literature were used to determine how the three bulk-fill insulation materials affect the different modes of heat transfer. The total thermal conductivity is the sum of the thermal conductivity of each mode of heat transfer, shown in Equation 2.11,

$$k_T = k_{sc} + k_{gc} + k_{cv} + k_r \tag{2.11}$$

where k_{sc} is the solid conduction, k_{gc} is the gas conduction, k_{cv} is the convection, and k_r is the radiation heat transfer. At high vacuum, gas conduction and convection are considered to be zero, while in this preliminary analysis, solid thermal conduction and radiation are assumed to be constant for all vacuum levels. Cryogenic thermal conductivity data for bulk-fill materials from 180 K down to 10 K is provided by Barrios [37]. Table 2.10 uses the data from Cryostat-100 and extrapolated data from the literature (see Rettelbach [39]) to calculate how each mode of heat transfer contributes to the total *k*-value. Although adding a bulk-fill insulation material to a vacuum system introduces solid conduction, the material's thermal performance is improved by reducing heat transfer due to radiation, convection, and gas conduction.

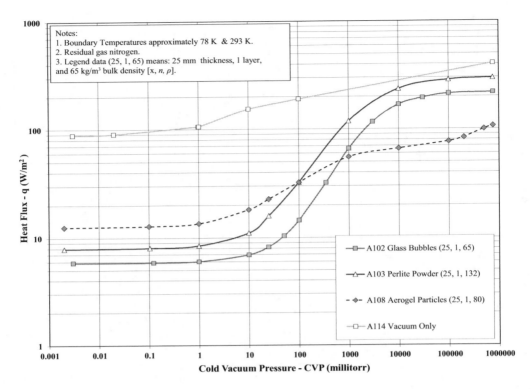

FIGURE 2.11 Variation of mean heat flux with cold vacuum pressure (1 millitorr = 0.13 Pa) for bulk-fill cryogenic insulation materials. The boundary temperatures are approximately 78 K and 293 K. The residual gas is nitrogen.

Opacified aerogel granules (manufactured by Cabot) and aerogel composite beads (manufactured by Aspen Aerogels) are some of the newer bulk-fill materials studied for cryogenic applications. Experiments are needed to examine evacuation and vacuum retention rates, flammability and fire compatibility, and the effects of different residual gases with a focus on nonvacuum cryogenic insulation systems. For example, a pipe test apparatus with a non-evacuated field joint assembly between two vacuum-jacketed sections provided data for laboratory-scale demonstrations and allows for practical solutions to old problems involving insulation with bulk-fill materials (see Johnson [24]). Aerogel-filled field joints were successfully implemented for the NASA Space Launch System launch tower servicing systems for both LH_2 and LO_2 piping [17]. In other examples, evacuated and sealed expansion packs of aerogel granules have been successfully demonstrated for external installation onto cryogenic pipelines for high thermal performance relative to conventional foam insulation (see Koravos [25]).

2.7 GLASS BUBBLE THERMAL INSULATION SYSTEMS

Research and development of the use of glass bubble bulk-fill material as part of a cryogenic tank insulation system began in the early 1970s by Dr. George Cunnington (Lockheed Palo Alto Research Laboratory) and Professor C. L. Tien (University of California, Berkeley) [35]. From the 1970s work through the CryoTestLab work in the 2000s, the goal was to provide a cost-effective, reliable, and higher performance thermal insulation system compared to perlite powder

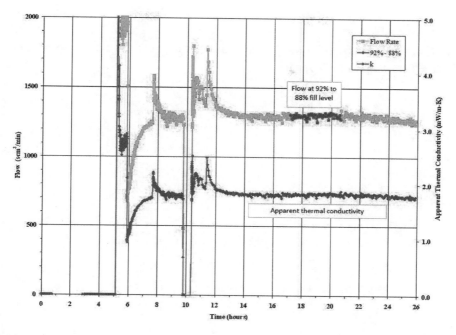

FIGURE 2.12   Example test of aerogel beads at 0.13 Pa (1 millitorr) CVP (A108) shows the nitrogen boil-off flow rate and the calculated k-value. Stabilization at the 90% level occurs at approximately 19 hours.

TABLE 2.10

Comparison of k-Values for Bulk-Fill Insulation Materials at High Vacuum Using Cryostat-100 and the Complementary Test Methods

Material	Test	WBT K	CBT K	T_{mean} K	k-value mW/m-K
	Cunnington & Tien	303	78	191	0.59
Glass bubbles	A102	293	78	186	0.70
	CESAT	293	77	185	0.69*
	CESAT	294	20	157	0.16*
	A103	293	78	186	0.94
Perlite powder	Fulk	304	20	162	0.65
	CESAT	296	20	158	0.53*
	Fulk	76	4	40	0.08
	Fulk	304	76	190	2.10
Aerogel beads	A108	293	78	186	1.71
	Fulk	76	20	48	0.20
	Barrios et al.	35	23	29	0.16

*Calculated k-value from tank testing.

[40]. Perlite is inexpensive, but the total cost of a vacuum-jacketed thermal insulation system is always relatively expensive no matter what filler is put into the annular space. Because 3M began large-scale production decades ago, and for many other applications besides an insulation material, the low-cost K1 product benefits from the economies of scale for common industrial uses.

TABLE 2.11

Calculation of the Heat Transfer Mode Contributions to the Total *k*-Value (*CVP* in millitorr [1 millitorr = 0.13 Pa], *k* in mW/m-K)

	CVP	k_T	k_{sc} [7]	k_r	k_{gc+cv}
	0.01	0.70	0.02	0.68	0.00
	0.1	0.70	0.02	0.68	0.00
	1	0.71	0.02	0.68	0.02
A102 Bubbles	10	0.83	0.02	0.68	0.13
	100	1.70	0.02	0.68	1.01
	1,000	7.76	0.02	0.68	7.06
	10,000	19.58	0.02	0.68	18.88
	100,000	25.13	0.02	0.68	24.43
	760,000	26.12	0.02	0.68	25.42
	CVP	k_T	k_{sc} [11]	k_r	k_{gc+cv}
	0.01	0.94	0.50	0.44	0.00
	0.1	0.95	0.50	0.44	0.01
	1	1.00	0.50	0.44	0.05
A103 Perlite powder	10	1.31	0.50	0.44	0.36
	100	3.83	0.50	0.44	2.88
	1,000	13.99	0.50	0.44	13.05
	10,000	27.85	0.50	0.44	26.90
	100,000	33.60	0.50	0.44	32.66
	760,000	*34.85*	*0.50*	*0.44*	*33.90*
	CVP	k_T	k_{sc}	k_r [12]	k_{gc+cv}
	0.01	1.71	1.43	0.28	0.00
	0.1	1.73	1.43	0.28	0.02
	1	1.83	1.43	0.28	0.12
A108 Aerogel beads	10	2.45	1.43	0.28	0.74
	100	4.33	1.43	0.28	2.62
	1,000	7.44	1.43	0.28	5.73
	10,000	8.89	1.43	0.28	7.18
	100,000	10.29	1.43	0.28	8.58
	760,000	14.34	1.43	0.28	12.63
	CVP	k_T	k_{sc}	k_r	k_{gc+cv}
	0.01	10.44	0.00	10.44	0.00
	0.1	10.66	0.00	10.44	0.22
A114 Vacuum only	1	12.54	0.00	10.44	2.10
	10	18.16	0.00	10.44	7.72
	100	22.44	0.00	10.44	12.00

Glass bubbles have use in both vacuum and non-vacuum systems. But in a system evacuated to moderate vacuum levels, glass bubbles have the best thermal performance of all bulk-fill materials. Glass bubbles have been successfully used in vacuum panels, complex cryostats, small dewars, tanker trailers, and large storage tanks.

For large-scale spherical LH$_2$ storage tanks (that is, above about 400-m^3 or 106,000-gallon capacity), the evacuated perlite powder system has been used for the last 60 years. Large-scale LH$_2$ tanks are site built and can therefore be of a spherical design. More recently, glass bubble thermal insulation systems have proved a cost-effective and reliable alternative to evacuated perlite powder systems [21]. The driver for this adoption has been substantial energy/product savings. Perlite powder is the main target for replacement with glass bubbles. Installations can be new construction or retrofits. The structural and load-carrying ability of panels that include glass bubble core materials have also been developed, including panels, piping, and tank applications of double-walled constructions to enable new lightweight energy-efficient designs.

2.7.1 MATERIAL TESTING AND THERMAL PERFORMANCE DATA

Extensive cryostat data on the thermal conductivity and total system thermal performance of glass bubble-based thermal insulation systems under cryogenic-vacuum conditions are available for the temperature range of 20 K to 300 K. Understanding the intricacies of the heat transfer mechanisms and their interactions through the vacuum pressure range is key. Glass bubbles are better than perlite powder through the entire vacuum pressure range. This performance increase is mainly through the additional radiation scattering afforded by the walls of the bubbles in combination with the low thermal conduction by virtue of the point contacts among adjoining spheres. A micrograph photo of K1 glass bubbles (average diameter of 65 microns) is given in Figure 2.13.

The vacuum sensitivity of glass bubbles vs perlite was investigated in detail and was shown to provide a very positive cost-efficiency benefit. Large vacuum-jacketed perlite tanks around the world are designed for high vacuum (below 0.1 millitorr) but typically operate at up to 10 to 50 millitorr (degraded or partial vacuum) in the real world. The performance benefit (boil-off reduction) is then not 35% (laboratory) but approximately 50% (field) or better. The data in Figure 2.14 are from tests using the Cryostat-100 insulation test instrument and include thermal performance for perlite powder, glass bubbles, and several different multilayer insulation systems for general reference.

In Figure 2.15 is provided a detailed evaluation of the thermal performance of glass bubbles versus perlite powder in the moderate vacuum range. This moderate vacuum range of approximately 1 millitorr to 100 millitorr of main interest for industrial applications, particularly those involving larger-scale LH2 storage tank systems. Thermal performance test data from extensive LH2 boiloff

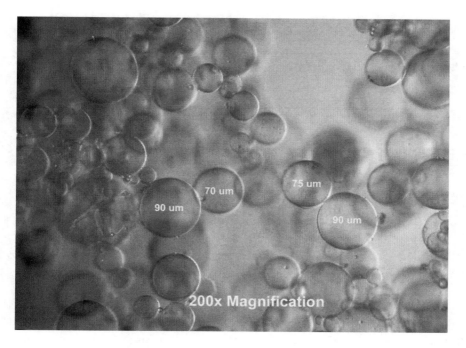

FIGURE 2.13   Micrograph photo of glass bubbles, type K1 by 3M. The average diameter is 65 microns.

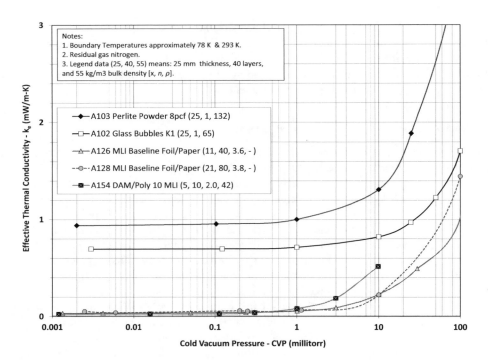

FIGURE 2.14 Effective thermal conductivity of perlite compared to bubbles under laboratory conditions using Cryostat-100. Bubbles have the lowest thermal conductivity for any evacuated bulk-fill material. Bubbles are also less sensitive to vacuum degradation compared to perlite.

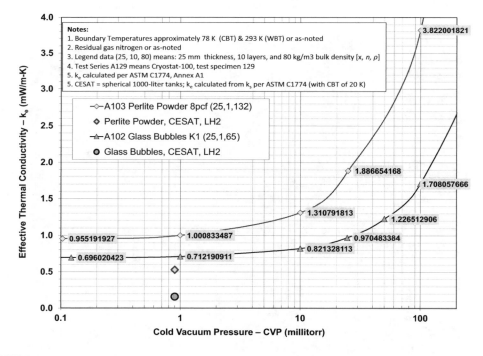

FIGURE 2.15 Thermal performance evaluation of glass bubbles compared to perlite powder in the moderate vacuum range of main interest for industrial applications.

testing at 20 K is also summarized in Figure 2.15. These data are from the CESAT project using two each 1000-liter spherical cryostats for the side-by-side testing of both glass bubbles and perlite powder under identical conditions.

Extensive mechanical testing was performed to verify by scientific material tests, prototype tank tests, and field tests that the bubbles do not break to any significant amount. The cases include vibration, compaction, installation, and thermal cycling [41–42]. This fact was established mainly for spherical tanks, with some work also applied for cylindrical vertical and horizontal tanks. Granular mechanics analysis and modeling of glass bubbles and systems was also performed in concert with the prototype and lab testing. Testing included vibration settling, compaction due to thermal cycling, moving wall tests, and vertical column tests [36]. The combined thermal-mechanical performance of glass bubbles for VJ tank insulation was demonstrated in the following cases: full-scale field demonstration on 50,000-gallon spherical liquid hydrogen tank at Stennis Space Center [20], research testing using two 1,000-liter spherical tanks (1/15th scale models of 850,000-gallon tanks at LC-39) with LH_2 and LN_2 [43]. and industry field demonstration on a pair of 6,000-gallon LN_2 tanks [44].

The field test facility is a 190-m^3 (50,000-gallon) LH_2 tank at Stennis Space Center with a long prior history of excellent vacuum maintenance and a top-performing track record with pristine, top-specification perlite powder that was removed and replaced with 3M K1 series glass bubbles. Three complete thermal cycles were completed over a six-year period with no vacuum problems and an average 46% reduction in boil-off as summarized by Fesmire [40].

The result of a 46% reduction in boil-off over six years of field demonstration surpasses the ~35% reduction (compared to perlite powder) found through years of extensive laboratory and sub-scale testing with liquid nitrogen (LN_2) and LH_2. Glass bubbles are far less sensitive to vacuum-level degradation compared to perlite powder, making the field result even better than the idealized laboratory result. A view of the loading process for the glass bubble insulation is shown in Figure 2.16.

FIGURE 2.16   Field demonstration in a 190-m^3 LH_2 storage sphere at NASA Stennis Space Center. Shown is the process of loading glass bubbles into the annular space of the VJ tank.

2.8   FIBERGLASS INSULATION SYSTEMS

While MLI systems can provide the ultimate thermal insulating capability, overall system design and operational factors usually limit complete utilization of their effectiveness. Fiberglass insulation composed of low outgassing micro-fibers provides effective high-performance capability in vacuum as well. Vacuum level requirements are considerably less strict for fiberglass, providing cost advantages both in manufacturing and life cycle for cryogenic vessels and piping. Installation around piping, structural supports and other complex geometries can be readily accomplished using fiberglass. Thermal performance tests were conducted to determine optimal placement and use of MLI (foil and paper type) and Cryolite micro-fiberglass blanket within cryogenic storage systems [45]. These tests indicated that it is preferential to use the blanket on the cold side of the MLI and that placing radiation shields within several blankets of Cryolite drastically improves thermal performance of the insulation system at higher vacuum levels. These results can be used to define or optimize future system design and construction techniques.

Evaluation of additional variations in MLI and Cryolite combinations was performed. A practical benefit of incorporating the Cryolite as part of an MLI-based high-vacuum system is twofold. First, the Cryolite layers can allow for better evacuation between layers. Second, the mechanical elasticity (spring effect) offers protection to the MLI layers to minimize compression and edge effects. The total thermal performance of the insulation system must be considered along with the mechanical performance advantages to determine the most effective system for a given vessel or piping application.

The MLI materials used in this study consist of layers of aluminum foil (7.2 microns thick, with emissivity of 0.03) separated by a micro-fiberglass paper spacer (Cryotherm 243, 12 g/m^2). The materials can be applied separately but are preferably collated and applied from a single roll (CRS-Wrap). The blanket material is a 25-mm-thick micro-fiberglass blanket of density 16 kg/m^3 (Cryolite). Photographs of the materials are shown in Figure 2.17. The removable cold mass assembly of Cryostat-2 is placed on a wrapping machine for precise control during installation of all materials and temperature sensors. A combination, or hybrid, system of fiberglass and aluminum foil layers is shown in Figure 2.18.

2.9   FOAM INSULATION SYSTEMS

Spray-on foam insulation materials were first developed in the 1960s and included applications for the Saturn V Moon rockets and subsequently the Space Shuttle. Other cryogenic tank insulation materials used in the past include rigid foams and cork. There are two main regions of interest for space launch and exploration applications: ambient pressure (non-vacuum) and high vacuum. Although the non-vacuum application is of primary interest, SOFI applications for in-space cryogenic storage are also of interest. In such cases where the cryogenic tanks are loaded on the launch pad, a composite system of materials that offers both non-vacuum and high-vacuum performance must be devised. Figure 2.22 illustrates the extreme thermal performance difference between these two cases. Even though aerogel blankets may be twice as good as SOFI at the ambient pressure, the performance of a typical multilayer insulation system is 100 times better. However, MLI is not practical for use in an open ambient, humid-air environment, while the aerogel blanket is fully hydrophobic and well-suited for such exposures. Therefore, the proper design and selection of insulation materials depend strongly on the environmental parameters.

Foam materials were sprayed at NASA's Marshall Space Flight Center in accordance with nominal Space Shuttle external tank flight specifications. Baseline (new condition) specimens were allowed to cure for approximately one month. Several 610- × 610-mm samples of the three types of foam, NCFI 24–124 (acreage foam used on external tanks), NCFI 27–68 (formerly proposed alternative acreage foam), and BX-265 (closeout foam), were packaged and shipped to Kennedy Space Center (KSC). The 610- × 610-mm samples of foam were then machined into 203-mm-diameter test specimens of 25 or 32 mm thickness. The measured densities of all specimens were comparable, ranging from 37 to 40 kg/m^3 prior to testing.

FIGURE 2.17 Photographs of micro-fiberglass spacer for MLI (Cryotherm). Left: aluminum foil (CRS-Wrap), Center: micro-fiberglass blanket material (Cryolite), and Right: MLI wrapping machine.

FIGURE 2.18 Cryolite and foil configuration.

Environmental exposure of the SOFI materials was an integral part of the performance testing. The aging and weathering simulations were performed at two exposure sites. The KSC vehicle assembly building (VAB) was used for the aging test (see Figure 2.23). The aging site simulates the area where the external tank is stored before the vehicle stacking and launch preparations begin. The conditions in this area are mild with ambient humidity levels and no direct sunlight. The corrosion beach site at KSC was used for the weathering test, as shown in Figure 2.24. The weathering

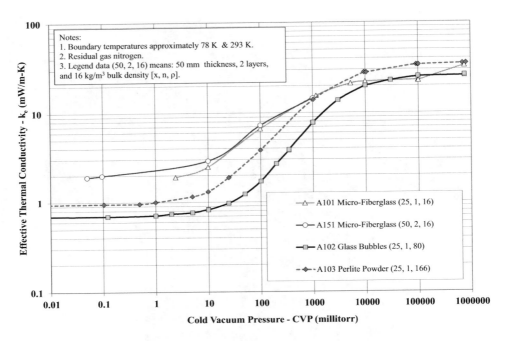

FIGURE 2.19 Cryolite micro-fiberglass Cryostat CS-100 thermal performance data. Cryogenic-vacuum thermal performance data for Cryolite fiberglass in comparison with bulk-fill insulation materials (perlite powder and glass bubbles) is presented in Figure 2.19. The cryostat data for fiberglass in comparison with a fiberglass and foil hybrid system is given in Figure 2.20. Finally, thermal conductivity (k) data for the fiberglass blanket is given in Figure 2.21 for the full range of vacuum and three different mean temperatures.

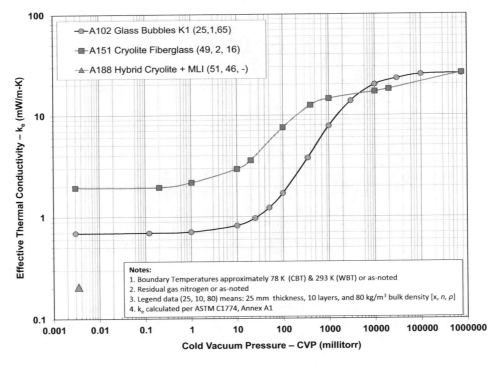

FIGURE 2.20 Cryostat CS-100 test data for Cryo-Lite fiberglass. Comparison with glass bubbles and hybrid Cryo-Lite/MLI system. Cryo-Lite micro-fiberglass, 16 kg/m³, one layer (25 mm) by Lydall Performance Materials.

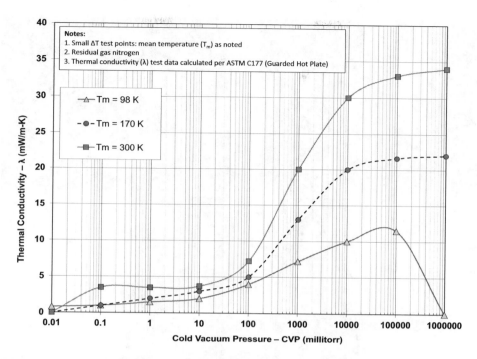

FIGURE 2.21 Guarded hot plate (C177) test data for Cryolite fiberglass. k-values (not k_e) given vs vacuum level for three different mean temperatures (T_m). Cryolite micro-fiberglass 16 kg/m³, one layer (25 mm) by Lydall Performance Materials.

site simulates the conditions that a mated external tank is exposed to while it is on the pad awaiting launch. The aging time for an external tank can be several years, whereas the typical weathering exposure is about one month. The maximum weathering exposure, as recorded in the Shuttle flight history, is about six months. The test specimen mounting enclosure, shown in Figure 2.22 (right), was designed to expose only the center portion of the top face of the material. Figure 2.22 (right) shows a SOFI test specimen type NCFI 24–124 after only one month of weathering. Exposure time durations are as follows: baseline, 2 weeks, 1 month, 3 months, 6 months, 12 months, 18 months, and 2 years.

In addition, a total of five 1-m-long cylindrical clamshell test articles were produced for testing by means of a cylindrical thermal performance test instrument (Cryostat-100). The specifications for these cylindrical test articles of approximately 25 mm thickness are given in Table 2.12. Photographs of one of the BX-265 test articles are shown in Figure 2.23. The test articles composed of SOFI material NCFI 24–124 include a specimen that is machined (shaved) and one that is in the net spray condition (rind).

The Cryostat-100 instrument uses the steady-state liquid nitrogen boil-off (evaporation rate) calorimeter method to determine effective thermal conductivity and heat flux. Preparations of the test article included installing temperature sensors through the thickness of the insulation at approximately 6.4-mm intervals. The two halves of the SOFI test article were secured tightly in place by using a series of four SST band clamps, as shown in Figure 2.24.

Cryogenic thermal performance testing included a total of over 100 tests of four baseline (new condition) test articles using Cryostat-100 (absolute effective thermal conductivity, k_e or k-value) [1]. A graphical summary of the results for three machined test articles is presented in Figure 2.25. The k_e for the baseline foam materials ranged from 21.1 mW/m-K at ambient pressure to approximately 7.5 mW/m-K at high vacuum. Of the two acreage foams, the alternate acreage foam NCFI 27–68 showed the lowest thermal conductivity overall, with a k_e of 20.7 mW/m-K at 760 torr and 7.3

FIGURE 2.22 Left: Weathering of SOFI test specimens exposure testing at the corrosion beach site of the Kennedy Space Center. Right: A one-month weathered specimen.

TABLE 2.12
Specifications of 1-m Long Cylindrical Clam Shell SOFI Test Articles

Ref. No.	Material	Outer Surface	Date Sprayed	T$_{FINAL}$ mm	Density kg/m^3
n/a	BX-265	Shaved	3/31/2005	24.1	44
A104	BX-265	Shaved	4/6/2005	26.7	42
A107	NCFI 24-124	Rind (partial)	5/11/2005	23.9	38
A105	NCFI 24-124	Shaved	6/9/2005	25.6	38
A106	NCFI 27-68	Shaved	7/6/2006	24.4	37

mW/m-K at 0.5 millitorr cold vacuum pressure. The closeout foam BX-265 recorded the lowest k_e at ambient pressure, while the acreage foam NCFI 27–68 showed the lowest k_e under vacuum conditions. Effective thermal conductivities for the three foams at ambient pressure were found to be in a close range from 20.9 to 21.1 mW/m-K.

The ambient thermal conductivities (λ), obtained using the heat flux meter in accordance with ASTM C518, ranged from 19 (baseline) to 32 mW/m-K (aged). The mean temperature (T_m) was approximately 297 K, and the pressure was atmospheric for all tests. It is important to recall that the ΔT for the ambient test method is only about 20 K, while the ΔT for the cryogenic test method is 215 K. The results for aging of BX-265 and NCFI 24–124 are given in Figure 2.26.

New information on the intrusion of moisture into SOFI under large temperature gradients was also produced through a related study [46]. Experimental data for a significant number of test

FIGURE 2.23   One of five 1-m-long cylindrical clamshell test articles were produced for testing using a cylindrical thermal conductivity test instrument (Cryostat-100). Machined thickness of approximately 25 mm is shown in the left view, while the precise longitudinal fit-up is shown in the right view.

specimens were produced and the overall trends are clear. The details of the cryogenic moisture uptake apparatus, and complete test results are reported elsewhere. SOFI materials were found to gain an extraordinary amount of water weight during a cryogenic propellant loading (cold soak) period. The cryogenic moisture uptake (CMU) results are expressed in terms of percentage weight gain for foam materials of similar density. Moisture uptake for the NCFI 24–124 specimens (acreage foam) averaged 30% for the baseline condition and 88% after three months of weathering. Moisture uptake continued to increase for both aging and weathering. The weight increase of SOFI was also found to be additive for three consecutive cryogenic thermal cycles, or simulated tanking operations, resulting in as much as 167% weight gain for the one-month weathered acreage foam, as shown by the example in Figure 2.27. The boundary conditions are 295 K and 78 K, with a 90% humid environment on the warm side. Further investigation and analysis are needed to understand the cryogenic moisture uptake phenomenon, determine the distribution and morphology of the moisture through the thickness of the foam, and enable improved engineering analysis of the propulsion performance of space launch vehicles.

The cryogenic thermal performance of cellular glass foam materials has also been investigated using different measurement techniques and methodologies [47]. Various approaches are needed depending on the size and shape of the test specimens. The machinability and stability of the completely closed cell material is also important for standardized, comparative testing among different laboratories and techniques for providing reference data.

FIGURE 2.24 Installation of a SOFI clamshell test article on the cold mass assembly of thermal conductivity instrument Cryostat-100.

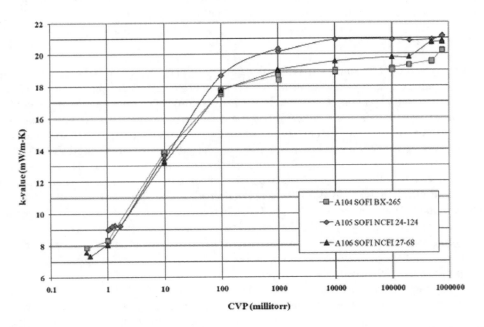

FIGURE 2.25 Variation of effective thermal conductivity (k_e) with cold vacuum pressure for three different SOFI materials. Boundary temperatures are approximately 293 K and 78 K. The residual gas is nitrogen.

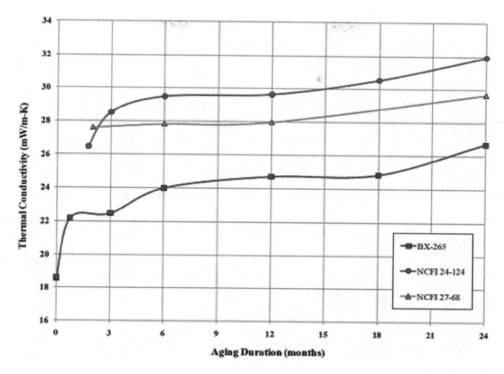

FIGURE 2.26 Variation in ambient thermal conductivity (λ) with aging duration for SOFI materials. The mean temperature (T_m) is approximately 297 K, and the pressure is atmospheric.

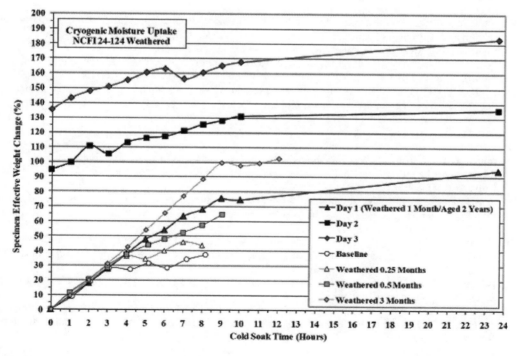

FIGURE 2.27 Example of weight gain due to cryogenic moisture uptake in spray-on foam insulation (9). Days 1–3 simulate consecutive days of launch vehicle cryogenic tank loading and draining operations.

REFERENCES

1. Fesmire J., Coffman B., Meneghelli B., and Heckle K., 2012. Spray-on foam insulations for launch vehicle cryogenic tanks. *Cryogenics*, Vol. 52, 251–261.
2. Fesmire J., 2015. Layered composite thermal insulation system for non-vacuum cryogenic applications. *Cryogenics*, Vol. 74, 154–165.
3. Scholtens B., Fesmire J., Sass J., and Augustynowicz S., 2008. Cryogenic thermal performance testing of bulk-fill & aerogel insulation materials, in *Advances in Cryogenic Engineering*, Vol. 53A, AIP, New York, pp. 152–159.
4. Fesmire J. and Johnson W., 2018. Cylindrical cryogenic calorimeter testing of six types of multilayer insulation systems. *Cryogenics*, Vol. 89, 58–75.
5. Augustynowicz S.D., Fesmire J.E., and Wikstrom J.P., 2000. *Cryogenic Insulation Systems*. 20th International Congress of Refrigeration, no. 2000–1147, IIR, Paris.
6. Fesmire J., 2015. Standardization in cryogenic insulation systems testing and performance data. *Physics Procedia*, Vol. 67, 1089–1097.
7. Fesmire J., Johnson W., Meneghelli B., and Coffman B., 2015. Cylindrical boiloff calorimeters for testing of thermal insulations, in *Advances in Cryogenic Engineering*, IOP Publishing Ltd, *IOP Conf. Ser.: Mater. Sci. Eng.*, 101 012056.
8. Fesmire J., Johnson W., Swanger A., Kelly A., and Meneghelli B., 2015. Flat plate boiloff calorimeters for testing of thermal insulation systems, in *Advances in Cryogenic Engineering*, IOP Publishing Ltd, *IOP Conf. Ser.: Mater. Sci. Eng.*, 101 012057.
9. Fesmire J., Bateman C., and Thomas J., 2020. Macroflash boiloff calorimetry instrument for the measurement of heat transmission through materials, in *Advances in Cryogenic Engineering*, IOP Publishing Ltd, *IOP Conf. Ser.: Mater. Sci. Eng.*, 756 012008.
10. Fesmire J., Augustynowicz S., and Scholtens B., 2008. Robust multilayer insulation for cryogenic systems, in *Advances in Cryogenic Engineering*, Vol. 53B, American Institute of Physics, New York, pp. 1359–1366.
11. Fesmire J.E., 2018. Aerogel-based insulation materials for cryogenic applications, in *International Cryogenic Engineering Conference*, Oxford University, United Kingdom, IOP Conference Series 502.
12. Fesmire J., Ancipink J., Swanger A., White S., and Yarbrough D., 2017. Thermal conductivity of aerogel blanket insulation under cryogenic-vacuum in different gas environments. in *Advances in Cryogenic Engineering*, IOP Publishing Ltd, *IOP Conf. Series: Mater. Sci. Eng.*, 278 012198.
13. National Institute of Standards (NIST), 2021. *Applied Chemicals and Materials Division*, Properties of solid materials from cryogenic to room temperatures, Boulder, CO, USA.
14. Flynn T., 2005. Cryogenic engineering. *Marcel Dekker*, pp. 467, 530.
15. Williams M. and Fesmire J., 2016. *Aerogel Hybrid Composite Materials: Designs and Testing for Multifunctional Applications*. NASA Tech Briefs Webinar.
16. Smith T.M., Williams M.K., Fesmire J.E., Sass J.P., and Weiser E.S., 2009. Fire and engineering properties of polyimide-aerogel hybrid foam composites. *Fire and Polymers V*, Vol. 10, 148–173, ACS.
17. Fesmire J., and Swanger A., 2019. *Advanced Cryogenic Insulation Systems*, 25th IIR International Congress of Refrigeration, Montreal, Canada, IIR No. 1732.
18. Demko J.A., Fesmire J.E., and Augustynowicz S.D., 2008. Design tool for cryogenic thermal insulation systems, in *Advances in Cryogenic Engineering*, Vol. 53A, AIP, New York, pp. 145–151.
19. Swanger A.M., 2018. *Large Scale Cryogenic Storage with Active Refrigeration*. Master's Thesis, University of Central Florida.
20. Sass J., Fesmire J., St. Cyr W., Lott J., Barrett T., and Baumgartner R., Glass bubbles insulation for liquid hydrogen storage tanks, in *Advances in Cryogenic Engineering AIP Conference Proceedings*, Vol. 1218, AIP, New York, pp. 772–779.
21. Fesmire J., Swanger A., Jacobson J., Notardonato W., 2022. Energy efficient large-scale storage of liquid hydrogen. in *Advances in Cryogenic Engineering, Cryogenic Engineering Conference,* IOP Publishing Ltd, *IOP Conf. Series: Mater. Sci. Eng.*
22. Adams J., et al., 2018. *Integrated Insulation System for Cryogenic Automotive Tanks (iCAT)*. United States: N. p., Web doi:10.2172/1457536.
23. Fesmire J., Augustynowicz S., and Nagy Z., 2004. Thermal performance testing of cryogenic piping systems, in *21st International Congress of Refrigeration, Washington, DC*, International Institute of Refrigeration, Paris.
24. Johnson W., Fesmire J. and Meneghelli B., Cryopumping field joint can testing, 2012, in *Advances in Cryogenic Engineering*, AIP Conference Proceedings, Vol. 1434, AIP, New York, pp. 15–27.

25. Koravos J., Miller T., Fesmire J. and Coffman B, 2010. Nanogel aerogel as a load bearing insulation material for cryogenic systems, in *Advances in Cryogenic Engineering, AIP Conference Proceedings,* Vol. 1218, AIP, New York, pp. 921–927.

26. Johnson W.L., Heckle K.W., and Hurd J., 2014. Thermal coupon testing of load-bearing multilayer insulation, in *AIP Conference Proceedings,* Vol. 1573, No. 1, AIP, New York, pp. 725–731.

27. Krenn A., 2012. Diagnosis of a poorly performing liquid hydrogen bulk storage sphere, in *Advances in Cryogenic Engineering, AIP Conference Proceedings,* Vol. 1434, AIP, New York, p. 376.

28. Fesmire J., 2006. Aerogel insulation systems for space launch applications, *Cryogenics,* Vol. 46, Issue 2--3, 111–117.

29. Fesmire J., and Sass J., 2008. Aerogel insulation applications for liquid hydrogen launch vehicle tanks. *Cryogenics,* Vol. 48, 223–231.

30. Coffman B.E., Fesmire J.E., Augustynowicz S.D., Gould G., White S., 2010. Aerogel blanket insulation materials for cryogenic applications, in *Advances in Cryogenic Engineering, AIP Conference Proceedings,* Vol. 1218, AIP, New York, pp. 913–920.

31. ASTM C1774, 2013. *Standard Guide for Thermal Performance Testing of Cryogenic Insulation Systems.* ASTM International, West Conshohocken, PA, USA.

32. Kropschot R.H., and Burgess R.W., 1963. Perlite for cryogenic insulation, in *Proceedings of the 1962 Cryogenic Engineering Conference,* ed. K.D. Timmerhaus, University of California, Los Angeles, CA, pp. 425–436.

33. Fesmire J.E., and Augustynowicz S.D., 2004. Thermal performance testing of glass microspheres under cryogenic-vacuum conditions, in *Advances in Cryogenic Engineering 49,* edited by J. Waynert et al., Plenum, New York, pp. 612–618.

34. Fesmire J.E., Augustynowicz S.D., and Rouanet S., 2002. Aerogel beads as cryogenic thermal insulation system, in *Advances in Cryogenic Engineering 47,* edited by S. Breon et al., Plenum, New York, pp. 1541–1548.

35. Cunnington G.R., and Tien C.L., 1977. Apparent thermal conductivity of uncoated microsphere cryogenic insulation, in *Advances in Cryogenic Engineering 22,* edited by K.D. Timmerhaus, Plenum, New York, pp. 263–271.

36. Fesmire J.E., Sass J.P., Nagy Z.F., Sojourner S.J., Morris D.L., and Augustynowicz S.D., 2008. Cost-efficient storage of cryogens, in *Advances in Cryogenic Engineering,* Vol. 53B, American Institute of Physics, New York, pp. 1383–1391.

37. Barrios M.N., Choi Y.S., and Van Sciver S.W., Thermal conductivity of powder insulations below 180K, in *Advances in Cryogenic Engineering,* to be published.

38. Fulk M.M., 1959. Evacuated Powder Insulation for Low Temperatures, in *Progress in Cryogenics 1,* Academic Press, New York, pp. 65–84.

39. Rettelbach T., Sauberlich J., Korder S., and Fricke J., 1995. Thermal conductivity of IR-opacified silica aerogel powders between 10K and 275K. *Journal of Physics D, Applied Physics,* Vol. 28, 581–587.

40. Fesmire J., 2017. Research and Development History of Glass Bubbles Bulk-Fill Thermal Insulation Systems for Large-Scale Cryogenic Liquid Hydrogen Storage Tanks, Cryogenics Test Laboratory, NASA Kennedy Space Center.

41. Fesmire J.E., Morris D.L., Augustynowicz S.D., Nagy Z.F., and Sojourner S.J., 2006. Vibration and thermal cycling effects. on bulk-fill Insulation materials for cryogenic tanks, in *Advances in Cryogenic Engineering,* Vol. 51B, American Institute of Physics, New York, pp. 1359–1366.

42. Werlink R.W., Fesmire J.E., and Sass J.P., 2012. Vibration considerations for cryogenic tanks using glass bubbles insulation. *Advances in Cryogenic Engineering, AIP Conference Proceedings,* Vol. 1434, 265–272.

43. Sass J.P., Fesmire J.E., Nagy Z.F., Sojourner S.J., Morris D.L. and Augustynowicz S.D., 2008. Thermal performance comparison of glass microsphere and perlite insulation systems for liquid hydrogen storage tanks, in *Advances in Cryogenic Engineering,* Vol. 53B, American Institute of Physics, New York, pp. 1375–1382.

44. Baumgartner R.G., Myers E.A., Fesmire J.E., Morris D.L., and Sokalski E.R., 2006. Demonstration of microsphere insulation in cryogenic vessels, in *Advances in Cryogenic Engineering,* Vol. 51B, American Institute of Physics, New York, pp. 1351–1358.

45. Kogan A., Fesmire J., Johnson W., and Minnick J., 2010. Cryogenic vacuum thermal insulation systems, in *Proceedings of the Twenty-Third International Cryogenic Engineering Conference and International Cryogenic Materials Conference,* Wroclaw University of Technology, Wroclaw, Poland.

46. Fesmire J., Williams M., Smith T., Coffman B., Sass J., and Meneghelli B., 2012. Cryogenic moisture uptake in foam insulation for space launch vehicles. *Journal of Spacecraft and Rockets*, Vol. 49, No. 2, 220–230.

47. Demko J., Fesmire J., Johnson W., and Swanger A., 2014. Cryogenic insulation standard data and methodologies, in *Advances in Cryogenic Engineering, AIP Conference Proceedings*, Vol. 1573, AIP, New York, pp. 463–470.

3 Multilayer Insulation Systems

James E. Fesmire, Quan-Sheng Shu, and Jonathan A. Demko

3.1 INTRODUCTION TO MULTILAYER INSULATION SYSTEMS

Multilayer insulation (MLI) systems are also known as superinsulation or evacuated reflective insulation. These systems are characterized by a number of radiation (or reflective) shields separated by low thermal conductivity spacers within an evacuated environment (about 1×10^{-5} torr). MLI systems are used when lower heat leakage rates than those obtained with other evacuated insulations, for example, perlite powder, glass bubbles, foams, aerogels, are required. Besides, MLI systems have a mass that is significantly lower, require much less insulation space between warm and cold boundaries, and are better suited for complicated geometries of cold masses compared to bulk-fill insulation systems. Therefore, MLI systems have a wide range of various applications, as partly shown in Figure 3.1: storage/transfer of cryogenic liquids, keeping the mass of superconducting (SC) devices cold, thermal protection in space exploration, and others.

As shown by the thermal performance data, a high vacuum level of less than 1×10^{-5} torr is needed for optimum MLI performance. Real-world systems operating at less than ideal high vacuum levels will have commensurately degraded thermal performance. For practical engineering conditions, there are inevitably several adverse issues that will seriously damage MLI performance. These unexpected issues include non-ideal layer density, cracks/slots/seams in the MLI system, penetration through the MLI system, and so on.

With the baseline of data established in comparison with other cryogenic insulation systems, this chapter first examines further details of MLI materials, MLI systems operating at low temperatures, and thermal radiation shielding for efficient use of cooling power. Theoretical calculations of MLI performance and comparisons with experimental data are introduced. Investigations of the impacts of unexpected engineering issues on MLI performance, as well as techniques to remedy them, are also discussed in detail.

3.1.1 WHAT IS THE BEST MLI?

The question of what the best MLI system is often asked in one form or another. Reducing the heat load into the system is the ultimate goal. An unacceptable heat load can cost too much money over time, cost too much time in operations, or render a system inoperable. Thus, there are many more questions to ask to determine the right thermo-economic approach and then the "best" MLI system. Material selection and design are crucial, but the two main factors in the end result are almost always level of vacuum (including the level of vacuum between the layers) and layer density. In turn, the layer density depends on the materials used, the design, and the installation. Experimental baseline data then become paramount for design and analysis of new cryogenic equipment that will rely on an MLI system of some kind.

3.1.2 ADVANTAGES AND APPLICATIONS OF MLI SYSTEMS

Heat flux values for properly designed and well-executed MLI systems can be less than 1 W/m² in high vacuum under those same conditions. Choosing a suitable thermal insulation system may be dictated by the value of the cryogenic fluid being isolated or by mass or thickness limitations imposed by the application. Also, an MLI system will typically have a mass that is significantly

DOI: 10.1201/9781003098188-3

lower, require much less insulation space between warm and cold boundaries, and be better suited to complicated geometries of cold masses in comparison with bulk-fill insulation systems. Standardization in methodologies for the testing and evaluation of MLI systems has included test apparatus, design approaches, and analytical techniques with a focus on the total cryogenic-vacuum system performance [1–4].

Uses of MLI systems generally fall into the following categories: (1) storage, transfer, and processes of cryogenic liquids (O_2, N_2, H_2, He, etc.); (2) keeping the internal masses cold in superconducting applications (SC magnets, SC cavities, SC cables etc.); (3) thermal protection in space cryogenic explorations; and (4) enabling low-temperature research for many laboratories (physics, biology, materials, chemistry, etc.). MLI systems are crucial to preserve the cryogen and enable the safe control of systems that range from fueling operations to scientific instruments to future transportation (as depicted by some examples in Figure 3.1). Understanding the nature and requirements of the specific application should guide the design, analysis, and testing of a given MLI system.

3.1.3 THERMAL PERFORMANCE TEST DATA

Over 100 different MLI systems have been tested, representing about 1,000 individual long-duration boil-off tests. From this data library of years of different research and projects, a select collection of 29 thermal performance data sets are presented herein, along with discussion of the key factors and sensitivities in MLI system design parameters. These data sets are the initial anchor for the international standard ASTM C740 *Standard Guide for Evacuated Reflective Insulation in Cryogenic Service*, first published in 2014, and provide a resource for comparison of different MLI system designs [5].

Comprehensive heat flux (q) and effective thermal conductivity (k_e) data for 26 different MLI systems are given for the full range vacuum-pressure (from high vacuum to ambient pressure). Six tested types are categorized as follows: Mylar/paper (MP), foil/paper (FP), Mylar/net (MN) and Mylar/net blanket (NB), Mylar/silk net (SN), Mylar/fabric (MF), and Mylar/discrete (DX). Vacuum-only (VO) data are also included for four different cold mass surface finishes. The Kaganer k-line, a composite of all 26 MLI systems, from 10 to 80 layers, is presented as a benchmark curve of k_e showing the relative trends across the vacuum pressure range. Benchmarking the results for heat flux, a baseline group of 13 MLI systems, from 20 to 60 layers, is also presented. This baseline group defines the Augustynowicz q-Band, which represents the mean heat flux +/–50%.

3.1.4 VACUUM-PRESSURE DEPENDENCY

Vacuum-pressure dependencies are important for new applications but also provide key insight for thermal modeling and analysis in highly mixed-mode heat transfer situations. MLI systems are used when heat flux values well below 10 W/m^2 are needed for an evacuated design. For the boundary conditions of 300 K and 78 K in high vacuum, heat flux values of around 0.3 W/m^2 have been achieved by a number of different MLI systems. For comparison among different systems, as well as for space and weight considerations, calculation of the effective thermal conductivity (k_e) of the system is important. Values of k_e of less than 1 mW/m-K are typical, and values as low as 0.02 mW/m-K are reported here (0.00014 Btu•in/h•ft2•°F or R-value 7,200). These systems are typically used in a high-vacuum environment (evacuated), but soft-vacuum or no-vacuum environments are also applicable to some designs or contingency situations. A welded metal, vacuum-jacketed (VJ) enclosure is often used to provide the vacuum environment.

High-performance cryogenic MLI systems are crucial for reducing the heat load in many different applications, including storage and transfer operations, cooling of scientific instruments, space launch and in-space propulsion, life-support, electrical power, and many other high energy-intensive systems. The question of what the best MLI system is starts with knowledge of the available

FIGURE 3.1   Cryogenic MLI systems are important for cryogenic storage and transfer, scientific instrument cooling, spacecraft propulsion, transportation of all kinds, and many other important applications.

thermal performance data. Material selection and design are important, but the two main factors are almost always level of vacuum and layer density. As these factors are highly interdependent, the experimental baseline data then become paramount for design and analysis of new cryogenic systems with the optimum thermo-economic performance.

3.2   MLI AND VACUUM

MLI systems are generally used in high-vacuum (HV) environments. The vacuum environment may be a vacuum vessel or jacket or the vacuum environment of space such as Earth orbit or the lunar surface. However, MLI systems can also be designed for soft-vacuum (SV) applications such as industrial products and commercial building equipment. No-vacuum (NV) applications can be important for loss-of-vacuum or contingency performance; cold boxes, terminations, or other purged enclosures; and space launch ground-hold operations. No-vacuum (ambient pressure) applications also include moderate temperature (non-cryogenic) thermal protection systems where radiation heat transfer is a significant amount of the total heat gain; examples include cold chain shipping and distribution of perishable products.

Another important consideration is loss of vacuum. This subject alone is wide ranging according to the system requirements and resultant effects. Loss of vacuum (LOV) may have to do with the degradation of vacuum over long periods of time. Vacuum degradation is normal and usually planned for by allowing for conservative margins in the system heat load. Vacuum degradation is often mitigated using adsorbents on the cold side and hydrogen getters on the warm side. Loss of vacuum in a sudden fashion must be adequately understood from the safety standpoint, including the sizing of pressure relief systems for the tanks, vessels, or cryostats. A further possibility could be sudden loss of vacuum (SLOV), which may invoke the addition of secondary insulation systems on the exterior or burst disk devices for pressure relief.

All the thermal data presented here are for steady-state conditions using laboratory apparatus [6–8]. Transient effects such as sudden loss of vacuum in a vacuum-jacketed system or the venting process during the ascent of a space launch vehicle are not addressed. However, because all vacuum-jacketed systems and even some space-based systems do not have perfect vacuum conditions, adequate description of the real-world thermal performance of MLI systems must necessarily include the heat flow as a function of the vacuum level [9, 10]. In addition, there are also specific design applications for partial vacuum conditions, including, for example, vacuum-jacketed composite tanks, vacuum-sealed panels, reflective foam composite systems, and the Mars surface environment [11–15].

3.3   MLI MATERIALS

3.3.1   System Variation with Different Reflectors and Spacers

System variations include different reflector materials (for example, aluminum foil and double aluminized Mylar), different spacer materials (for example, polyester fabric, Dacron net, micro-fiberglass paper, and discrete spacers), number of layers from one 1 to 80, layer densities from 0.5 to 5 layers per millimeter, and installation techniques such as layer by layer, blankets (multilayer assemblies), sub-blankets, seaming, butt joining, spiral wrapping, and roll wrapping. For example, photographs of blanket-type MLI test specimens being installed on a cylindrical calorimeter test apparatus are shown in Figure 3.2.

Typical spacer and reflector materials used for the MLI systems tested are listed in Table 3.1. Reflector materials include aluminum foil (Foil), 7 micron thickness, and double-aluminized Mylar (Mylar), 6 micron thickness with 400 angstroms coating thickness. Spacer materials include micro-fiberglass paper (Lydall Cryotherm-243, 25 micron thickness) [12?], polyester non-woven fabric (Fabric), Dacron polyester net (Net), silk net, and discrete molded plastic spacers. All materials used are available from commercial sources.

By definition, MLI systems have multiple layers of reflectors separated by low thermal conductivity spacer material in any number of different combinations. Reflector materials depend on the low emissivity characteristic of clean, polished metal surfaces. The metal can be a sheet of foil, or it can be a coating or deposition onto an appropriate nonmetal. The two most commonly used

TABLE 3.1

Typical Spacer and Reflector Materials Used in Experimental MLI Systems*

Category	Material	Description	Thickness, mm
Reflector	Mylar	Double-aluminized Mylar; 400 angstroms aluminum	0.006 (¼ mil)
	Foil	Aluminum foil; type 1145–0; 99.45% purity	0.007 (7 micron)
	Paper	Micro-fiberglass paper; Lydall Cryotherm-243	0.025 (25 micron)
	Fabric	Polyester non-woven fabric	0.08
Spacer	Net	Polyester netting; Dacron B4A and B2A	0.02
	Silk net	Silk netting	0.02
	Discrete	Discrete molded plastic	1.8

*Gold-coated Mylar in special space missions, crinkled Mylar without spacer, and Mylar with small pin-holes for pumping are also studied.

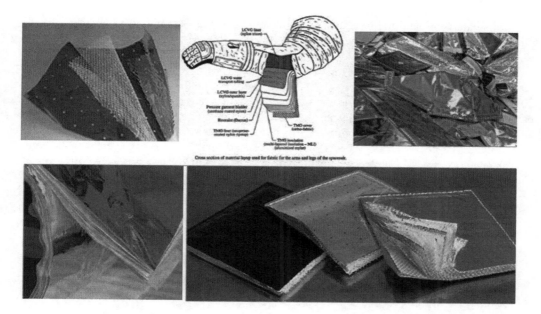

FIGURE 3.2 Samples of various MLI materials and the combination of MLI materials for spacesuits.

reflector materials are (1) thin aluminum foil and (2) vapor-deposited aluminum on a polymeric film. The polymeric film is commonly composed of polyester (such as Mylar) or polyimide (such as Kapton). Some MLI designs do not include spacer materials but instead rely on reducing thermal contact between layers by crinkling the reflector material before installation or reflectors that are metalized only on one side.

Because the reflector materials must be thin and highly reflective, the foils are usually high-purity metals with high thermal conductivity. Metals such as gold and silver can be used, but the usual choice is aluminum because of cost. The most commonly used aluminum foil is 1145–0. This material has 99.45% purity, is soft, and can be obtained in thin sheets. Metalized non-metals include aluminized polyester film. Various sheet materials may be vapor deposited (in vacuum) with aluminum or gold to a thickness sufficient for optical opacity.

Spacer sheets used between the reflectors need to have a very low thermal conduction in the direction perpendicular to the sheet. Typical examples of such spacers are silk net, polyester net, polyester non-woven cloth, fiberglass woven cloth, fiberglass mats, fiberglass paper, and rayon fiber paper. Fibers or strands only a few micrometers in diameter or less are desirable in order to reduce thickness, reduce thermal conduction, and present many points of contact.

3.3.2 CLASSICAL THERMAL PERFORMANCE OF MLI SYSTEMS

Prior to detailed discussions of various design, performance, and technical-economic aspects of MLI systems, two classical performance curves of MLI systems are first introduced with examples, as illustrated in Figures 3.3 and 3.4.

First, an MLI system of 40 layers of Al-foil and paper with boundary temperatures of 78 K and 293 K and nitrogen as the residual gas was tested, and the heat flux through the MLI as a function of vacuum levels is presented in Figure 3.3 by Fesmire (which is introduced in the calorimeter chapter of the book). These data are typical for such a system of layers of multilayer insulation. The optimum performance (lowest heat flux) occurs at a CVP well below 0.01 millitorr and begins to degrade rapidly above 1 millitorr.

In this data summary, the heat flux ranges from about 200–300 W/m^2 at no vacuum (10^4–10^6 millitorr), about 40–100 W/m^2 at soft vacuum (10–10^3 millitorr), and about 0.6–0.8 W/m^2 at high vacuum (10^{-3}–10^{-1} millitorr).

In the second example, at a fixed high vacuum level of 1×10^{-5} torr, the heat flux through the MLI blanket as a function of the numbers of reflection layers was studied. The heat flux values through 30, 60, and 90 reflection layers and with different cold surface preparations (Cu surface, black-painted Cu, and Al-taped Cu) were tested between 77 and 293 K, as shown in Figure 3.4 by QS Shu [16]. Relevant data reported are summarized as follows: the heat flux from 277 K to a black-painted fin through 30 layers of MLI is 0.64 W/m^2, 60 layers is 0.47 W/m^2, and 90 layers is 0.38 W/m^2. In the case without any MLI layers, the heat flux from 297 K to the black fin is 24.7 W/m^2, to a plain Cu fin is 13.9 W/m^2, and to an Al-taped fin is 4.8 W/m^2.

FIGURE 3.3 LEFT: Preparation of MLI test specimens on the Cryostat-100 cold mass assembly: layer joining A134. Right: Completed test specimen A138 for insertion to the vacuum chamber [1].

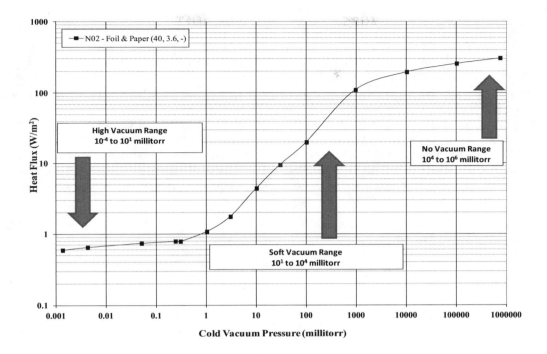

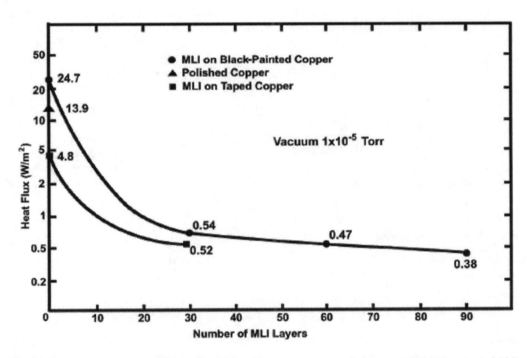

FIGURE 3.4 Upper: Variation of heat flux with cold vacuum pressure for an example MLI system of 40 layers foil and paper [1]. Lower: Heat flux as a function of the number of MLI layers and surface preparation at high vacuum range of 10^{-5} torr [16].

3.4 CALCULATION OF MLI THERMAL PERFORMANCE

The calculation models introduced here are based on data from vertical cylindrical calorimeters. The heat flow rate through the insulation test specimen and into the test chamber of the cylindrical cold mass is directly proportional to the LN_2 boil-off mass flow rate by the heat of vaporization, as already discussed. The value of k_e is determined from Fourier's law of heat conduction through a cylindrical wall, as already discussed.

While k_e is helpful for general comparison among different thermal insulation systems, the detailed performance analysis of highly evacuated MLI systems, because they are based in radiative heat transfer effects, should ultimately rely upon the calculation of heat flux (q) rather than k_e. Although calculation of the theoretical thermal performance of MLI is a highly specialized subject in its own right, two well-known approaches are briefly summarized as follows. Further details and literature references are given in ASTM C740 [5].

3.4.1 LOCKHEED EQUATIONS

In the first example approach, the generalized form of the Lockheed equations gives an empirical form for heat flux as follows [17]:

$$q = \frac{C_s \times \bar{N}^{2.63}(T_h - T_c) \times (T_h + T_c)}{2 \times (N+1)} + \frac{C_R \times \varepsilon \times (T_h^{4.67} - T_c^{4.67})}{N} + \frac{C_G \times P \times (T_h^{0.52} - T_c^{0.52})}{N} \qquad (3.1)$$

All three modes of heat transfer are accounted for by the leading coefficients: solid conduction (C_S), radiation (C_R), and gaseous conduction (C_G). The coefficient C_S is a function of the spacer material and its surface contact resistance with the facing radiation shields. C_R is a function of the radiation shield material and its perforation size and pattern, if any. C_G is a function of the residual gas pressure between the layers. Even at high vacuum levels, some gas molecules do exist between the layers of radiation shields and spacers, necessitating a term for gaseous conduction.

For the specific case of silk net spacers and unperforated double-aluminized Mylar radiation shields, the coefficient are as follows: $C_S = 8.95 \times 10^{-8}$, C_R, = 5.39×10^{-10}, and $C_G = 1.46 \times 10^{-4}$. It is important to note that the various Lockheed equations are primarily based on data from MLI systems composed of double-aluminized Mylar radiation shields with silk net spacers and tested using a flat plate, not cylindrical, boil-off calorimeter [17]. Due to the ($N + 1$) term in the denominator of the radiation term, the Lockheed equations should not be used for a layer-by-layer method of analysis.

3.4.2 EQUATION BY McINTOSH

Another well-known approach to theoretical calculation of MLI system performance is the physics-based equation developed by McIntosh [18]. The general form is given as follows:

$$q = \sigma \left(T_h^4 - T_c^4\right) / \left[(1/\varepsilon_h + 1/\varepsilon_c - 1)\right] + C_G P \, \alpha \, \left(T_h - T_c\right) + C_S fk \left(T_h - T_c\right)/x \qquad (3.2)$$

This equation applies to a total MLI system rather than a layer-by-layer approach. The McIntosh equation, like the Lockheed equations, has three terms: one for radiation between shields, one for solid conduction (k) through the spacer material and surface conductance, and one for gaseous conduction due to residual gas molecules between the layers. The term f is the relative density of the spacer compared to the solid form of the material. The gaseous conduction term includes the residual gas pressure (P) within the layers and the accommodation coefficient (α) for energy transfer onto a specific gas-surface combination. The radiation term includes the Stefan-Boltzmann (σ) constant and the effective emissivity (ε) of the hot and cold surfaces.

3.4.3 Hybrid Approach by Augustynowicz

The use of these or other equations available in the literature requires adequate understanding of all three heat transfer modes as well as the testing methodologies used and the influences of installation for a given application. It should be noted that the McIntosh equation must be solved on every layer, whereas the Lockheed equation is solved for the bulk system. Alternatively, a hybrid type approach is given by Augustynowicz, where solution of the temperature distribution through the MLI blanket is determined by an energy balance for groupings of layers within the total N layers of the system [19].

Because radiation heat transfer within an MLI system produces a nonlinear temperature gradient, k_e will vary approximately as the third power of the mean temperature. Thus, k_e can be properly used for comparison of performance of different MLI systems only when the boundary temperatures are the same [9, 20]. For an evacuated MLI system with the idealized assumptions of "perfect" vacuum between layers and zero solid conduction between layers, the effective thermal conductivity can be expressed as follows:

$$k_e = \left(N / x\right)^{-1} / \left[h_c + \sigma_e \left(T_h^2 - T_c^2\right) \left(T_h - T_c\right) / \left(2 - e\right)\right] \tag{3.3}$$

In any case of thermal performance calculation, the total insulation thickness must be carefully defined.

Whenever k_e is used to describe the thermal performance of an MLI system, a statement indicating the method used in making the thickness measurement and the accuracy of such measurement is needed. In some cases, an estimate of a range of thicknesses for a given installation may be needed. If a system specific analysis is required, the inside diameter of the vacuum can or vacuum jacket can be used to establish a thickness for determining an overall *system* thermal conductivity (k_s) including the vacuum enclosure, any annular gap, and the MLI system itself. Technical consensus standards ASTM C740 and C1774 provide further details on the terminology, calculations, and data reporting requirements [5, 9].

3.4.4 Empirical Equation by CERN Large Hadron Collider

Besides the previous three calculation equations, there are other empirical calculation formulas for MLI performance, which were developed in several leading institutes for their own facilities [20–21]. One of the representative efforts is the CERN Large Hadron Collider (LHC) team's empirical equation of MLI performance for their about 40 km of SC accelerators, which is the single largest user of MLI systems in the world. Modeling the thermal heat exchange through MLI can be quite complex and material (reflective and insulating layers) dependent, as well as strongly influenced by the application of the blanket in the cryostats. A simplified engineering model considers two main contributions, one accounting for radiation, proportional to the difference of temperatures at the fourth power, and a second accounting for the residual solid conduction across the blanket thickness proportional to the temperature difference between the warm and the cold surface and to the average temperature [20]:

$$q_{\text{MLI}} = \left[\frac{\beta}{N+1} \cdot \left(T_1^4 - T_2^4\right)\right] + \frac{\alpha}{N+1} \cdot \frac{T_1 + T_2}{2} \cdot \left(T_1 - T_2\right) \tag{3.4}$$

where β and α are corrective factors that should be obtained experimentally. T_1 and T_2 represent the temperatures of cold and warm surfaces. Values obtained for the MLI system of the thermal shield of the LHC cryostats are $\beta = 3.741 \times 10^{-9}$ and $\alpha = 1.401 \times 10^{-4}$.

3.5   ENERGY SAVING: MLI WITH INTERMEDIATE SHIELDS

There are two strategic tasks equally crucial to energy savings in keeping liquid gases or cold masses cold. The first is to choose the best MLI for the applications. The other is to optimize the MLI system both thermally and mechanically. From a thermal management point of view, the task is to use the least work to intercept and remove the heat from the ambient environment to the interior.

3.5.1   BASIC PRINCIPLES AND TYPICAL CONFIGURATIONS

Overall system efficiency is measured by the coefficient of performance (COP). In cryogenics, the Carnot efficiency is the theoretical maximum efficiency where $COP_c = T_c/(T_h - T_c)$. So, $Q_c/W_{in} = T_c/(T_h - T_c)$, where Q_c is heat moved, W_{in} is external work needed, T_h is high T, and T_c is low T. Based on Carnot's theorem, to remove the same amount of heat, the higher the temperature level of the heat being removed, the lower the external watts of the cooling device consumed. Therefore, intermediate temperature (T) shields are important approaches to improve thermal efficiency in real MLI systems.

To demonstrate how the shields greatly reduce energy consumption, six typical configurations are chosen and illustrated in Figure 3.5. The 77-K level can be an LN_2 vessel, shields for an LHe vessel, SC cold mass, LH_2 vessel, and so on; the 20-K level can be a LH_2 vessel and shields for the SC cold mass, special space device, and so on; and the 4-K level can be an LHe vessel, SC mass, and space device. These configurations are:

1. Only 40 MLI layers on 4 K surface (LHe vessel or SC magnet) without shield
2. 40 MLI layers on 80 K intermediate shields but no MLI on 4 K surface
3. 40 MLI layers on 80 K shield, 20 layers on 20 K shield but bare 4 K surface
4. Same as case 3 on 80 K and 20 K shields but an additional 10 layers on 4 K surface

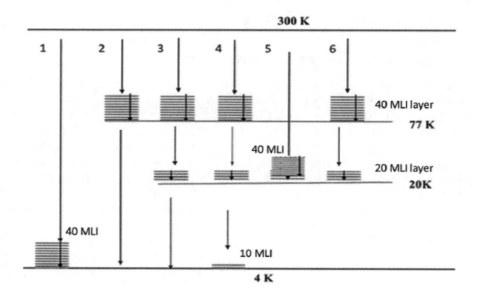

FIGURE 3.5   Six typical cases for demonstration of MLI system with intermediate shields [21].

5. The cold surface is 20 K (LH$_2$ vessel or 20 K mass), only 40 MLI on 20 K surface but no shield

6. 40 MLI layers on the 80 K intermediate shields and 20 MLI layers on 20 K surface

3.5.2 DEMONSTRATION OF ENERGY SAVING BY INTERMEDIATE SHIELDS

Based on the six cases (configurations), the heat loads per m^2 from each warm T_h surface to the facing cold surfaces T_c are calculated or taken from experimental data. These heat load data are placed in the corresponding cells with bold in Table 3.2. Then, the minimum work needed to remove the heat loads is calculated using the Carnot formula. For easy comparison, the minimum work data are also placed in the same cell (occupied by the corresponding heat loads) but using italics underneath the heat load data.

An example on how to use this table is as follows: In case 1, the heat load from 300 K through 40 MLI layers reaching 4.2 K is 0.62 W, and *43.6 W* of refrigeration power is needed to remove that heat. In case 4, the heat load at 77 K is 0.6 W, and to remove it from the 77 K level, a refrigeration power of *1.67 W* is needed. In another case, the heat load from 77 K to 20 K through 20 MLI layers is 0.025 W, and to remove that heat, we need *0.35 W*. Finally, the heat from 20 K to 4 K through ten MLI layers is only 0.0002 W (0.2 mW), and only *0.014 W* of refrigeration power is needed to remove it. The integrated total minimum power for each case is given in the last column of the table.

3.5.3 DESIGN METHODOLOGY OF INTERMEDIATE SHIELDS WITH MLI SYSTEMS

Regarding how many intermediate shields to employ and what temperatures to allocate between T_h and T_c, it depends on the cooling resources available; the weight, volume, and shape of the cold

TABLE 3.2

The Heat Loads to 1-m^2 of Cold Surface (in Bold) and Minimum Refrigeration Work Needed to Remove the Heat Loads to Ambient (in Italic)

Case	Description	Heat Loads, W (in Bold) Min Work, W (in Italic)						Total Min Work, W
		300–4 K	300–80 K	300–20 K	80–20 K	80–4 K	20–4 K	
1	Only 40 MLI layers on 4 K	**0.61** *43.6*						*43.6*
2	40 layers on 80 K shield		**0.60** *1.67*			**0.195** *13.93*		*15.6*
3	40 layers on 80 K, 20 layers on 20 K		**0.60** *1.67*		**0.025** *0.35*		**0.00064** *0.046*	*2.066*
4	40 layers on 80 K, 20 layers on 20 K, 10 layers on 4 K		**0.60** *1.67*		**0.025** *0.35*		**0.0002** *0.014*	*2.034*
5	Only 40 layers on 20 K			**0.61** *8.7*				*8.7*
6	40 layers on 80 K, 20 layers on 20 K		**0.60** *1.67*		**0.025** *0.35*			*2.02*

Source: For Six Typical Configurations [21]

*The Carlo coefficient: from 80 K to 300 K: 0.36, 20 K to 300 K: 0.07, 4 K to 300 K: 0.014.

mass; and the operational cost vs hardware cost. Shield design will be discussed in Chapter 6. The various cooling methods for shields are briefly listed here for reference:

- Passive conductive shield-foils connected to cold vapor exit neck
- Passive conductive shields cooled by cold vapor exiting pipe
- Conductive shields cooled by an internal LN_2 vessel
- Active forward/return cryogenic liquid and vapor-cooled shields
- Multistage cryocooler actively cooled shields

3.6   THERMAL PERFORMANCE OF MLI SYSTEMS

Extensive experimental data from cryostat testing are available for MLI systems tested down to 77 K (that is, with liquid nitrogen as the boil-off medium).

3.6.1   DESCRIPTION OF MLI TEST SPECIMENS

Cryogenic thermal performance testing with Cryostat-100 included hundreds of tests of 68 different MLI test specimens. Presented here is a select group of test series for 26 different MLI systems plus three vacuum-only systems with an emphasis on baseline combinations of materials. Not selected in this group were any test specimens with specialized attachment hardware, mechanical connections, or engineered closures, nor those without a documented installation process. Test specimens with unusual or experimental installation methods were not selected. Specifically, hybrid-type systems with foam layers or aerogel blanket materials were not included in the selected group.

A guide and nomenclature for the different MLI test specimens are given in Table 3.3. The different lay-up configurations or installation techniques used, including rolled, spiral, layered, blanket, or fabricated, are also listed. Three additional test series cover the vacuum-only cases with three different cold mass surface coverings/finishes: stainless steel, black chrome, and copper sleeve. Additionally, one of the Mylar/paper test specimens provides the results for a single layer of aluminum film, giving a total of four useful VO cases.

TABLE 3.3
Guide to Different MLI Systems

Category	Code	Lay-up*	Reflector	Spacer	x	n	z
					mm	*layers*	*layers/mm*
Mylar and paper	MP	L or B	Mylar	Paper	2–19	2–40	0.7–2.1
Foil and paper	FP	R or S	Foil	Paper	11–21	40–80	2.3–3.8
Mylar and net	MN	L	Mylar	Poly net	5–16	10–40	2.1–2.6
Mylar and net	NB	B	Mylar	Poly net	23–64	40–60	0.9–2.6
Mylar and fabric	MF	L or B	Mylar	Poly fabric	6–15	10–40	1.6–5.1
Mylar and silk net	SN	L	Mylar	Silk net	8–24	10–20	0.9–1.4
Mylar and discrete	MX	F	Mylar	Discrete	8–39	5–20	0.5–0.6
Vacuum only	VO	n/a	As indicated	n/a	n/a	n/a	n/a

*Lay-up configurations:

Rolled (R) = continuously rolled in individual layers, collated layer pairs, or multiple layer pairs

Spiral (S) = wrapped in continuous spiral fashion in layer pairs

Layered (L) = layer by layer; stack-up of individual layers or layer pairs

Blanket (B) = pre-assembled blanket or stack-up of sub-blankets; seams may be interleaved, overlapped, or folded over

Fabricated (F) = fabricated system engineered for a particular end-use geometry (usually flat plate or cylindrical)

The cold boundary temperature was approximately 78 K, and the residual gas was nitrogen in all cases. The effective area for heat transfer is calculated from the mean diameter of the test specimen and the effective length (0.58 m), as previously discussed. The test results are divided into the following four tabular sections for clarity:

- Mylar/paper and foil/paper types
- Mylar/net and Mylar/net blanket types
- Mylar/fabric and Mylar/silk net types
- Mylar/discrete type and vacuum only

3.6.2 CRYOSTAT TEST DATA FOR SELECT MLI SYSTEMS

Accordingly, these results are presented in tabular form, in their entirety, in the four tables of Appendix B. One example is given in Table 3.4. The warm boundary temperature was approximately 293 K or 300 K, as noted. In a few special cases, elevated warm boundary temperatures of 305 K, 325 K, or 350 K were applied; these elevated temperatures are only provided in the tables and not used in any of the subsequent figures. The warmer boundary temperatures are additionally indicated in color code. Note that the warm boundary temperature is taken as the outer surface of the MLI system, not the vacuum chamber wall temperature.

The data in Table 3.4 include information on the MLI system physical design: total thickness (x), number of layers (n), and layer density (z). In addition to the columnar entries, the physical design data (x, n, z) are also given in the name designation for each MLI system. The installation method and details are noted as well. Where available, the MLI system bulk density (ρ) is given. The bulk density is the as-installed density and ranges from 24 to 145 kg/m³. Although highly dependent on layer density, a typical bulk density for an MLI system is shown to be roughly 50 kg/m³.

Two 40-layer systems of Mylar/fabric, MF-2 and MF-3, are identical except for their layer densities. In this case, an 80% increase in layer density (from 2.8 to 5.1 layers/mm) resulted in a 280% increase in heat flux (from 0.71 to 2.7 W/m²). Examination of the two foil/paper systems, FP-1, and FP-3, shows that a doubling of the number of layers, from 40 to 80, gives a substantial 34% reduction in the heat leak (from 0.59 to 0.39 W/m²). This 80-layer system also shows a strong thermal performance advantage through the degraded vacuum region with CVP higher than 1×10^{-4} torr.

Graphical summaries of the results are expressed in terms of the variation of k_e with CVP. These data sets are given in their entireties in Appendix B. The four data sets are as follows:

- Figure B.1: Mylar/paper and foil/paper types
- Figure B.2: Mylar/net and blanket (NB) types
- Figure B.3: Mylar/fabric and Mylar/silk net types
- Figure B.4: Mylar/discrete type and vacuum only

One of the four figures is given here as Figure 3.6 for the variation of effective thermal conductivity (k_e) with cold vacuum pressure (CVP). These plots show the entire pressure range for rudimentary comparison among different MLI systems. The *Kaganer k-line*, a benchmark curve of all 26 MLI systems, is given in Figure 3.7 and shows the relative trend across the vacuum pressure range. This benchmark curve is a composite or mean of the effective thermal conductivity at each of 10 points of cold vacuum pressure (that is, nine decades of vacuum level).

Graphical summaries of the heat flux results are presented in Figures 3.7 and 3.8. Figure 3.8 presents the results for the variation of heat flux (q) with cold vacuum pressure (CVP) for a baseline group of 13 different MLI systems with numbers of layers (n) from 20 to 60. This group was chosen from the full group of 26 systems, minus those with fewer than 20 layers or more than 60 layers. This baseline group defines the *Augustynowicz q-band*, which represents the mean heat flux +/- 50%. This benchmark curve is a composite or mean of the heat flux at each of 10 points of cold vacuum pressure (that is, nine decades of vacuum level).

TABLE 3.4

MLI System Thermal Data Summary: Mylar/Fabric (MF) and Mylar/Silk Net (SN) types; Each Name Designation Includes Total Thickness (x), Number of Layers (n), and Layer Density (z) as (x, n, z)

Cryostat Test Series (MLI System)	CVP μ	WBT K	Flow sccm	Q W	q W/m²	k_e mW/m-K	x mm	n	z lyrs/mm
MF1-A145 Mylar/fabric (6, 10, 1.6)	0.006	294	68	0.282	0.893	0.026	6.4	10	1.57
Layer by layer	0.1	293	146	0.604	1.91	0.057			
Apr-10	0.3	293	171	0.707	2.24	0.066			
	1	293	277	1.149	3.64	0.108			
	10	293	1456	6.037	19.1	0.568			
	99	293	7684	31.862	101	3.00			
MF2-A133 Mylar/fabric G1 blanket (8, 40, 5.1)	0.005	299	206	0.854	2.68	0.095	7.8	40	5.12
One 40-layer blanket	0.3	300	258	1.070	3.36	0.118			
Bulk density = 146 kg/m³ Sep-09	1.0	300	310	1.285	4.04	0.142			
MF3-A134 Mylar/fabric G2 blanket (15, 40, 2.8)	0.006	300	56	0.233	0.707	0.046	14.5	40	2.75
Two 20-layer sub-blankets	0.1	300	98	0.406	1.23	0.081			
Bulk density = 82 kg/m³	0.3	300	115	0.477	1.44	0.095			
Sep-09	1	300	153	0.634	1.92	0.126			
SN1-A177 Mylar/silk net double (24, 20, 0.9)	0.006	293	29	0.120	0.348	0.038	23.5	20	0.85
Layer by layer	0.1	293	58	0.242	0.700	0.077			
Staggered seams with 1" overlap typical	1	293	146	0.605	1.75	0.191			
Bulk density = 20 kg/m³	760000	284	16647	69.027	200	22.8			
Mar-14	0.004	305	34	0.140	0.405	0.042			
	0.004	326	45	0.186	0.538	0.051			
SN2-A178 Mylar/silk net single (15, 20, 1.4)	0.003	293	27	0.114	0.344	0.023	14.7	20	1.36
Layer by layer	0.1	293	58	0.240	0.725	0.050			
Staggered seams with 1" overlap typical	1.0	293	138	0.572	1.73	0.118			
Bulk density = 24 kg/m³ Mar-14									
SN3-A179 Mylar/silk net single (8, 10, 1.3)	0.001	292	41	0.171	0.538	0.019	7.7	10	1.30
Layer-by-layer	0.002	292	42	0.172	0.542	0.019			
Staggered seams with 1" overlap typical	0.1	294	59	0.246	0.774	0.028			
Bulk density = 25 kg/m³ Apr-14	1.0	293	222	0.921	2.89	0.103			

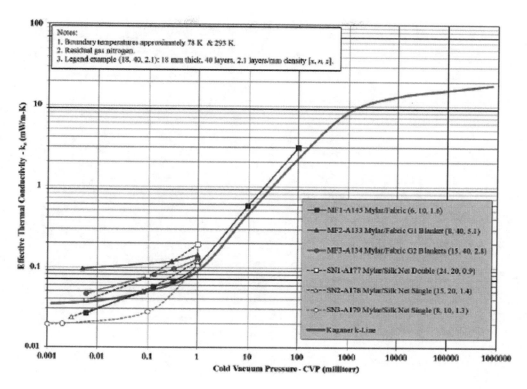

FIGURE 3.6 Variation of k_e with CVP: Mylar/fabric and Mylar/silk net types.

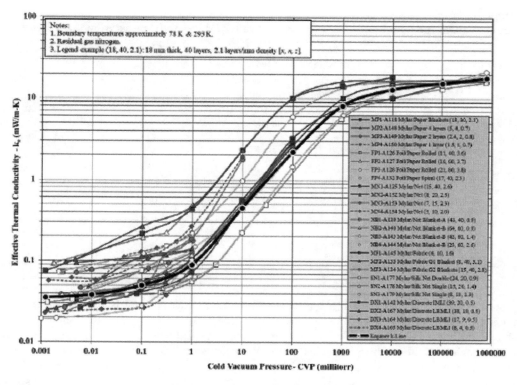

FIGURE 3.7 Variation of k_e with CVP: Kaganer k-line composite result for full range of different MLI systems.

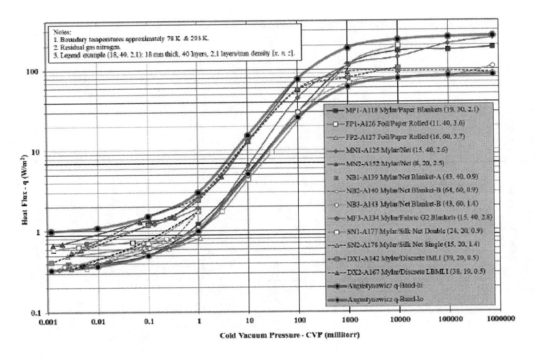

FIGURE 3.8 Variation of heat flux (q) with CVP: Augustynowicz q-band of 13 MLI systems from 20–60 layers.

TABLE 3.5
Tabular Values for Benchmark MLI Curves: Kaganer k-Line and Augustynowicz q-band

Kaganer k-line	CVP μ	k_e mW/m-K	Augustynowicz q-band	CVP μ	q_{high} W/m²	q_{mid} W/m²	q_{low} W/m²
Composite of 26 MLI systems	0.001	0.035	Baseline 13 MLI systems	0.001	1.00	0.665	0.330
From 10 to 80 layers; all types	0.01	0.038	From 20 to 60 layers; all types	0.01	1.10	0.735	0.370
CBT = 78 K	0.1	0.050	CBT = 78 K	0.1	1.52	1.01	0.500
WBT = 293 K	1	0.090	WBT = 293 K	1	3.00	2.00	1.00
Residual gas nitrogen	10	0.450	Residual gas nitrogen	10	15.5	10.3	5.10
	100	2.20		100	77.0	51.5	26.0
	1000	8.20		1000	184	123	62.0
	10000	13.0		10000	232	155	78.0
	100000	15.5		100000	250	167	84.0
	760000	18.0		760000	257	172	86.0

For reference and use in engineering analysis or thermal data comparisons of different MLI systems, the values for the two benchmark curves, the Kaganer k-line and the Augustynowicz q-band, are given in Table 3.5. These values represent idealized laboratory cases and not the real system implementation, which will yield higher heat loads depending on the support structures, penetrations, installation quality factor, and other factors. The benchmark curves provide a starting point for design, analysis, or comparison with other evacuated MLI systems. Where overall size or volume is a primary consideration, the Kaganer k-line should be used, as the k_e takes into account the thickness. This benchmark curve can also be used for broader, order-of-magnitude comparisons across different thermal insulation systems. For more focused comparisons with MLI systems in the typical range ($n = 20$ to 60 layers), the Augustynowicz q-band should be used.

3.6.3 SUPPORTING CRYOSTAT TEST DATA FOR OTHER MLI SYSTEMS

The designation numbers (A123 or C123, for example) for these data sets also provide cross-reference to any related publications that may include a portion or select subset of test points for a specific application. For example, further details with system-specific application notes on FP-type systems [18], MN-type systems [19, 21], SN-type systems [22], MX-type systems [23, 24], and others are available in the literature. For cross-comparison with the data presented here, detailed experimental investigations on specific types of MLI systems, tested under similar conditions and geometry, are given by Scurlock [25], Bapat [26], Jacob [27], and others. Experimental work by Ohmori [28] has shown the differences in thermal performance between vertical and horizontal configurations of MLI systems. Thermal performance data for only a few layers of MLI are given by Shu [29]. Finally, for extension to a variety of other layered system concepts, Johnson [30–32] reports similar cryostat test data on seams, aerogel blanket, and hybrid stack-ups.

3.7 DISCUSSION OF MLI THERMAL PERFORMANCE

Insulations of the type described previously are generally used when lower heat leakage rates than those obtained with other evacuated insulations are required. For example, typical MLI systems can provide ten times lower heat leakage rates than evacuated perlite systems.

3.7.1 GENERAL PERFORMANCE CONSIDERATIONS

Well-executed, highest-performance MLI systems can provide heat leakage rates nearly ten times lower than typical MLI systems. Other techniques such as vapor-cooled shielding in combination with MLI systems may also be employed for the lowest possible heat leakage rate. For maximum effectiveness of the installed (total) system, any connecting, mechanical, or structural elements must be designed in conjunction with the design of the thermal insulation [4, 22].

MLI systems are readily suited to cylindrical geometries and are therefore commonly designed as thermal insulation for piping or tanks. MLI systems are also designed for a wide range of blankets, panels, and other thermal protective elements and configurations [33–35].

The type and amount (vacuum pressure) of residual gas has a strong influence on the resulting thermal performance of MLI systems. The vacuum level, if known, is usually measured at the warm boundary or vacuum enclosure. The vacuum levels between layers are generally unknown and can have significant effects on the thermal performance. Understanding and applying all the available information from the heating, purging, evacuation, and vacuum monitoring steps can help to account for residual gas effects and explain the overall thermal performance results.

3.7.2 DETAILED PERFORMANCE CONSIDERATIONS

The number of layers (n) for MLI systems can be from 1 to 100 or more. If size and weight are not an issue, then more layers are generally better for reducing the heat flux. However, sagging in thicker blankets can result in additional compression between the layers and give diminishing returns or even reduced thermal performance. The increased entrapment of residual gas molecules between layers is also likely to degrade performance [36, 37].

The layer density (z) is crucial for estimating the thermal performance of MLI systems. The optimum layer density must be considered in light of the performance targets as well as the practicality of installation techniques. The optimum layer density often varies with different combinations of reflector and spacer materials. In addition, vacuum pressure and layer density are tied together. Lower layer density blankets transition to degraded performance (higher heat flux) at lower pressures in correspondence to the Knudsen number.

The cold boundary temperatures typically range from 111 K for liquefied natural gas to 4 K for liquid helium. Liquid nitrogen at a normal boiling point of 77 K is of course right in the middle of this range and offers a popular test condition for many MLI systems. While the overall change in ΔT is not extremely large when changing the cold boundary temperature from 77 K to 4 K (only 223 K vs 296 K for a corresponding WBT of 300 K), the influence of lower temperature can have profound effects on the cold vacuum pressure and heat transfer mechanisms. Testing and analysis are needed to understand the influence of different CBT. Extrapolations of test data are often unavoidable, but such performance predictions should be taken with precaution and only heat flux data used for the final design performance calculation.

3.7.3 EFFECTS OF SYSTEM REQUIREMENTS

System requirements can vary greatly with regard to boundary temperatures, temperature matching among conductive layers, vapor shields, and so forth. The effects of different boundary temperatures are important and are discussed later. The effects of helium purge gas for an MLI system, sometimes used for space launch vehicles or spacecraft, are shown in Figure 3.9 [5]. The main point here is that the heat flow through an MLI system backfilled with helium gas at 760 torr is about three times higher compared to nitrogen or air.

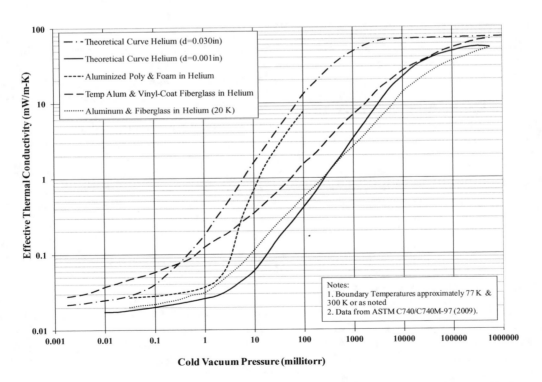

FIGURE 3.9 Variation of effective thermal conductivity with cold vacuum pressure for helium residual gas. (Note: 1 millitorr = 0.133 Pa.)

3.8 EFFECT OF NUMBER OF LAYERS AND LAYER DENSITY

The general radiation theory described by Equation 3.5, when properly applied in a layer-by-layer model, gives the following relation:

$$q \propto \frac{\epsilon \sigma \left(T_H^4 - T_c^4 \right)}{N+1} \tag{3.5}$$

Thus, between two boundary temperatures, the heat flux times the number of layers should be relatively constant, or at least comparable. Figure 3.10 shows the high vacuum data between 293 K and 78 K for heat flux vs number of layers and is not very instructive due to the rectangular hyperbola shape that would be expected. However, Figure 3.11, which plots $q \times N$ vs the number of layers, is quite instructive as to the thermal performance trends that can occur when a large number of layers are used.

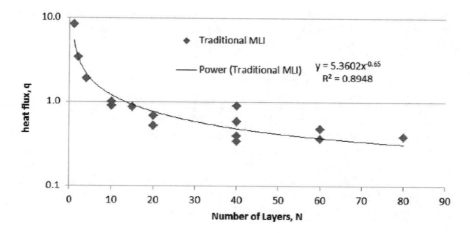

FIGURE 3.10   Variation of heat flux q with the number of layers in an MLI system.

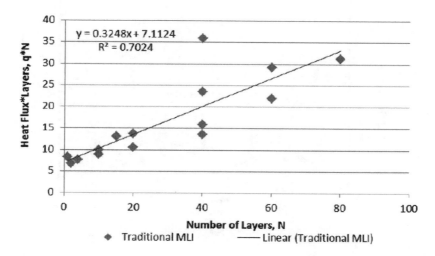

FIGURE 3.11   Variation of the quantity $q \times N$ with the number of layers in an MLI system.

3.8.1 Layer Density Estimation and Analysis

The layer density for a given system is determined by a number of practical factors, including space available, support of the layers, interface with structures, and so on, as well as the type of materials used and the required installation method. The layer density can have a dramatic effect due to solid conduction and quality of evacuation of the total system. Johnson provides theory and analysis on layer density effects [38]. The concept of an optimal layer density is further supported by recent experimental work [39]. Thus, the best layer density for a given MLI system design is not only determined by design and material thicknesses but also by the practical terms of installation, evacuation, and means of mechanical support.

Understanding layer density and how to achieve a target density, or a certain variable density through the thickness, in the final rendered and installed case is paramount in achieving the necessary thermal performance for an MLI system. It is obvious that the highest vacuum level is important but not necessarily obvious that vacuum level, layer density, materials, and installation are all intertwined.

The efficiency of MLI systems in accelerator cryostats also depends on the workmanship of MLI handling. In higher packing density, the solid conduction contribution increases due to enhanced contacts (areas and pressure), and the residual gas thermal conduction also increases due to pumping capacity decreases between MLI layers. There is an optional packing density, which is allocated between 15 and 25 layers per cm, as shown in Figure 3.12.

3.8.2 Practical Rules for Installation

MLI efficiency also depends on practical implementation rules. The ideal packing density should be preserved after installation of the blankets. Using too many layers in a reduced space between warm and cold surfaces is a typical mistake, which would cause unexpected heat leakage through the MLI system. A good rule is to leave a minimum gap of at least a few cm between the warm surface and the outermost MLI layer, as in experimental data shown by Shu et al. [40].

When mounted in horizontal accelerator cryostats, MLIs tend to be compressed by their weight on the top part of the thermal shields, whereas it tends to be looser at the bottom. A good rule is to check that an acceptably correct packing remains after installation. Also, when preassembled MLI

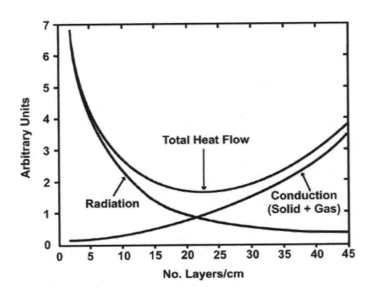

FIGURE 3.12   Total heat flow as a function of the MLI packing density.

blankets are to be mounted around circular geometries, one should include in the design the larger circumferential dimension of the outermost layers with respect to the innermost ones.

The choice of the ideal number of MLI layers is not a straightforward one and is very much dependent on the materials employed (film material and thickness, single or double-alienization, alienization thickness, type of spacers, etc.). In the CERN LHC accelerator, an optimal and cost-effective solution was chosen for the MLI blankets of the thermal shields (~65 K) [20]: 30 reflective layers (double-aluminized-400 Å polyethylene film) interleaved with polyester net. An additional ten-layer MLI blanket is mounted on the cold mass, essentially to reduce heat loads from residual helium gas conduction in case of degraded vacuum from helium leaks.

3.9 COMPARISON OF DATA TO THERMAL MODEL

In general, there are severe limitations on any approach to obtain a generalized model for an MLI system with its combination of three modes of heat transfer and generally very low rates of heat flow. The inherent complexity of such systems, intertwined among installation technique, layer density, and materials, does not easily lend itself to modeling except within specific design bounds of materials and environmental conditions. However, the modeling is essential, particularly in light of the time and expense required in cryogenic-vacuum testing and experimentation. The numerous thermal performance data sets given here and summarized in the benchmark data curves together with the appropriate modeling approach provide a starting point for the design and implementation of new MLI systems for the primary enabling work in cryogenic systems of all kinds.

The thermal performance data given for the wide range of MLI systems can help determine the level of validity of a given theoretical model. For example, the Lockheed equation given in Equation 3.1 is applied. By merging a Dacron polyester net spacer term from Hedayat (instead of the silk netting-based conduction term of the original Lockheed development) [38] with the non-perforated radiation shield term from Lockheed, an updated form of the Lockheed equation is given as follows:

$$q = \frac{\left(2.4E-4*\left(0.017+7E-6*\left(800-T_{avg}\right)+2.28e-2*\ln\left(T_{avg}\right)\right)\right)\bar{N}^{2.63}\left(T_h-T_c\right)}{N+1}$$

$$+\frac{5.39E-10*\int*\left(T_h^{4.67}-T_c^{4.67}\right)}{N}+\frac{1.46E4*P*\left(T_h^{0.52}-T_c^{0.52}\right)}{N} \tag{3.7}$$

The new equation in comparison with the Lockheed equation is represented in Figure 3.13. The scale factor is the ratio of Q_{test}/Q_{model}, and thus 1 is the idealized result. In general, the Lockheed equation provides a lower model value, while the new equation provides a higher value. Both are generally off by up to a factor of 3. Recent results suggest that another factor of 2 should be added when applying calorimeter data to tank applications due to the larger numbers of seams and penetrations [32, 35, 41].

In addition, Figure 3.14 provides the variation of the scale factor with the layer density for an MLI system. This analysis includes data from the following 12 Cryostat-100 data sets: A118, A125–126, A139, A145, A148–150, A152–154, and A159. Further testing and analysis are being performed to understand the effect of the warm boundary temperature on the scale factor; preliminary results suggest an inverse correlation.

3.10 MLI PERFORMANCE BELOW 77 K

Many thermal performance data in the literature are based on a cold boundary temperature of 77 K based on using liquid nitrogen as the boil-off test medium. The use of MLI for liquid hydrogen

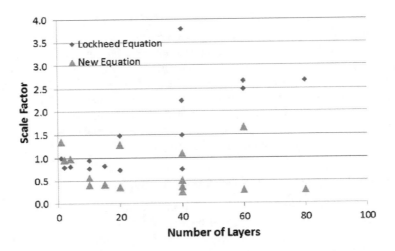

FIGURE 3.13   Variation of the scale factor with the number of layers in an MLI system.

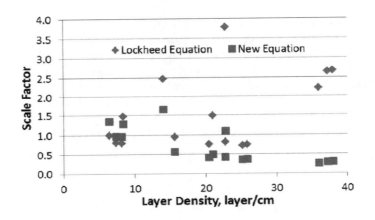

FIGURE 3.14   Variation of the scale factor with the layer density in an MLI system.

(20 K), liquid helium (4 K), and others can mean that there are significantly lower cold boundary temperatures involved. But testing under these conditions is usually far more complicated and expensive. Therefore, both analytical and novel experimental methods are often employed to cover these extreme low temperature requirements.

There are many applications in which the temperatures of cold surfaces to be thermally protected are much lower than 77 K, such as 1.8 K–20 K for SC applications, cryogenic storage/transmission of LH_2 and LHe, and some space explorations. In order to greatly save cooling power (energy), as discussed in Section 3.4, intermediate T shields are employed to block the thermal radiation from the 300 K environment to the cold surfaces. These shields can be 77 K and LN_2 cooled or in the T range from 20 K–65 K when cooled by cryocoolers and return cryogens. Therefore, the MLI thermal performance between these T pairs becomes very useful, and these data have been experimentally determined in many laboratories around the world. Although these MLI performance data are measured in the specific T arrangements, the technical data have been widely referenced for many similar applications.

3.10.1 MLI Performance for 77 K to 4.2 K

Shu et al. at Fermilab reported the performance of a normal MLI system between 77 K and 4.2 K [42–43]. Thirty MLI layers (300 Å single-aluminized crinkled Mylar (SACM), NRC-2, without spacer layers) are wrapped on a copper flat plate cooled to 4.2 K by a soldered siphon LHe pipe. A copper plate of 2.24 m^2 (area, A) is surrounded by a copper box, which is cooled by two LN$_2$ siphon pipes. The test apparatus and data-taking procedure are discussed in detail, as shown in Figure 8.14 of Chapter 8 in this book. The accuracy of the wet test meter used for the boil-off is ±0.2%. The temperatures at different points in the MLI blanket and of the fin, guard vessel, and guard shield were also measured and recorded during the tests. After the thermal equivalent was reached, the heat load from the 77 K copper box to the 4.2 K black-painted fin through the 30-layer MLI blankets was measured, Q_0 was 29.9 mW, and the corresponding heat flux, $Q_0/A = 13.2$ mW/m^2 at vacuum pressure lower than 1×10^{-6} torr.

3.10.2 MLI Performance for 65 K to 6 K

Ohmori et al. reported investigations of MLI performance below 77 K [44–45]. In a recent study, the MLI sample is wound on the surface of an oxygen-free high thermal conductivity drum. The inner drum is cooled by the second stage of a cryocooler at around 6 K, and the outer drum is cooled by the first stage at around 65 K. The MLI is made from 9-μm-thick double-aluminized Mylar (DAM), and very light nonwoven polyester fabric is fused to one side of the DAM. The mass of a single film of the MLI is 19.6 g/m^2. Three-mm-diameter evacuation holes are punched on the MLI. The heat flux of 15–16 W at boundary T of 298 K–65 K with 50 layers of MLI and 25–36 mW between 65 K and 6 K with 20 layers were obtained, respectively. MLI thermal performance measured by calorimeter is given in Table 3.6 (*N/h* is the layer density). Also, Nicol et al. [46–48] have presented several test results of MLI performance, among them heat flux of about 50–80 mW/m^2 between 77 K and 20 K with five and ten MLI layers. They also stated that the number of layers becomes important when vacuum pressure is below 10^{-6} torr, and 600-Å -thick aluminum coatings offer no significant improvement over 350 Å .

3.10.3 MLI Performance Test for 260 K–19 K

In order to demonstrate the application of MLI and other technologies on a relative scale for upper stages, Johnson et al. report that NASA is developing the Structural Heat Intercept, Insulation, and Vibration Evaluation Rig (SHIIVER) [47–48]. One of the objectives of SHIIVER is to demonstrate the benefits of MLI as applied to upper-stage hydrogen and oxygen applications. The test assembly is also cooled by cryocoolers and relies on a calibrated conduction rod to determine heat flow, shown in Chapter 5 of this book. The test results are shown in Table 3.7.

TABLE 3.6
Thermal Performance of MLI Sample

MU sample	N [layers]	N/h [layers/mm]	T_h [K]	T_c [K]	[W/m^2]
Around 70 K drum	50	15.2	298	64	1.5
		18.1	298	68	1.6
Around 4 K drum	20	11.5	64	5.5	0.025
		15.3	68	6.4	0.034

Source: KFP-9B08 [45]

TABLE 3.7

Test Results for 50- and 30-Layer MLI Blankets

Coupon Number	Configuration	Q_{total} W	Q_{net} W	Thickness, cm	Layer Density. Lay/cm	q_{net} W/m²
1	50 layers	0.937	0.931	2.7	18.5	0.670
2	30 layers	0.923	0.923	1.9	15.8	0.674

Source: Calibrations [48]

TABLE 3.8

Heat Flux vs Layer Number for Type A MLI at Density 20/cm and vs Packing Density for Type B MLI at 37 Layers

Layer Number	$T_c = 80$ K [W/m²]	$T_c = 5$ K [W/m²]	Packing Density	$T_c = 80$ K [W/m²]	$T_c = 5$ K [W/m²]
10	0.95	0.97	30/6 mm⁻¹	2.52	-
20	0.49	0.51	30/8.5 mm⁻¹	1.48	1.44
30	0.43	0.44	30/15 mm⁻¹	0.40	0.51

Source: All $T_h = 300$ K [49]

3.10.4 OTHER EXPERIMENTAL STUDIES DOWN TO 4 K

Mazzone et al. carried out measurement of MLI from 300 K to 77 K and directly from 300 K to 4 K at CERN [49]. In-situ measurement of the thermal conductivity is obtained for two types of MLI blankets with different numbers of layers and layer densities: type A: 400 Å (Al) and type B: 150 Å coating; measured data are shown in Table 3.8.

Venturi et al. preliminarily introduce the function of a single MLI layer and single AL foil between 20 K and 60 K to 4.2 K [50]. Considering a future circular collider (FCC) of about 80–100 km in circumference, a predicted shield of *T* range from 20 K to 60 K will be considered. Results show that below 50 K, the heat flux to 4.2 K is almost constant for both configurations, and the benefit of putting one foil on the test vessel decreases the heat flux to 20 mW/m².

3.11 CHALLENGES AND REMEDIES IN REAL MLI SYSTEMS

Multilayer insulation has been studied for a long time, and many aspects of its performance can be predictable. However, the efficiency of real MLI systems also depends on practical implementation and workmanship. The ideal packing density should be well examined after installation of the blankets. MLI systems tend to be compressed by their weight on the top part of the thermal shields and cold masses. Differential thermal contractions between the MLI and its support can also cause deformations of insulation structures when cooled down.

There can be serious problems and damage associated with cracks, slots, and penetrations in practical MLI systems, which require better understanding, explanation, and remedy. Reducing the effects of cracks and penetrations on MLI performance is critical for economics, thermal performance, and mission assurance.

3.11.1 GREATLY UNEXPECTED HEAT FLUXES THROUGH CRACKS/SLOTS

Shu et al. reported experimental studies of the unexpected effects of cracks/slots (in different sizes and shapes) on the overall heat loads through MLI blankets of 2.24 m² both at 300 K–77 K and at

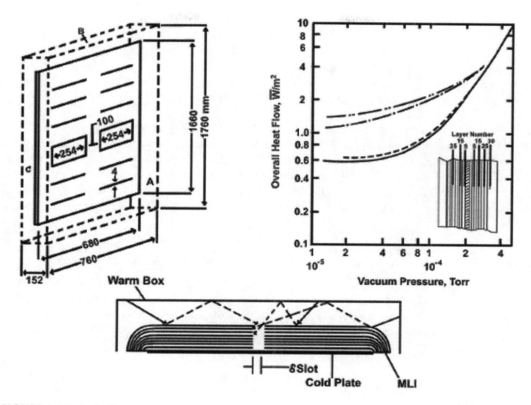

FIGURE 3.15 Left: Sketch of cracks/slots in the MLI blanket for testing between 300 K and 77 K and 77 K and 4.2 K. Right: Heat flux vs vacuum level (300–77 K), without cracks, one-dimensional sharp cut, 4-mm slots, square holes. Lower: Radiation flux re-reflected by the shiny surface of the MLI blanket, which then enters the slot to enhance the effect of the black cavity. [51]

77 K–4.2 K [51–53], as illustrated in Figure 3.15. In order to study the effect of slot depth on heat flux, cracks/slots of various widths and lengths were cut in the MLI blanket. An adjustable frame of Micarta bars was used to cut accurately sized cracks in the fluffy MLI blanket. The gap between the bars was exactly the width desired. The bars were clamped to the test fin at each end, taking special care not to damage the MLI on the back side; they must be held firmly while cutting so that the layers of insulation will not be pulled and torn instead of cut by the blade.

The dependence of the heat load on crack width, geometry, properties of the cold surface under the crack, depth of the crack, and overall vacuum pressure is systematically reported in detail in [51, 52]. Generally, serious damage on overall performance of the MLI blanket by cracks and slots is beyond expectation. In one tested group, fourteen 6- or 4-mm-wide, 245-mm-long slots were cut in both a 30-layer and a 90-layer MLI blanket.

As one testing example, the area, A, of the fin (MLI blanket) is 2.24 m², and the total surface area of slots (14 slots of 4 mm width × 254 mm length), ΔA (using the optically measured widths), is 0.0114 m². The testing results are:

1. The heat flux, Q_0, from a 277-K copper box to a black-painted fin through 30 layers of MLI without cracks is 1.4 W (equal to 0.625 W/m²).
2. The total heat flux through the MLI blanket with slots, $Q_{c,b} = 2.74$ W, and the total heat flux increase due to the slots, $Q_{o,b} = 1.34$ W (equal to 1.34 W/0.0114 m² = 117.5 W/m²).
3. The heat flux from the box to the black fin without MLI is 61.3 W (equal to 27.4 W/m²).

Based on this experimental data, the conclusions can be summarized as follows:

- The ratio of the heat flux increments due to unit slot area to the heat flux through per-unit area of 30 MLI layers is (117.5/0.625 Wm^{-2}) = 188.
- The ratio of the heat flux increments due to unit slot area to the heat flux directly to the per-unit area of the black-painted surface without MLI is (117.5/27.4 Wm^{-2}) = 4.3.

The experimental data prove that cracks/slots can cause great unexpected damage, as much as 188 times larger per unit area than an MLI blanket without cracks/slots. These results are in good agreement with the prediction of the enhanced black cavity model. Studies also show that the increments of heat flux due to slots are not sensitive to cold surface emissivity. A systematic investigation was performed in the following aspects: (1) effects of various cracks/slots on heat loads with different numbers of layers and (2) temperature distribution and thermal conductivity among cracks/slots [51].

3.11.2   Shu's Enhanced Black Cavity Model Theory for MLI with Cracks/Slots

To theoretically explain why the effects are so serious on MLI performance, the so-called enhanced black cavity model (EBCM) was developed by Q. S. Shu [51]. The results calculated using EBCM are consistent with series testing data. The math of the enhanced black cavity model is introduced in the details of the cited reference, and the core concepts are briefly discussed in the following with Figure 3.15 (Lower):

1. A crack/slot in an MLI blanket was considered first as a black cavity in a low-emissivity enclosure, which absorbed all the incident radiant energy in the sandwich structure of the MLI edges.
2. The multi-reflection of radiation fluxes between the shiny MLI and inner surface of the vacuum jacket greatly enhances the radiant flux into the black crack.
3. The temperature distribution in the region near the crack increases the heat transfer.
4. The relative dimensions of a crack also influence its absorptivity; this is considered a cavity effect here.
5. If there are n slots, the total radiation absorbed can be calculated by Equation 3.8 in the first order of approximation.

$$Q_S = \eta \sum_{i=1}^{n} \varepsilon_2 \alpha T_2^4 \left[S_i \left(F_{S_i-A} + F_{S_i-B} + F_{S_i-C} \right) \right] \tag{3.8}$$

where n is the number of slots; η is the total enhancement factor (determined experimentally); α is the Stefan-Boltzmann constant 5.67×10^{-8} Wm^{-2} K^{-4}; ε_2 is the emissivity of the warm box; T_2 is the temperature of the warm box; S_i is the area of slot i; F_{S_i-A}, F_{S_i-B}, and F_{S_i-C} are the view factors from slot i to the warm walls A, B, and C, respectively. η can be experimentally determined by measuring the Qs for a particular geometric enclosure of the MLI blanket. In the particular case in Figure 3.15, η is 3.8, while F_{S_i-A}, F_{S_i-B}, and F_{S_i-C} are calculated.

3.11.3   Patch-Covering Technique for Remedy of MLI Performance

To eliminate the most performance degradation of MLI blankets by cracks/slots, the patch covering technique (PCT) is the most effective remedy method. In the PCT, MLI patch layers are employed to cover the cracks, slots, and seams. The issues with the patch covering technique are: how many MLI layers to employ, where and how the MLI layers are to be inserted in the MLI depth to cover

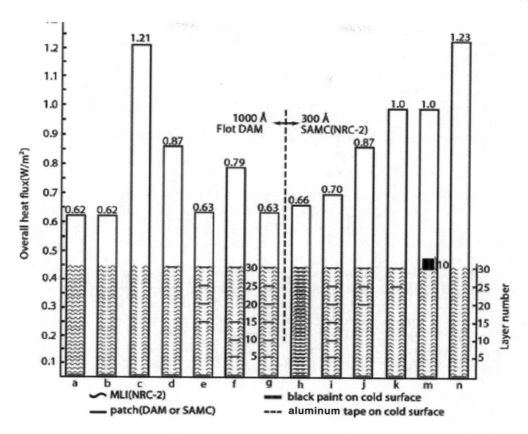

FIGURE 3.16 Graphic summary of experimental patch study. A: No cracks; B: one-dimensional slits (sharp knife cut); C and N: no patch; D to M: various patch distributions. [51]

the cracks/slots, and which MLI materials to use. To practically answer these questions, a systematic investigation to reduce the heat load to a 77 K and 4.2 K surface through cracks/slots in a MLI blanket was conducted by Shu et al. at Fermilab [40–42, 51].

Typical comparisons of testing results are illustrated in Figure 3.16, which summarizes 15 tests with MLI constructed using SACM material. The test specimen is a 30-SACM layer blanket. The Y-axis presents the overall heat flux in each column. The columns on the X-axis represent the test results: column height is the heat flux, and the configuration indicates the number of patches used and their locations. Two background tests are performed: first, the heat loads through the MLI blanket without any cracks/slots are measured as 0.62 W/m², shown in column a in Figure 3.16; the column height represents the heat load value. Second, 14 one-dimensional cuts (254 mm length) are made with a sharp knife through the MLI blanket, and the heat load is measured as 0.62 W/m², shown in column b (equal to that without cuts). The data prove that properly sharp cuts have no negative impact on MLI blanket performance, and the normal practice of properly cutting MLI blankets for better evacuation is acceptable.

Then the 14 one-dimensional cuts were enlarged to 4 mm width with a special tool, and the measured heat loads of the MLI blanket greatly increased to 1.21 W/m², as indicated in column c in Figure 3.16. Data from patches of 300-Å SACM (NRC-2) are shown on the left side of the dashed vertical line, and data of the 1,000-Å DAM patches are on the right side. The main conclusions are summarized as follows:

1. Four to five patches inserted on several top layers will totally recover MLI thermal performance, as shown in columns e and g. Therefore, too many patches deeply inserted into the MLI blanket are not cost effective, as shown in column f.
2. 1,000-Å DAM patch materials are better than 300-Å SACM.
3. For engineering convenience, covering cracks/slots with 10 to 15 patches on the top is also effective, as shown in column m. In addition, various patches overlapping on cracks/slots are very effective.
4. Columns e and f show patch cover on the warm side is better than on the cold side.

3.11.4 Engineering Remedy for MLI with Many Joins/Seams

Johnson et al. introduced engineering remedy tests of MLI blankets with many seams for space exploration. The total number of MLI layers is variable but can be on the order of 50 for cryogenic propellant tanks of long-duration missions. Performance degradation in MLI systems due to joints and seams in insulation blankets has been recognized since the introduction of MLI. For large tanks, seam numbers increase as tank dimensions exceed the roll widths available for insulation materials. Early studies focused more on quantifying the effect of seams in an MLI blanket than understanding the mechanisms. Shu, as part of the superconducting supercollider effort, developed a theoretical model of seam behavior and validated it against a series of tests between liquid nitrogen and room temperature. Recently, Johnson and Fesmire examined joint/seam issues with a test apparatus at Kennedy Space Center (KSC) and not only experimentally verified the work by Hinckley but also extended this research into LH_2 temperature to study a broader range of seam configurations [47–48], as illustrated in Figure 3.17.

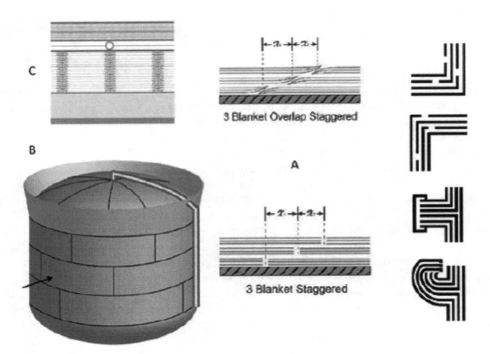

FIGURE 3.17 Configurations of three MLI blankets staggered, B: 3D of propellant tank, C: a practical MLI blanket, D: optional blanket seams at the edge/corner and flat. [48]

3.11.5 MLI CONFIGURATION OF JOINTS/SEAMS AND TESTING RESULTS (300 K TO 20 K)

Two sets of coupons were fabricated for testing: one 50-layer set and one 20-layer set. Both were made of two sub-blankets. As expected on the 50-layer test, the layer-by-layer interleaved joint had the lowest heat leak. The overlap joint had a slightly better performance than the straight and staggered butt joints. Surprisingly, staggering the butt joint did not decrease the heat load, and increasing the stagger distance did not help. In fact, the test with the largest stagger was worse than the straight butt joint, although this may be due to damage incurred by repeated handling rather than the joint itself. Even for the worst-performing seam, the results are only 5% more heat leak than the best-performing seam.

Results with the 20-layer testing are a bit less conclusive; however, similar results are seen. The overlap seam still performs very well in this case, performing nearly identically to the baseline interleaved blanket. The poor performance of the offset butt joint is yet unexplained, being 20% worse than the interleaved blanket. The full butt joint slightly outperforms the offset butt joint and is within 15% of the interleaved blanket. However, labor intensiveness and time consumption must be considered. The theoretical butt seam heat load from Hinckley is on the order of 0.094 W/m for a 20-layer blanket and 0.050 W/m for a 50-layer blanket similar to those tested in this campaign. Those values are on the same order of magnitude of the 0.15 W/m and 0.06 W/m measured.

3.11.6 PATCH-COVERING METHOD FOR 4 K SURFACES

Due to requirements for large superconducting magnet development, the patch-covering technique for MLI was extended to be verified between 77 K and 4.2 K [42]. Shu et al. conducted the experimental investigation with the same device and cutting slots as shown in Figure 3.15. The total measured heat load without cracks is $Q_0 = 29.9$ mW/m^2 and with cracks is $Q_c = 32.42$ mW. The heat load through slots, defined as $Q_S = Q_C - Q_0$, was 2.52 mW/m^2. Since the total area of the slots is $A_S = 0.0114$ mm^2, the heat flux through the slots is $Qs/As = 221$ mW/m^2. The heat flux through the slots was about seven to eight times that through an MLI blanket without slots. Then, the slots on layers 15, 20, 25, and 30 were covered with patches, and the total heat load was 30.3 mW, which is approximately equal to the heat load through MLI without slots. The patch material was 1,000-Å double-aluminized flat 6.35-µm Mylar. Compared to MLI between 300 K and 77 K, where the heat flux through cracks is nearly 200 times that of a blanket without slots, the crack/slot effect from 77 K to 4.2 K is much less than that from 300 K to 77 K. One of the reasons is that the heat transfer in a MLI system between 300 K and 77 K at high vacuum is primarily by radiation, while solid conduction/residual gas plays a more important role between 77 K and 4.2 K.

3.12 EXPERIMENTAL STUDY OF HEAT TRANSFER MECHANISMS

Besides heat fluxes obtained by measuring and calculation, experimental study of the temperature distributions and local equivalent thermal conductivity among the layers will help in understanding

TABLE 3.9
Test Results

Coupon Number	Configuration	Q_{total}, W	Q_{net}, W	Q_{seam}, W	% change
2	Single seam	0.93	0.92	0.135	
3	Double seam	1.06	1.06	0.27	14.6%

Source: For Single and Double Seams [48]

the heat transfer mechanism in MLI systems. This information not only extends theoretical knowledge but also gives more alternatives for engineering practices. As a practical example, the T distribution data inside an MLI blanket can be used to monitor the states of thermal assembly, thermal equilibrium, and vacuum levels deep inside vacuum jackets.

3.12.1   Eight Experiments for T Distributions

Long copper-constantan thermocouples (diameter 0.07 mm) are thermal-anchored to a 300-K warm box and 77-K cold plate and placed on layers 1, 5, 10, 15, 20, 25, and 30 for the 30-layer MLI blanket as presented in Figure 3.15 (with/without slots, respectively). When the thermal equilibrium is reached (24 hours after filling LN_2), the heat loads (boiloff) and temperatures are recorded at the highest vacuum reached. Then, boiloff and temperatures are taken again following the new vacuum pressure step up to the new thermal equilibrium levels. A total of eight experiments were conducted with blankets, which have total layers 5, 10, 20, and 30, by Shu et al. [40, 51].

3.12.2   Temperature (T) Distributions

The typical T distributions in different MLI layers for four blankets as functions of both location depth in MLI (layer number) and vacuum level are illustrated in Figure 3.18. For each group of T distribution, the curve with high vacuum is located on the northwest side, and the curve with low vacuum tends to be on the southeast side.

Shown in Figure 3.19 (Left) is the temperature of the last layer (warmest) and first layer (coldest) in MLI blankets between 77 K and 277 K as a function of the total number of MLI layers. All possible temperatures in MLI blankets of from 5 to 90 layers fall into the cross-hatched area. The temperature of the first layer (cold) in blankets with fewer layers is much higher. So, the coldest layer can hang on the cold mass with blankets with more layers, but the coldest layer needs to be kept floating from the cold mass for blankets with fewer layers.

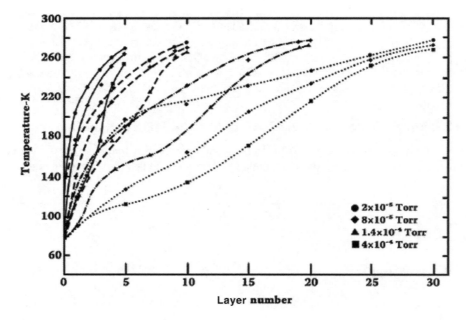

FIGURE 3.18   Temperature distribution MLI as a function of the vacuum level. •, 2×10^{-5} torr; ♦, 8×10^{-5} torr; ▲, 1.4×10^{-4} torr; ■, 4×10^{-4} torr [40].

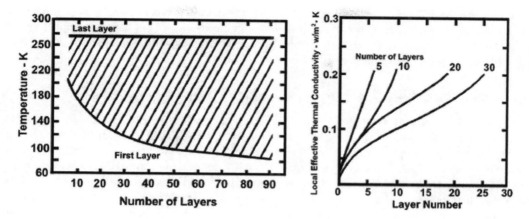

FIGURE 3.19 Left: Temperature of the first (coldest) and last (warmest) layers as a function of total number of layers of different blankets. Right: Local equivalent thermal conductivities in four different MLI blankets. [40]

3.12.3 Calculation of Local Equivalent Thermal Conductivity

All of the effects influencing the temperature distribution in MLI blankets will cause the effective thermal conductivity to be a function of depth in the MLI blanket. The total effective thermal conductivity can be defined as

$$K(N) = \frac{Q(N)}{D(N)} \frac{N}{\Delta T(N)} \tag{3.9}$$

Defining K_{ij} as the effective thermal conductivity between the ith and jth layers

$$K_{ij} = \frac{Q_{ij}}{D_{ij} A} \frac{N_{ij}}{\Delta T_{ij}} \tag{3.10}$$

where A (m$_2$) is the heat transfer area, D (N) is the total layer density for an N-layer system (m^{-1}), D_{ij} is the layer density between layers i and j (m^{-1}), K (N) is the effective thermal conductivity of an N-layer system (W m^{-1} K^{-1}), $K'(N) = K(N) D(N)$ (W m^{-2} K^{-1}), and K_{ij} is the effective thermal conductivity between layers i and j (W m^{-1} K^{-1}).

The local equivalent thermal conductivities in four different MLI blankets are calculated as in Figure 3.19 (Right).

3.12.4 Local Equivalent Thermal Conductivity w/o Slots for 77 K to 4.2 K

In order to understand the heat transfer mechanism in MLI between 77 K and 4.2 K, Shu et al. conducted a series of tests with similar methodology and the same test apparatuses but different thermocouples. Chromium/gold 0.07% iron type thermocouples were used [42]. In addition, 14 slots, 254 mm long by 4 mm wide, were cut in the 30-layer MLI blanket and vacuum at 5 × 10^{-8} torr. The local equivalent thermal conductivity calculated from the temperature distribution is interesting: It has a broad maximum near the center of the blanket. This is in contrast to an MLI blanket between room temperature and 77 K, where the equivalent thermal conductivity always increases with increasing temperature.

The temperature difference between the last layer and the warm box is ~7 K for a 30-layer blanket between 77 K and 4.2 K, which is larger than that in the 300 K to 77 K case. Therefore, contact of the last layer with the warm box will cause a more serious heat load increase from 77 K to 4.2 K, probably due to solid conduction contribution in the MLI [42].

3.13   MLI COMPOSITES, HYBRIDS, AND STRUCTURAL ATTACHMENTS

For high-performance MLI systems (that is, $q < 1$ W/m^2 or $k_e < 0.1$ mW/m-K), the amount of energy coming through the supports can contribute 50% or more of the total heat leak [53]. Superconducting power cables, for example, require $k_e < 1$ mW/m-K [3]. Laboratory or idealized constructions of MLI can easily provide this level of performance. Some examples of ideal MLI systems with k-values in the range of 0.05 mW/m-K are given by Ohmori [44] and Kaganer [54]. However, the reality of design, fabrication, and operation can reduce the thermal effectiveness by one or two orders of magnitude. Refrigeration systems must, therefore, be increased to compensate for the additional heat load due to the reduced insulation effectiveness.

3.13.1   IDEAL MLI VS PRACTICAL MLI

Practical experimental comparisons between ideal MLI and actual MLI, including rigid vs flexible piping, have been previously reported. For similar 60-layer systems of foil and paper, the increased heat transfer was about 80% for the rigid piping installation and about 50% more from rigid to flexible [55]. Even ideal MLI performance can be degraded by the weight of the MLI itself and the resulting contact heat transfer between layers. A non-dimensional contact pressure parameter has been proposed by Ohmori to account for this additional heat transfer [56]. In the case of flexible piping, the weight of the inner line pressing against the outer line, combined with the bending that compresses the layers of materials, increases solid conduction while inhibiting full evacuation between the layers. Localized damage due to spacer structures and bending of piping has been shown to increase heat transfer by approximately 40% [57]. The total (installed) system thermal conductivity, or k_s, has also been defined as a practical measure of the overall efficiency of the insulation as installed in a complete system [58].

The first and foremost operational problem with MLI is the vacuum pumping. Many closely spaced layers make proper evacuation below 0.1 millitorr between all layers difficult to achieve. A degraded vacuum to 13.3 Pa (100 millitorr) will cause an increase in heat flux by more than two orders of magnitude, from 1 to 100 W/m^2 [57]. The support structures and methods of attaching, taping, or securing the MLI blankets can cause edges to be compressed, which will hinder the vacuum pumping process. Studies of variable-density MLI systems have shown that lower layer density nearer the cold mass will provide a significant performance benefit [59].

A loss of vacuum can mean a major loss of product or a facility shutdown. A sudden loss of vacuum leading to overpressurization of a system could have catastrophic consequences, including injury to personnel and major equipment damage. The possibility for these events in new high-performance systems must of course be minimized by engineering and technical standards.

3.13.2   ADDITIONAL CONSIDERATIONS OF MLI SYSTEMS

Combining thermal, mechanical, and operational considerations, the insulation system design could include supports composed of insulation materials. Robust MLI systems, therefore, meet all of the following criteria to a reasonable extent.

Additional considerations of MLI systems:

- High-vacuum environment is required
- Evacuation must include reaching high vacuum between all layers

- System design must allow for coverage of complex shapes
- System design must consider structural supports and other mechanical obstacles
- Installation must not introduce any significant heat paths nor prevent proper evacuation
- Layers must stay put during installation, evacuation, operation, and maintenance
- Evacuation or release of vacuum must not damage the layers
- Effect of degraded vacuum levels on thermal performance must be considered

For more robust MLI systems, including both thermal and mechanical considerations, layered composite insulation (LCI) systems have been developed.

3.13.3 LAYERED COMPOSITE INSULATION SYSTEMS

Like MLI systems, layered composite insulation (LCI) systems incorporate alternating layers of reflectors and spacers [8, 10]. LCI system designs have larger interlayer spacing to reduce vulnerability to compression (and consequent heat leak) caused by installation and use. This arrangement also allows the spacer layers to keep their loftiness, which is a key part of the very low thermal conductivity. The overall density of an LCI system is typically less than one layer per mm.

An LCI system includes radiation shield layers, powder layers (aerogel or fumed silica), and carrier layers (non-woven fabric or fiberglass paper). The layers are put together by a continuous roll-wrap process. The powder layer can be deposited on the surface of the carrier layer or within the carrier layer itself. LCI products can be produced in forms such as multiple layer rolls, blankets, and cylindrical sleeve packages. The products can be tailored for a specific application. Optional edge strips can be used to set a gauge of layer thickness or provide mechanical load capability. A single layer or many layers can be used. The optional outer wrapper material can be used if improved handling ability or an extra measure of powder containment is desired. A stack-up including 5, 10, or 15 layers for installation within a tank or piping annular space is a typical installation. Various configurations of LCI, including radiation, powder, and carrier layers, were tested for cryogenic thermal performance.

The insulation test materials were horizontally roll-wrapped onto the cylindrical cold mass for Cryostat-200 (comparative, cylindrical apparatus) tests. The cold mass for Cryostat-200, before and after insulation wrapping, is shown in Figure 3.20. The materials were horizontally roll-wrapped onto a copper sleeve for Cryostat-1 tests. Cryostat-100 test article preparation was accomplished by positioning the materials layer by layer onto the vertically oriented cold mass. Standard MLI or superinsulation (SI) constructions are composed of a reflective shield (aluminum foil 0.00724 mm thick) and spacer (fiberglass paper 0.061 mm thick), double-aluminized Mylar with paper spacer, or double-aluminized Mylar with bonded non-woven polyester spacer (Cryolam). The installed thickness for most test articles was from 20 to 25 mm.

3.13.4 THERMAL TEST RESULTS OF LCI SYSTEMS

Over 400 tests of 30 different MLI and LCI systems were performed. Preliminary results using Cryostat-1 have been previously reported [14]. Further testing was completed using Cryostat-2 and Cryostat-100. Values for k_e of LCI systems are shown in comparison with three popular MLI systems. Performance is comparable at high vacuum even though the LCI systems have only half the layers of MLI systems. LCI combinations utilizing Mylar were found to have an advantage for the full vacuum range.

A compilation of selected results from Cryostat-100 testing is given in Figure 3.21. Test article descriptions are presented in Table 3.10. The LCI systems shown by the solid lines are clearly superior in the soft vacuum range. The optimized LCI system C130 is also shown to be comparable to the benchmark MLI system C123 in the high vacuum range. The Kaganer line, representing ideal MLI, is for an aluminum foil and fiberglass spacer system (40 layers, 1.5 layers per mm, 293 K and 90 K

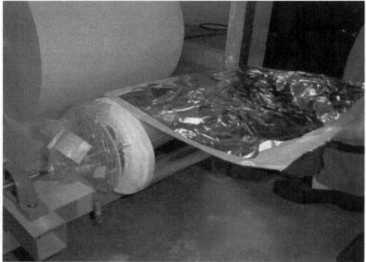

FIGURE 3.20   Cryostat-2 cold mass shown before (upper) and after (lower) roll-wrapping with insulation materials.

boundary temperatures). For comparison, layered systems of aerogel blankets and aerogel blankets plus MLI are also presented. As expected, the layered aerogel is best at ambient pressure, while the aerogel and MLI combination system gives a dramatic advantage at higher vacuum levels. The higher-density prototype LCI system (A110) is not the best thermal performer but offers a high level of mechanical load-carrying capability if that feature is of importance in the overall system design.

3.13.5   APPLICATION AND DISCUSSION OF LCI SYSTEMS

A robust MLI design, like the example LCI systems, must consider the total system design, including coverage of complex shapes, structural supports, and other mechanical obstacles. The high vacuum requirement is further defined to mean that evacuation or release of vacuum must not damage

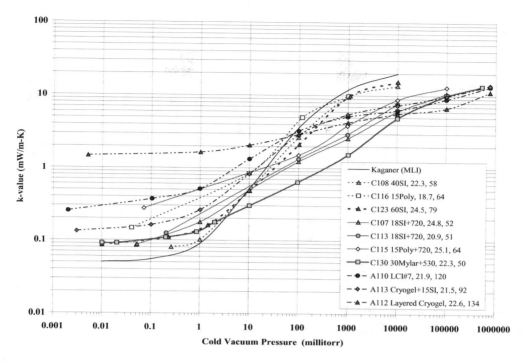

FIGURE 3.21 Test results reported in k_e for LCI, MLI, and other layered insulation systems. Boundary temperatures are approximately 78 K and 293 K; the residual gas is nitrogen (1 millitorr = 0.133 Pa).

the layers and that evacuation must include reaching high vacuum between all layers. Finally, installation methods for robust MLI systems must be worked out to prevent additional heat leaks and to allow complete evacuation of the system.

Cryogenic thermal performance of LCI systems are approximately six times better than that of MLI systems at soft vacuum (1 torr) and comparable to MLI systems in high-vacuum environments. For the benchmark MLI system C123 compared to the optimized LCI system C130, the k_e data are summarized as follows: 0.086 vs 0.091 mW/m-K at HV and 10.0 vs 1.6 mW/m-K at SV. The LCI thermal performance is excellent at ambient pressure but not as good as aerogel blankets at this condition. LCI systems consisted of half the number of layers of corresponding MLI systems but with overall thicknesses kept approximately the same. Presenting the results in k_e rather than heat flux gives the best comparison between different systems, as small variations in thickness and boundary temperatures are normalized.

LCI systems can be used for double-walled piping or tank constructions where degraded vacuum or loss of vacuum is a concern. Designed as HV systems, LCI would provide a back-up level of performance to prevent loss of product or to reduce system maintenance. With the expense required to produce and maintain a high vacuum level, industry in many cases is looking for an efficient, low-cost system with high performance in the soft vacuum range [13, 14].

3.14 DEMONSTRATION OF SUCCESSFUL MLI SYSTEMS

Discussions in this section only focus on MLI systems for space explorations and SC applications. MLI systems for storage vessels, tanks pipes, and cryostats are presented in Chapter 5. Robust MLI technology can enable new technology in other energy-intensive areas such as hydrogen transportation and superconducting power transmission.

TABLE 3.10

Description of Cryostat Insulation Test Articles from Cryostat-100 Testing

Test Series No.	Insulation System Type	Description of Insulation System	Total Thickness (mm)	Installed Density (kg/m³)
C107	LCI	18 layers foil + paper + fumed silica 720	24.8	52
C108	MLI	40 layers foil + paper at 1.8 layers/mm (40 SI)	22.3	58
C113	LCI	18 layers foil + paper + fumed silica 720	20.9	51
C115	LCI	15 layers foil + polyester fabric + fumed silica 720	25.1	64
C116	MLI	15 layers foil + polyester fabric	18.7	51
C123	MLI	60 layers foil + paper (60 SI)	24.5	79
C130	LCI	30 layers Mylar + paper + fumed silica 530	22.3	50
A110	LCI	Prototype system using aerogel blankets	21.9	120
A112	Material	6 layers of Cryogel aerogel blanket	22.6	134
A113	Material	Cryogel with 15 layers foil + paper	21.5	92

3.14.1 MLI Systems for Space Exploration

New aerospace and space exploration initiatives are requiring cryogenic insulation systems that will perform well under the full range of vacuum levels: Earth (no vacuum), Mars (soft vacuum), and Moon (high vacuum). Earth-to-Moon missions requiring cryogenic tanks will require launch pad hold times in the ambient pressure environment and then long-term storage in the high-vacuum environment of space. Robust MLI, working in concert with load-supporting insulation technology and active thermal systems, will help enable truly mass-efficient spacecraft designs by replacing the complex, heavy support structures with an insulation system that does its thermal job, carries the mechanical loads, and provides vibration damping in operational environments.

The space exploration arena consists of space launch to low Earth orbit (LEO) and beyond. The world's existing large launch vehicles can loft payloads of about 25 tons to LEO, or 9 tons to Earth escape [60]. For the NASA Artemis program and Space Launch System (SLS) vehicle, needing about 160 tons in LEO is several times the capability of any existing launch vehicle [64]. For all programs, presently and in the future, LH_2 and LO_2 as propellants have been employed to launch vehicles. Therefore, cryogenic storage tanks of LH_2 and LO_2 are essential.

For a short period of time, launch and subsequent on-orbit payload delivery (typically <24 hours), these launch vehicles and cryogenic upper stages include spray-on foam insulation (SOFI) and thin, limited-layer MLI. Longer-duration storage and utilization of hydrogen and oxygen cryogenic fluids are routinely demonstrated with every space shuttle flight via the power reactant storage and distribution (PRSD) tanks. These tanks store supercritical cryogenic hydrogen and oxygen, the respective gasses from various usages.

For the long-duration space applications, cryogenic propellant tanks are all insulated with multi-layer insulation to minimize radiative heat loads in the space vacuum environment. MLI with advanced insulation structures is employed, as shown in Figure 3.22. The total number of MLI layers is variable but can be 50 layers or more for long-duration missions. Performance degradation in MLI systems due to joints, seams, attachments, and other design details has been recognized as a serious concern and should be considered in the design.

3.14.2 MLI Systems for Space Science Missions and Payload Applications

If an instrument in space is to receive and deliver signals, its white noise must be less than the signals. Because white noise is proportional to T^3, the T of the system of sensors must be kept very low. MLI

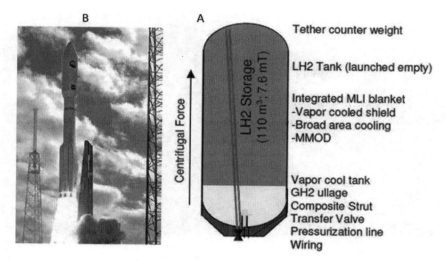

FIGURE 3.22 A: The LH$_2$ propellant module for long duration. B: Atlas V 551 launch of the Pluto New Horizon spacecraft [60]. Source: *ULA*

blankets provide passive thermal control to a variety of spacecraft, launch vehicles, and instruments in vacuum. Films of different polymeric substrates such as Teflon or Kapton are coated with aluminum by the vacuum deposited aluminum (VDA) process. Spitzer demonstrated an equivalent average cryogen boil-off rate of 0.05% per day, achieving an operational mission duration total of over five years on-orbit. Blankets for the Hubble Space Telescope's (HST's) external surfaces consist of a 17-layer MLI stack of 5-mil (127-μm)-thick VDA-Teflon outer layers and 15-layer 0.3-mil (7.62-μm) VDA-Kapton inner layers, as illustrated in Figure 3.23 (Right) [61]. Bay areas 5 and 8 had small amounts of damage (less than 1 mm) after years of service before being repaired. The left photos show two astronauts dressed in MLI-insulated spacesuits working on a spacecraft, the Hubble Space Telescope.

3.14.3 MLI SYSTEMS FOR SUPERCONDUCTING ACCELERATORS

The LHC superconducting magnet (cold mass) cooled in superfluid helium at 1.9 K is contained inside a stainless-steel helium vessel with an external diameter of 0.6 m and length 15 m and housed by a cryostat [62]. The LHC has about 2,000 dipole magnets in the 27-km-long tunnel, and each magnet's cold mass is 30,000 kg (2,000 kg/m length). The external diameter of the cryostat outer vessel is about 1.0 m. The cold mass is supported inside the cryostat by three support posts. Per unit length of cold mass, the lateral cold surface area is about 28 m^2 per magnet to be protected by MLI blanket (Figure 3.24). The 30–40-layer MLI blanket with layer density of 15–25 layers/cm is wrapped on the intermediate *T* shield, which is cooled by cryogen to 60–80 K. A 10-layer MLI is allocated on the 1.8-K cold mass.

The X-ray free electron laser is another large SC accelerator project, with 1,000 m of SC accelerating cavities. Its 12-m-long cryomodule houses eight superconducting cavities and is operated at 2 K. The cryomodules provide mechanical support and thermal insulation to the RF cavities. In their final design, 5 K shields are omitted, but a 30–40-layer MLI blanket on the 55 K shield and 10 MLI layers on the 2 K assembly were kept.

3.14.4 MLI SYSTEMS FOR FUSION PROJECTS

The JT-60SA superconducting tokamak was a testing fusion project prior to the International Thermonuclear Experimental Reactor (ITER) project, and its SC magnets must be thermally

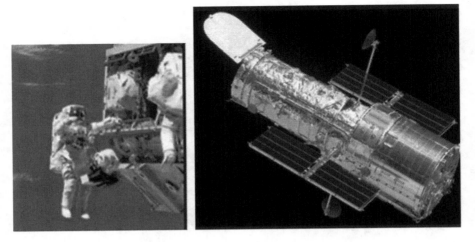

FIGURE 3.23   Right: the outer surfaces of HST are covered by MLI blankets [61]. Left: Astronaut working with spacecraft. Source: *NASA*

FIGURE 3.24   LHC dipole magnet being connected in tunnel (left), and preproduced MLI blankets being installed on the magnet cold mass (right) [62]. Source: *Courtesy CERN*

isolated from warm ambient *T* to achieve economical operation. Its huge refrigerator, with equivalent cooling power of 9 kW at 4.5 K, produces the helium for the SC coils. The JT-60SA cryostat is composed of the cryostat base, cryostat vessel body cylindrical section (CVBCS), and cryostat top lid. All these sections are thermally insulated by MLI blankets. Regarding ITER, its cryogenic system is introduced in Chapter 13.

REFERENCES

1. Fesmire, J., and Johnson, W., 2018. Cylindrical cryogenic calorimeter testing of six types of multilayer insulation systems. *Cryogenics*, Vol. 89, 58–75.
2. Fesmire, J.E., 2018. *Aerogel-Based Insulation Materials for Cryogenic Applications*. International Cryogenic Engineering Conference, Oxford University, UK.
3. Demko, J.A., Fesmire, J.E., and Augustynowicz, S.D., 2008. Design tool for cryogenic thermal insulation systems. *Advances in Cryogenic Engineering*, Vol. 53A, AIP145–151.

4. Johnson, W.L., Kelly, A.O., and Fesmire, J.E., 2014. Thermal degradation of multilayer insulation due to the presence of penetrations. *Advances in Cryogenic Engineering, AIP AIP Conf. Proc.*, Vol. 1573, 471.

5. ASTM C740—Standard Guide for Evacuated Reflective Cryogenic Insulation. ASTM International, West Conshohocken, PA, USA (2013).

6. Adams, J., et al., 2018. *Integrated Insulation System for Cryogenic Automotive Tanks (iCAT)*. United States: N. p., Web doi:10.2172/1457536.

7. Johnson, W.L., and Fesmire, J.E., 2012. Thermal performance of low layer density multilayer insulation using liquid nitrogen. *Advances in Cryogenic Engineering*, Vol. 34.

8. Fesmire, J.E., Tomsik, T.M., Bonner, T., Oliveira, J.M., Conyers, H.J., Johnson, W.L., and Notardonato, W.U., 2014. Integrated heat exchanger design for a cryogenic storage tank. *Advances in Cryogenic Engineering, AIP Conf. Proc.*, Vol. 1573, 1365–1372.

9. Swanger, A., Jumper, K. et al., 2015. Modification of liquid hydrogen tank for integrated refrigeration and storage. *IOP Conf. Series: Materials Science and Engineering*, Vol. 101.

10. Fesmire, J.E., and Augustynowicz, S.D., 2002. Thermal performance of cryogenic piping multilayer Insulation in actual field conditions. *Cryogenics 2002, IIR proceedings*, 2003-2679.

11. Fesmire J., 2015. Layered composite thermal insulation system for non-vacuum cryogenic applications. *Cryogenics*, Vol. 74, 154–165.

12. Fesmire, J., 2016. Boiloff calorimetry for the measurement of very low heat flows, Part 1: Short history of heat measurement in cryogenics. *Cold Facts, CSA*, Vol. 32, No. 4.

13. Sun, P. et al., 2009. Experimental study of the influences of degraded vacuum on multilayer insulation blankets. *Cryogenics*, Vol. 49, No. 12, 719–726.

14. Augustynowicz, S.D., and Fesmire, J.E., 2000. Cryogenic insulation system for soft vacuum. *Advances in Cryogenic Engineering*, Vol. 45, Plenum Press, New York, pp. 1691–1698.

15. Riesco, M. et al., 2010. Venting and high vacuum performance of low-density multilayer insulation. *AIP Conference Proceedings*, Vol. 1218, No. 1, 796–803.

16. Shu, Qs., Fast, R. et al., 1986. An experimental study of heat transfer in multilayer insulation systems from room temperature to 77 K. *Advances in Cryogenic Engineering*, Vol. 31, Plenum Press, New York, pp. 455–463.

17. Cunnington, G. et al., 1971. Thermal performance of multilayer insulations, interim report. LMSC-A903316/NASA CR-72605, Lockheed Missile and Space Company, Sunnyvale, CA.

18. McIntosh, G.E., 1993. Layer-by-layer MLI calculation using a separated mode equation. *Advances in Cryogenic Engineering*, Vol. 39B, Plenum Press, New York, pp. 1683–1690.

19. Augustynowicz, S.D., Demko, J. et al., 1993. Analysis of multi-layer insulation between 80 K and 300 K. *Advances in Cryogenic Engineering*, Vol. 39B, pp. 1675–1682.

20. Parma, V., 2013. Cryostat Design, 2013 CERN School, Superconductors for Accelerators, May 2013.

21. Shu QS, Private Tech Note, AMAC-TN-04-01, Virginia 2004.

22. Fesmire, J.E., Augustynowicz, S.D. et al., 2008. Robust multilayer insulation for cryogenic systems. *Advances in Cryogenic Engineering*, Vol. 53B, *AIP Conference Proceedings*, Vol. 985, 1359.

23. Fesmire, J.E., Augustynowicz, S.D., and Darve, C., 2003. Performance characterization of perforated MLI blanket. *Proceedings of ICEC 19*, New Delhi, 2006-0162.

24. Knoll, D., Willen, D., Fesmire, J. et al., 2012. Evaluating cryostat performance for naval applications. *Advances in Cryogenic Engineering, AIP Conference Proceedings*, Vol. 1434, No. 265, pp. 39–46.

25. Scurlock, R.G., and Saull, B., 1976. Development of multilayer insulations with thermal conductivities below 0.1 0.1 μW cm−1 K−1. *Cryogenics*, Vol. 16, No. 5, 303–311.

26. Bapat, S.L., Narayankhedkar, K.G., and Lukose, T.P., 1990. Experimental investigations of multilayer insulation. *Cryogenics*, Vol. 30, No. 8, 711–719.

27. Jacob, S., Kasthurirengan, S. et al., 1992. Investigations into the thermal performance of multilayer insulation (300–77 K) Part 1: Calorimetric studies. *Cryogenics*, Vol. 32, No. 12.

28. Ohmori, T., 2005. Thermal performance of multilayer insulation around a horizontal cylinder. *Cryogenics*, Vol. 45, No. 12, 725–732.

29. Shu, Q.S., Fast, R.W. et al., 1986. Heat flux from 277 to 77 K through a few layers of multilayer insulation. *Cryogenics*, Vol. 26, No. 12, 671–677.

30. Johnson, W.L., and Fesmire, J.E., 2010. Cryogenic testing of different seam concepts for multilayer insulation systems. *Advances in Cryogenic Engineering*, Vol. 1218, 905–907.

31. Johnson, W.L., and Fesmire, J.E., 2015. Demonstration of hybrid multilayer insulation for fixed thickness applications. *IOP Conf. Series: Materials Science and Engineering*, Vol. 101.

32. Johnson, W.L., Fesmire, J.E., and Demko, J.A., 2010. Analysis and testing of multilayer and aerogel insulation configurations. *Advances in Cryogenic Engineering*, Vol. 1218.

33. Johnson, W.L., Fesmire, J.E. et al., 2015. Thermal performance testing of cryogenic multilayer insulation with silk net spacers. *IOP Conf. Series: Materials Science and Engineering,* Vol. 101.
34. Dye, S., Kopelove, A., and Mills, G.L., 2014. Load responsive multilayer insulation performance testing. *AIP Conference Proceedings,* Vol. 1573, No. 1, 487–492.
35. Johnson, W.L., Heckle, K.W., and Hurd, J., 2014. Thermal coupon testing of load-bearing multilayer insulation. *AIP Conference Proceedings,* Vol. 1573, No. 1, 725–731.
36. Johnson, W., 2010. Thermal performance of cryogenic multilayer insulation at various layer spacing. Thesis for Master of Science, Dept of Mechanical, Materials, and Aerospace Engineering, University of Central Florida, Orlando, FL.
37. Johnson, W.L., 2010. Optimization of layer densities for multilayered insulation systems. *Advances in Cryogenic Engineering, AIP Conference Proceedings,* Vol. 1218, 804–811.
38. Johnson, W.L., and Fesmire, J.E., 2012. Thermal performance of low layer density multilayer insulation using liquid nitrogen. *Advances in Cryogenic Engineering,* Vol. 57A, *AIP Conference Proceedings,* 1434 39.
39. Hedayat, A., Hastings, L. et al., 2002. Analytical modeling for variable density multilayer insulation for cryogenic storage. *Advances in Cryogenic Engineering,* Vol. 47.
40. Shu, Q.S. et al., 1986. Heat flux from 277 to 77 K through a few layers of multilayer insulation. *Cryogenics,* Vol, 26, 86/120671-07.
41. Shu, Q.S., Cheng, G., Yu, K., Hull, J.R., Demko, J.A., Britcher, C.P., Fesmire, J.E., and Augustynowicz, S.D., 2004. Low thermal loss cryogenic transfer line with magnetic suspension. *Advances in Cryogenic Engineering,* Vol. 49, 1869–1876.
42. Shu, Q.S., Fast, R.W., and Hart, H.L. 1988. Crack covering patch technique to reduce the heat flux from 77 K to 4.2 K through multilayer insulation. *Adv. Cryo. Engr.,* Vol. 33.
43. Leung, E.M.W., Fast, R.W., Hart, H.L., and Heim, J.R., 1980. Techniques for reducing radiation heat transfer between 77 and 4.2 K. *Adv. Cryo. Eng,* Vol. 25, 489–497.
44. Ohmori, T. et al., June 1987. Thermal performance of candidate SSC magnet thermal insulation systems. SSC-N-459, Superconducting Super Collider Laboratory.
45. Ohmori, T. et al., 2014. Test apparatus utilizing Gifford–McMahon cryocooler to measure the thermal performance of MLI objectives background. *ICEC 2014.*
46. Boroski, W., Nicol, T. et al., 1991. Design of the multilayer insulation system for the superconducting super collider 50mm dipole cryostat. *ITSSC—1991.*
47. Johnson, W. et al., 2020. Performance of MLI seams between 293 K and 20 K. *Advances in Cryogenic Engineering, IOP Conf. Series: Mater. Sci. Eng.,* 755 012152.
48. Johnson, W. et al., 2020. Testing of SHIIVER MLI coupons for heat load predictions. *2020 IOP Conf. Ser.: Mater. Sci. Eng.,* 755 012151.
49. Mazzone, L. et al., 2002. (MLI measurement) CERN LHC/2002–18 (ERC).
50. Venturi, V. et al., 2020. Qualification of a vertical cryostat for MLI performances tests between 20 K and 60 K to 4.2 K. *IOP Conf. Ser.: Mater. Sci. Eng.,* 755, 012154.
51. Shu, Q.S. et al., 1987. A systematic study to reduce the effects of cracks in multilayer insulation, Part 1: Theory. *Cryogenics,* Vol. 27, No. 298.
52. Shu, Q.S. et al., 1987. Theory and technique for reducing the effect of cracks in multilayer insulation from room temperature to 77 K. *Advances in Cryogenic Engineering,* Vol. 33.
53. Shu, Q.S., Demko, J., and Fesmire, J., 2015. Developments in advanced and energy saving thermal isolations for cryogenic applications. *Advances in Cryogenic Engineering, IOP Conf. Series: Materials Science and Engineering,* Vol. 101 012014.
54. Kaganer, M.G., 1969. *Thermal Insulation in Cryogenic Engineering,* Israel Program for Scientific Translations, Jerusalem.
55. Fesmire, J.E., Augustynowicz, S.D., and Demko, J.A., 2002. Thermal insulation performance of flexible piping for use in HTS power cables. *Adv in Cryogenic Engineering,* Vol. 47, *AIP Conference Proceedings,* 613 1525.
56. Ohmori, T., 2006. Thermal performance of multilayer insulation around a horizontal cylinder. *Cryogenics,* Vol. 45, 725–732.
57. Fesmire, J.E., Augustynowicz, S.D., and Demko, J.A., 2002. Overall thermal performance of flexible piping under simulated bending conditions. *Adv in Cryo Eng,* Vol. 47, *AIP Conference Proceedings,* 613 1533.
58. Fesmire, J.E., and Augustynowicz, S.D., 2002. Thermal performance of cryogenic piping multilayer insulation in actual field conditions. *Cryogenics 2002IIR Proceedings,* Praha Czech Republic, pp. 94–97.

59. Hastings, L.J., Hedayat, A. et al., 2004. Analytical modeling and test correlation of variable density multilayer insulation for cryogenic storage. NASA/TM-2004–213175, May.
60. McLean, C., Mustafi, S., Walls, L. (KSC) et al., 2011. Simple robust cryogenic propellant depot for near term applications. 978-1-4244-7351-9/11/-2011 IEEE.
61. Ward, M. et al., 2019. MLI impact phenomenology observed on the HST Bay 5 MLI panel. *First Int'l. Orbital Debris Conf.* https://www.semanticscholar.org/paper/MLI-Impact-Phenomenology-Observed-on-the-HST-Bay-5-Ward-Anz-meador/a1107a2473fc6687c771719c5b6f6e6eaab446ad
62. Lebrun, P., 2004. Design of a cryostat for superconducting accelerator magnet: The LHC main dipole case, CERN Report.
63. Meyer Tool & Mfg., 2010. The use of multilayer insulation blankets on 77K and on 4K surfaces, Oak Lawn, IL. https://www.mtm-inc.com/using-multilayer-insulation-blankets-on-77k-and-on-4k-surfaces.html
64. Fesmire, J., and Swanger, A., 2021. Overview of the New LH2 Sphere at NASA Kennedy Space Center, Department of Energy and NASA, Advances in Liquid Hydrogen Storage Workshop, virtual.

4 Thermally Efficient Support Structures for Cryogenics

Jonathan A. Demko, Quan-Sheng Shu, and James E. Fesmire

4.1 INTRODUCTION

Thermally efficient support structures are essential for all cryogenic applications because all vessels containing cryogenic liquids (liquefied gases) and other cold masses must be isolated from ambient temperature through supports. The preservation of cryogenic liquids and keeping cold mass temperature unchanged within a few Kelvins of absolute zero is a great challenge. In terms of mechanical support, if the cold mass of a superconducting (SC) magnet is about 15 tons, it is equivalent to the weight of that about 18.7 m^3 of LN$_2$, 214 m^3 of LH$_2$, or 124 m^3 of LHe. It is crucial to minimize the heat flows through the support structures and simultaneously to have strong supports against the weight, impacts of transportation, and thermal expansion-contraction of the inner and outer vacuum enclosures.

The support structures of cryogenic vessels for LH$_2$, LHe, LN$_2$, and so on are an integral part of the thermal insulation system. For good thermal performance to be achieved, the vessel design must orchestrate a fine balance among supports, piping, and insulation. The basic approach for cryogenic storage vessel design was outlined in two main textbooks by Barron and Fynn in the 1980s [1–2].

In recent decades, particularly since the 6-km superconducting accelerator was successfully operated at 4.5 K (Fermilab, 1982), many superconductivity projects with cold masses of tens of kilometers in length, hundreds of tons in weight, and huge complex machines, such as X-ray free electron laser (XFEL), Large Hadron Collider (LHC), continuous beam electron accelerator facility (CEBAF), spallation neutron source, and International Thermonuclear Experimental Reactor have been in operation or under development [3–7]. In the meantime, there are many advanced ultra-large, sophisticated, and unique cryogenic applications that have been completed and are being developed around the world, such as a 5,000-m^3 LH$_2$ tank for space launch, ship carriers of LNG and LH$_2$, special cryostats for space exploration, SC MRIs, and various physics/chemistry research. Obviously, the design and implementation of highly efficient support structures are equally crucial to high-performance thermal insulation for all cryogenic applications.

In this chapter, we will mainly introduce the advanced designs, new materials and structures, and representative examples of cryogenic supports while also briefly discussing and summarizing basic design considerations and practical analyses and data as well as principal formulas. The focus will be on the following subjects concerning thermal management technologies:

- Basic design considerations and materials
- Thermal optimization for minimizing heat leak
- Ring-shaped supports for cryogenic pipes and pipe complexes
- Rod-tube supports for cryogenic vessels and SC cold mass
- Post-shaped supports for long cylindrical heavy-weight cold mass
- Supports for thin/long cold mass with very large warm bore
- Contact-free supports with magnetic levitation

DOI: 10.1201/9781003098188-4

4.2 BASIC DESIGN AND MECHANICAL CONSIDERATIONS

4.2.1 GENERAL CONSIDERATIONS

Cryogenic devices typically consist of a warm outer vessel and an inner vessel that holds a cryogenic liquid or isolated cold mass from the environment. These are separated by a different insulation system. The necessary penetrations to the inner vessel from the outer vessel include various pipes, supports, and other functional insertions. Support members are a crucial component to the thermal management of cryogenic systems. In all applications, these must provide a capability to effectively transmit the required loads with minimal thermal impact on the refrigeration system. The thermal requirement can be met using different advanced options and sophisticated approaches for structural supports within the insulating space. In general, there are several typical support structures: (1) tension/compression rods, tubes, and chains of high-strength steel; (2) tension/compression posts, tubes, or saddle bands of metal and G-10 type composites; (3) flexible cables or chains of metal; and (4) compression multiple-contact supports (disks, rings etc.). These successful developments and results based on collective efforts over decades by national laboratories, industry, and universities are simplified and depicted in Figure 4.1 and will be discussed in detail. In general, each of the structures or any combination of them can be used either for cryogenic vessels or various cold masses.

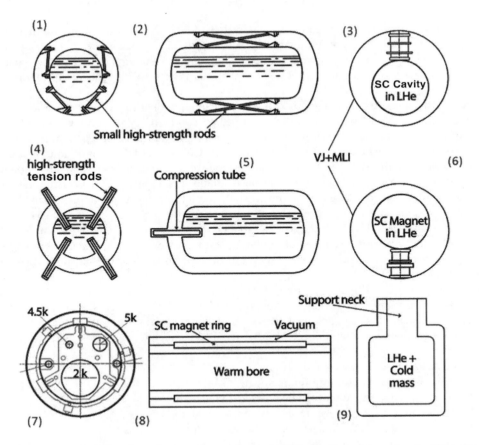

FIGURE 4.1 Simplified sketches of cryogenic supports, which thermally isolate various cold masses from outer vessels at environment temperatures. 1, 2, 4, and 5: Tension and compression rods. 3 and 6: Tension and compression posts. 7: Ring-shaped support. 8: Horizontal tension bars. 9: Neck tube support.

In addition, supports must ensure position stability, allowing for the thermal contractions that occur when the device reaches thermal equilibrium in its operating mode. This may be particularly true for accelerator SC magnets and superconducting radio-frequency (SRF) cavity systems. Supports in tension are generally preferred over supports in compression. In tension configurations, the support can be designed for minimum cross-sectional area and maximum heat conduction length and thus minimize the heat leak to the inner vessel. As SC equipment becomes more widely used, electrical isolation is required as well. The following list summarizes requirements of supports in cryogenic systems and the relative importance of the requirement. Not all requirements will be found in every situation.

- Thermal isolation—high priority, all situations
- Electrical isolation—high priority, some situations
- Mechanical loads—high priority, all situations
- Static loads—high priority, all situations
- Dynamic or shock loads (transporting the system, 3G)—high priority, some situations
- Light weight—low priority, some situations
- Low volume—medium priority, some situations
- Vibration isolation (spacecraft, magnets with cryocoolers)—high priority some situations.

4.2.2 MECHANICAL CONSIDERATIONS

The focus of this work is heat management, but the mechanical load carrying requirement is in conflict with the minimization of heat transfer. A brief discussion of mechanical load constraints is provided. Mechanical loads come in three basic forms: tension and compression, bending, and torque or twisting. Tension and compression loads are taken along a structural member. Lateral loads are also present in applications such as superconducting accelerator magnets, as described in Nicol [8] and Nieman et al. [9], which can introduce bending and twisting.

Structural supports for cryogenic equipment must sustain the required loading under tension or compression as well as thermal stresses. This discussion is not intended to be comprehensive of the mechanics of materials and stress calculations but only to provide some background for stress considerations needed in the design of supports. The designer should consult more comprehensive references such as Beer [10] and Shigley [11] for more complete discussions.

The ability to safely carry a load is in terms of the stress, σ, which is a normalized load force, F, per unit area, A, in SI units of MPa. There are several types of stresses, axial, shear, bending, and thermal, that result from thermal contractions and expansions of the materials. A normal stress is illustrated in Figure 4.2A and B is given by Equation 4.1 for an axial load in tension or compression. Equation 4.1 assumes the stress is uniformly distributed over the area, which frequently is not the case.

$$\sigma = \frac{F}{A} \tag{4.1}$$

Equation 4.1 also applies for shearing stresses as illustrated in Figure 4.2C for a simple pinned connection.

A transversely loaded beam will have a shear force, F_{shear}, (Figure 4.2D) and a bending moment, M, in N·m, as illustrated in Figure 4.2E. The normal stress due to bending can be calculated according to the elastic flexure formula:

$$\sigma_{bend} = \frac{Mc}{I} \tag{4.2}$$

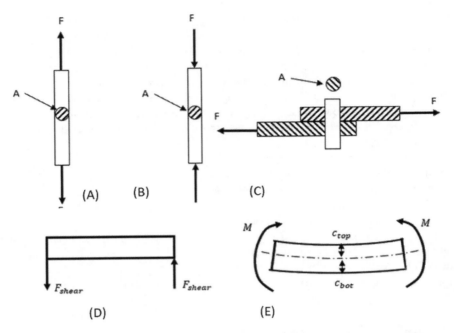

FIGURE 4.2 Illustration of various forces and stresses: A: load under tension, B: load under compression, C: shear stresses in a pinned connection, D: internal shear force, and E: bending moment in a beam.

where the distance, c, in meters, is measured from the neutral axis to either the top or bottom surface, and the moment of inertia or the second moment of the cross-section, I, in m⁴, is determined with respect to the centroidal axis perpendicular to the plane of the couple, M, given in N-m. The neutral axis is the line of zero fiber stress in a member subject to bending [12]. For the couple shown, the normal stress on the top surface will be in compression, a negative stress, and the bottom surface will be in tension, a positive stress. In addition to the normal stresses, longitudinal shearing stresses must exist in any member subjected to transverse loading.

The longitudinal shear force across the thickness of the cross-section per unit length of the beam is the shear flow denoted by the symbol q. This can be calculated by:

$$q = \frac{VQ}{I}$$

where V is the resultant vertical shear force at the cross-section being considered. Q represents the first moment of the area about the centroidal axis of the entire cross-section, as described in Byers and Snyder [12]. The parameters describing a the cross-section of a rectangular beam are provided in Figure 4.3. The maximum shearing stress on a horizontal face for a narrow rectangular beam is given by Equation 4.3, where the shear stress τ_{max} is in Pa. Thus,

$$\tau_{max} = \frac{3}{2}\frac{F_{shear}}{A} = \frac{3}{2}\frac{F_{shear}}{bh} \tag{4.3}$$

A torque, T (N·m), applied to a support will experience a maximum shear stress according to the elastic torsion formula:

$$\tau_{max} = \frac{TR}{J} \tag{4.4}$$

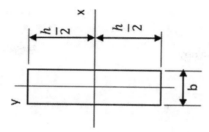

FIGURE 4.3 Cross-section of a narrow rectangular beam.

where R (m) is the outer radius of a circular shaft, and J (m^4) is the polar moment of inertia. For a circular cross-section, this is given by:

$$J = \frac{1}{2}\pi R^4 \tag{4.5}$$

These are basic stress relations that can be applied in many situations. The mechanical load determines the cross-section needed for a support. Many other analytical calculations for stresses can be found in Young [13].

The heat load from a support can be a significant fraction of the total heat load to a system. Assuming that the support is accurately modeled by one-dimensional heat conduction, the heat transfer for a support of length, L (m), and uniform cross-section, A (m^2), can be determined using an average thermal conductivity of the support rod material, $\overline{k}_t$ (W/m·K):

$$\overline{k}_t = \frac{\int_{T_C}^{T_H} k(T)dT}{(T_H - T_C)} \tag{4.6}$$

The heat transfer, $\dot{Q}$ (W), can be calculated using the conduction integral, as presented in Equation 4.6.

$$\dot{Q} = \frac{\overline{k}_t A(T_H - T_C)}{L} \tag{4.7}$$

The temperatures T_H (K) and T_C (K) are the temperatures at the hot and cold ends of the support. Normally, T_H is the ambient or room temperature, but in more complicated system designs, there may be intermediate temperatures due to the incorporation of thermal shields.

For the simple case of axial loading only, the support member must be able to carry the weight of the cold component and any accelerations. This sets the design cross-sectional area for a tension member as:

$$A = \frac{F}{\sigma_a} = \frac{f_{safety}}{\sigma_y}F \tag{4.8}$$

where F is the design load of the support member, σ_a is the allowable stress for the support material, σ_y is the yield strength, and f_{safety} is the unitless factor of safety for the design. This area can be substituted into the heat conduction equation to provide:

$$\dot{Q} = \frac{\overline{k}_t \left(T_H - T_C\right) F f_{safety}}{\sigma_y L} = \frac{F f_{safety} \left(T_H - T_C\right)}{L\left(\sigma_y / \overline{k}_t\right)} \tag{4.9}$$

where $\left(\sigma_y / \overline{k}_t\right)$ is the strength to conductivity ratio, which is only a function of the support material. In order to minimize conduction heat leak along the member and provide the most thermally efficient design, a material with as large a value of this ratio as possible should be used. In many cases, support designs are not simple rods, as will be discussed later. But the use of this parameter still applies in guiding the selection of candidate materials.

4.3 MATERIALS

Many different structural materials are available for use as structural components. The optimum selection will depend on several factors, not the least of which is the ability to carry the loading and remain within a heat load budget. Detained lists of materials are available in the appendix.

One factor that must be considered is the ductility of the material. Many materials become brittle when they are cooled below their glass transition temperature. Regarding metals, body-centered cubic (bcc) structure metals such as copper, nickel, copper-nickel alloys, aluminum and its alloys and austenitic stainless steels, titanium, and zirconium remain ductile at low temperatures. In general, face-centered cubic (fcc) metals remain ductile. Materials such as polytetrafluoroethylene (Teflon), also remain ductile at low temperatures.

Materials that become brittle include metals such as iron, carbon and low-alloy steels, molybdenum, niobium and zinc. In general, most body-centered cubic metals and plastics become brittle at low temperatures. The commonly available polyvinyl chloride (PVC) pipe is commonly used to replace piping at room temperature, but PVC is very brittle at low temperatures and should be avoided.

High vacuum is often used as part of the thermal isolation, so metals must provide the capability to maintain high vacuum inside the vessel. Outgassing of hydrogen trapped in stainless steels during the formation process is usually accounted for by the use of getters in the vacuum space. In some applications, corrosion resistance is also needed.

Many composites are used due to the ability to tailor thermal and strength properties. These tend to be more porous than metals and have the potential to outgas. Fiberglass composites, such as G-10 and G-11, are widely used because of their strength, low thermal conductivity, very low outgassing, and for many applications the dielectric strength. Kevlar straps have been applied in some situations because of their high strength.

The ratio of strength to thermal conductivity, σ_y / k (N-K/W-m), is one important parameter to qualify whether the material is suitable for cryogenic support structures. Some values for typical materials are provided in Table 4.1. The thermal conductivities of some materials vary greatly with temperature change but not linearly. Therefore, as introduced in conduction section of Chapter 1, the thermal conductivity integral

$K = \int_{T_1}^{T_2} k\left(T\right) dT$ and the mean thermal conductivity $\overline{k}$ over T_1 to T_2 are used in ratios for

applied temperature regions.

TABLE 4.1

Strength to Conductivity Ratios for Typical Materials used in Support Members. The mean thermal conductivity between 90 K and 300 K is used in the reference for optimal design for liquid oxygen temperature applications

	Yield Strength, σ_y (MPa)	$\overline{k}_{avg}$ (90–300 K), (W/m·K)	σ_y / k (N-K/W-m)
Teflon	20.7	1.522	13.6
Nylon	75.1	1.522	48.4
Mylar	275.7	0.962	286.6
Dacron fibers	206.8	0.962	630.7
Kel-F fibers	206.8	0.381	542.8
Glass fibers	896.1	4.88	183.6
304 SS	627.3	12.41	50.5
316 SS	820.3	12.41	66.1
347 SS	882.3	12.41	71.1
1100-H16 aluminum	135.8	260.1	0.52
2024–0 aluminum	75.8	88.4	0.86
5056–0 aluminum	137.9	114.9	1.20
K Monel (45%)	641.0	18.63	34.4
Hastelloy C (annealed)	379.1	12.36	30.7
Inconel cold-drawn	310.2	13.35	23.2

Source: From Barron [2]

4.4 THERMAL OPTIMIZATION

4.4.1 MATHEMATICAL ANALYSES FOR OPTIMIZATION

Two Heat Intercept Stations at Intermediate Temperatures. Allocation of heat intercept stations at an intermediate temperature at the cryogenic support structures between the cold mass and environment T is an essential and efficient way both to minimize the heat reaching the cold mass and reduce the refrigeration power needed. The temperatures and distances of the stations along the support structure would be decided by the cooling sources available in practice and results from the math analyses. A simple but very instructive analysis was performed by McAshan for the superconducting supercollider (SSC) cryogenic system [14]. It assumes a uniform cross-section stainless-steel member with heat stations. The optimization is based on the fact that the power needed to remove a heat load at a given temperature decreases as the temperature increases. In this memo, McAshan performs an optimization of the refrigeration requirements along a stainless-steel (UNS S304000) and Ultem-2100 support assuming a uniform cross-section and variable thermal conductivity with heat stations at two locations along the post. The uniform cross-section simplifies the analysis but is a reasonable assumption since it is the mechanical load that determines the area of the support. The optimization clearly shows that refrigeration requirements can be minimized for the 4 K level with heat stations around 80 K and 20 K.

The situation is illustrated in for a case with two heat stations. The cross-sectional area of the support, A, is assumed to be constant. This would most likely be the case since the area would be determined by the stresses in the member. The total length of the support is L, which is broken into segments L_1, L_2, and L_3, which separate the heat stations at the different temperatures. The

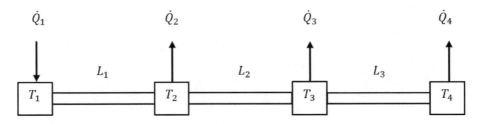

FIGURE 4.4   Uniform support between room temperature, T_1, and liquid helium temperature, T_4, with two heat stations at intermediate temperatures, T_2 and T_3.

thermal optimization is performed by minimizing the Carnot refrigeration power, P_{Carnot}, needed for this member.

The warm end of the support is assumed to be at room temperature, T_1, so the heat load, $\dot{Q}_1$, enters the support. The refrigeration loads that need to be removed by the cryogenic refrigeration are $\dot{Q}_2, \dot{Q}_3$, and $\dot{Q}_4$ at heat station temperatures T_2, T_3, and T_4. To optimize the support geometry, a minimum of the Carnot refrigeration power is sought. The Carnot power is given by

$$P_{Carnot} = \eta_2 \dot{Q}_2 + \eta_3 \dot{Q}_3 + \eta_4 \dot{Q}_4 \tag{4.9}$$

where the Carnot efficiencies, defined as the ideal input power to the refrigerator per unit of refrigeration power required, are given by:

$$\eta_i = \frac{T_1 - T_i}{T_i} = \frac{300 - T_i}{T_i} \tag{4.10}$$

The heat transfer between any two heat stations can be expressed in terms of the thermal conductivity integral, $\bar{k}(T)$, and are given by:

$$\dot{Q}_4 = \frac{A}{L_3}\left(\bar{k}(T_3) - \bar{k}(T_4)\right) = \frac{A}{(L - L_1 - L_2)}\left(\bar{k}(T_3) - \bar{k}(T_4)\right) \tag{4.11}$$

$$\dot{Q}_3 = \frac{A}{L_2}\left(\bar{k}(T_2) - \bar{k}(T_3)\right) - \dot{Q}_4 \tag{4.12}$$

$$\dot{Q}_2 = \frac{A}{L_1}\left(\bar{k}(T_1) - \bar{k}(T_2)\right) - \dot{Q}_3 - \dot{Q}_4 \tag{4.13}$$

According to the McAshan memo, the optimization can be performed by expressing the Carnot power as:

$$P_{Carnot} = \frac{A}{L}\left(\frac{D_1}{X_1} + \frac{D_2}{X_2} + \frac{D_3}{X_3}\right) = \frac{A}{L}\left(\frac{D_1}{X_1} + \frac{D_2}{X_2} + \frac{D_3}{(1 - X_1 - X_2)}\right) \tag{4.14}$$

where the terms in this expression are given by:

$$D_1 = \eta_2 \left(\bar{k}(T_1) - \bar{k}(T_2) \right)$$

$$D_2 = (\eta_3 - \eta_2) \left(\bar{k}(T_2) - \bar{k}(T_3) \right)$$

$$D_3 = (\eta_4 - \eta_3) \left(\bar{k}(T_3) - \bar{k}(T_4) \right)$$

$$X_1 = \frac{L_1}{L}$$

$$X_2 = \frac{L_2}{L}$$

$$X_3 = \frac{L_3}{L} = 1 - \frac{L_1}{L} - \frac{L_2}{L} = 1 - X_1 - X_2$$

The minimum Carnot power is found with respect to the dimensionless locations of the heat stations along the support X_1 and X_2 to be:

$$X_1 = \left(1 + \left(\frac{D_2}{D_1} \right)^{0.5} + \left(\frac{D_3}{D_1} \right)^{0.5} \right)^{-1} \tag{4.15}$$

$$X_2 = X_1 \left(\frac{D_2}{D_1} \right)^{0.5} \tag{4.16}$$

Example Results of Single Heat Reception Station. The equations can be applied to the case of a single heat station on the support by letting $T_2 = T_3$ and $D_2 = 0$. The remaining step of the optimization is to determine the heat station temperatures which minimize the Carnot power. For an example, assume that the warm end of a support is at "room temperature" 300 K and supports a cold mass at 4.5 K. It is made from a 4-inch IPS schedule 40 stainless steel UNS S304000 pipe with a length of 0.1 m and has a cross-section of 0.001065 m². The thermal conductivity, k, and the thermal conductivity integral, $\bar{k}$, of UNS S304000 from NIST curve fits are shown in Figure 4.5.

The solution shows that the distance along the support at which a heat station is located depends on the temperature of the heat station. The Carnot power at the optimized locations for this support is provided in Figure 4.6. The first stage or warm heat station was assumed to be at temperatures of 70 K, 80 K, and 90 K.

Example Results of Two Heat Station Intercepts. At each of these temperatures, which are near liquid nitrogen temperatures, the second stage or cold heat station was varied from 10 to 30 K. It is clearly seen from the curves that the minimum Carnot power falls between 17 K and 23 K. The results for this support without heat stations and with heat stations at temperatures of 80 K and 20 K are compared in Table 4.2. Without heat stations, the total heat load of 32.3 W must be removed at the low temperature end, which is 4.5 K. Removing the conducted heat from the support at the intermediate temperatures of 80 K and 20 K decreases the heat load to 4.5 K from 32.3 W to 1.4 W. The Carnot efficiency for refrigeration at that temperature is $\eta_{4.5\,K} = 65.7$, resulting in a significant

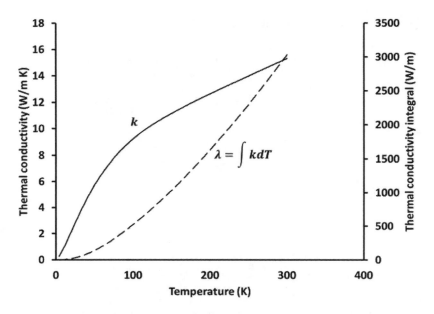

FIGURE 4.5 Thermal conductivity and thermal conductivity integral of UNS S304000. Source: *https://trc. nist.gov/cryogenics/materials/304Stainless/304Stainless_rev.htm, last accessed April 14, 2021*

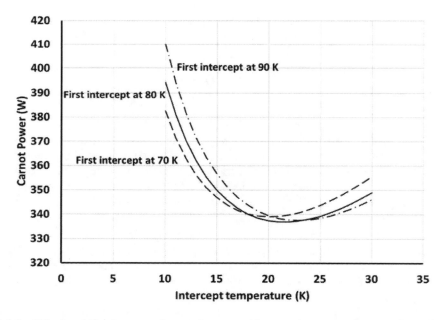

FIGURE 4.6 Calculated Carnot power for varying second heat station temperatures at the optimum locations along the support.

drop in the Carnot power required to remove the heat at the cold end. At higher heat station temperatures, the Carnot efficiencies are $\eta_{80\,K} = 2.75$ and $\eta_{20\,K} = 14$. Accounting for these terms, it is seen that the ideal amount of power required for refrigeration drops from 2,123 W for support without heat stations to 337.5 W with two heat stations.

TABLE 4.2

Comparison of Carnot Power for an UNS S304000 Support without Heat Stations and with Optimized Heat Stations When High *T* Is 300 K Calculated Using the Approach of McAshan

Temperature	No Stations		Two Stations		
	$\dot{Q}$	Carnot Power	$\dot{Q}_i$	L_i / L	Carnot Power
80 K	–		48.94 W	0.483	
20 K	–	2,123 W	9.16 W	0.343	337.5 W
4.5 K	32.3 W		1.14 W	0.174	

4.4.2 Thermal Optimization with Computing Codes

The mathematical analyses for thermal optimization are practically satisfactory for most cryogenic support with rather simple geometric shapes. Of course, the experimental verifications are normally conducted to double-check the analytical results for final design. However, the mathematical analyses approaches will face the burden of accuracy when the geometrical shapes become sophisticated. Currently, computing code, such as ANSYS and SINDA, is utilized for simulation of thermal optimization with more accurate results. As an example, the simulation with ANSYS code is demonstrated for a sophisticated support structure of a large, multi-channel cryogenic pipe in Section 4.5.

4.5 SUPPORTS FOR PIPES AND PIPE COMPLEXES

4.5.1 Ring Supports for Cryogenic Fluid Transfer Pipes

Cryogenic fluid transfer pipes are frequently and most efficiently achieved using double-walled vacuum-jacketed (VJ) insulated lines. The typical construction for this type of pipe starts with a stainless-steel inner pipe that is sized properly for the flow rate. Surrounding that is typically a multilayer insulation (MLI) blanket and then a containment pipe, which maintains a vacuum around the cold inner pipe, isolating it from the warm ambient space. Between the cold inner pipe (or pipes in multichannel transfer line) and warm outer pipe, some type of ring shape support is periodically used to maintain a gap between all pipes both to avoid compressing the MLI blanket and to prevent pipes of different temperatures from coming into contact with each other.

In order to minimize the heat leak from the outer pipe to the cold pipe through the supports, the supports are designed to minimize the cross-section areas and to lengthen the path of the heat flows with low thermal conductivity materials (such as G-10) while the mechanical performance is satisfied. To reach these goals, various beautiful patterns of ring-shaped supports have been designed and implemented. An improved pipe spacer design that also minimizes contact area at warm and cold surfaces and a simple three-point contact design are shown in Figure 4.7A–B. In some applications, the supply and return of helium are conducted with double-walled insulated lines inside of other double-walled vacuum jacketed lines, such as a double line with length around 6.5 km at the Tevatron, Fermilab. A classic ring shape support for it is illustrated in Figure 4.7C [15].

Heat transfer into a vacuum-insulated cryogenic transfer line can be divided into two main categories: (1) heat transfer through MLI and (2) heat transfer from supports. The warm vacuum pressure, with no cryogen inside the line, may be in the range of 0.1 to 1 Pa, while the cold vacuum pressure will be below 1×10^{-3} Pa. The total heat leak can be calculated as

$$Q_{pipe} = Q_{insulation} + Q_{supports} \tag{4.17}$$

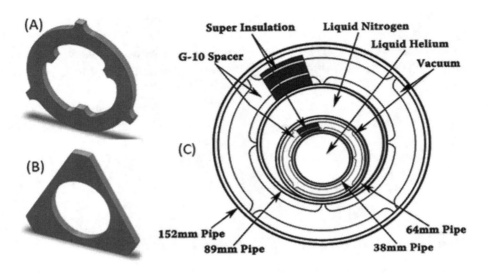

FIGURE 4.7 Designs of pipe spacers for VJ. A: Improved spacer design; minimizes contact area, B: simple three-point contact design, C: Fermilab double line with length around 6.5 km [15].

In this section, we only discuss the thermal load from support rings, $Q_{supports}$. Since the supports pass through a vacuum space, gas conduction and convection on the face of the support ring are negligible. The thermal solution for a penetration (support) can be obtained assuming one-dimensional heat conduction with consideration of the area integral of the cross-section. Thus, the major mode of heat transfer associated with penetration is generally heat conduction, which can be determined mainly with the approaches introduced in Section 4.4. In addition, Figure 4.7 also indicates that the contacts between support to cold and warm surfaces provide an extra-thermal resistive mechanism, which blocks the heat flows [16]. Thermal contact resistance is attributable to several factors, the most notable being that contact between two surfaces is made only at a few discrete locations rather than over the entire surface area. There is a large temperature drop between two contact surfaces.

4.5.2 Thermal Simulation of Ring Support Designs

Single-Channel Cryogenic Pipe. Deng et al. reported ANSYS simulation results of a triangle support with three uniform circle holes to reduce the area of cross-section and increase the length of the heat path, as shown in Figure 4.8 [17]. The test pipe section is made from SS-304 with one support in center, inner pipe 21.9 mm at 77 K, and outer pipe 88.9 mm and 1 m length, which is wrapped MLI with the layer density of 25/cm. The highest temperature is 291 K, where the triangle support contacts the warm outer pipe, which is close to the ambient temperature. Three points, T_{s1}, T_{s2}, and T_{s3}, at the triangle support are 275 K, 270 K, and 275 K, respectively. Temperature T_{s2} is less than T_{s1} and T_{s3} because triangle support always has only two points to contact with the outer pipe, which bears the weight of the inner pipe. The temperature contour shows a secondary heat leak by radiation from outer pipe walls. A test assembly of validation was set up, and good agreement between test data and simulation results was obtained (about 5%). The heat leakages of multilayer insulation, a support, and the overall leakage are 1.02W/m, 0.44 W, and 1.46 W/m from experimental data, respectively.

Multi-Channel Cryogenic Pipe

Currently, helium is transported in multi-channel cryogenic pipes, which transfer helium at different states, such as liquid, gaseous, supercritical, and superfluid, to meet the cooling requirements of SC

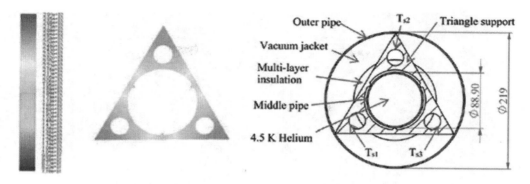

FIGURE 4.8   The temperature contour of support (left) and simplified triangle support [17].

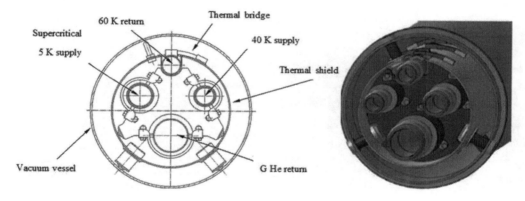

FIGURE 4.9   Sketch of an XFEL process He process line (L_1) and its 3-D model with weak fixed support [18].

magnets and SRF cavities. The most typical four-channel cryogenic transfer lines consist of two pipes (supply and return of the cooling cryogens, about 5 K) and of two pipes at warmer temperatures for supply and return of helium for thermal shield cooling. These have temperatures of 40–80 K. Duda et al. [18] conducted a comprehensive entropy analysis of a support system, shown in Figure 4.10, for a multi-channel transfer pipe 100 m in length with the cross-section identical to the XATL (Accelerator Module Test Facility) [18]. The tested components were used in the construction of the XFEL. The number of process pipes, their cross-sections, and the parameters of the transferred medium are dictated by the specific requirements of the devices in which the cryogen is used. Although they have different shapes or principles of operation, such components of cryogenic transfer lines as supports, vacuum barriers, and compensation bellows have very similar functions. Figure 4.9 shows all types of supports used in one- and multi-channel cryogenic transfer lines. The MLI is intentionally omitted in order to show the structures.

The minimum entropy production method is used to optimize the multi-channel transfer line L_1 and another modified L_2 (sketch not shown). The source temperatures are shown in the sketch, and the thermal shield is thermally fixed to a 60 K return pipe.

The L_1 sliding support is designed to connect three process pipes with temperatures of 4.5 K, 5 K, and 45 K. The advantages of this concept are its simplicity of design and assembly. However, from the perspective of thermodynamics, a disadvantageous heat influx occurs between the pipe supplying thermal shields at 45 K and the pipes with a temperature of near 5 K. Using the calculated

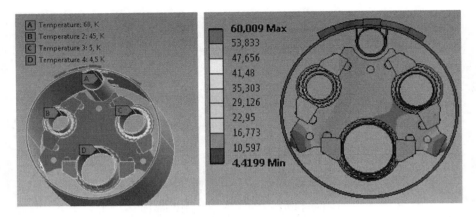

FIGURE 4.10   Boundary conditions set for the process pipe sliding support in the L_1 line (left). Temperature contour for the process pipe sliding support in the L_1 line (right) [18].

TABLE 4.3

Heat Transfer and Entropy Flow Calculated for the Sliding Support

5 K Supply		4.5 K Return		45 K Supply		60 K Return		
$\dot{Q}$	$\dot{S}_{\Delta T}$	$\dot{Q}$	$\dot{S}_{\Delta T}$	$\dot{Q}$	$\dot{S}_{\Delta T}$	$\dot{Q}$	$\dot{S}_{\Delta T}$	$\Sigma \dot{S}_{\Delta T}$
(W)	(W/K)	(W)	(W/K)	(W)	(W/K)	(W)	(W/K)	(W/K)
0.24	0.051	0.53	0.12	−0.19	−0.004	−0.58	−0.008	**0.152**

Source: Of Line L_1 [18]

TABLE 4.4

Heat Transfer and Entropy Flow Calculated for the Sliding Support System

5 K Supply		4.5 K Return		45 K Supply		60 K Return		
$\dot{Q}$	$\dot{S}_{\Delta T}$	$\dot{Q}$	$\dot{S}_{\Delta T}$	$\dot{Q}$	$\dot{S}_{\Delta T}$	$\dot{Q}$	$\dot{S}_{\Delta T}$	$\Sigma \dot{S}_{\Delta T}$
(W)	(W/K)	(W)	(W/K)	(W)	(W/K)	(W)	(W/K)	(W/K)
−0.002	−0.003	0.45	0.10	0.376	0.007	−0.826	−0.011	**0.095**

Source: Of the L_2 Transfer Line [18]

values of temperature and heat transfer to individual process pipes, the corresponding flows of entropy between the different pipes were calculated. Table 4.3 includes the values of both heat transfer and entropy flows. The sliding support used in this model thermally connects pipes with a temperature of approximately 5 K with pipes for thermal shield cooling, which have temperatures of 45 K and 60 K. This connection results in heat transfer at the support location from the 45 K and 60 K pipes to the 5 K pipes. The entropy flow out of the higher-temperature pipes is negative, and the entropy flow into the 5 K pipe is positive.

A second design, L_2, consists of supporting the thermal shield supply pipe directly against the thermal shield of the transfer line with the use of a dedicated support. Values of both heat transfer and entropy flows for this design are in Table 4.4. In order to further reduce heat flux, the 5 K supply pipe was supported to the vapor return pipe (4.5 K), while the vapor return pipe was connected in another location and with another support to the thermal shield. This arrangement is seen to have lower total entropy generation than the L_1 arrangement and is more energy efficient.

4.5.3 OTHER ADVANCED SUPPORTS FOR CRYOGENIC PIPES

Spiral Support for Flexible Transfer Pipes. Spiral-wire-beads support is designed for narrow and long cryogenic transfer pipes. If both inner and outer pipes are corrugated tubes in order to remain flexible, the beads are omitted to reduce the friction while the transfer pipe is bending or curving, reported by Knoll in Figure 4.11 [19]. The insulation space is evacuated and wrapped with MLI to minimize radiation and gas convection. The conductive heat flow will also be restricted by the high thermal resistance between loose contacts between wires and pipes. This type of support is currently utilized in high temperature superconducting (HTS) power cable systems.

Sophisticated Shape Support of Cryogenic Pipes. When cryomodules of SC magnets or SRF cavities are installed in an accelerator tunnel and the cryogenic system is distanced away or on the ground, the transfer lines will be very long, and the heat loads from ambient temperature to the cryogenic liquid will be large. Then the performance of the transfer lines is crucial for the efficient long-term operation of the cryogenic system. An example of a sophisticated design for a cryogenic transfer pipe is shown in Figure 4.12. Hosoyama, Nakai et al. [20] have reported many support

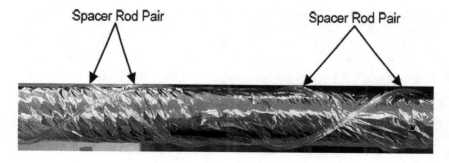

FIGURE 4.11 Spiral-wire support for a narrow flexible transfer line [19].

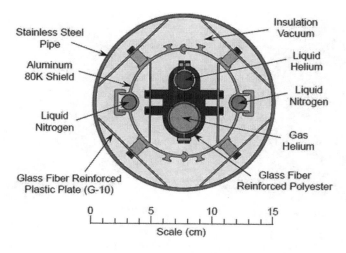

FIGURE 4.12 Support assembly of sophisticated design for cryogenic transfer pipe [20].

assemblies of sophisticated design for cryogenic transfer pipes, as one of the examples illustrated in Figure 4.11 shows [20]. The liquid helium line (supply line) and gas helium line (return line) are thermally shielded by liquid nitrogen supply and return lines to decrease the heat load from room temperature. These lines are supported by low thermal conducting materials with long thermal paths and added thermal contact resistance in a vacuum insulation pipe to minimize conduction heat load. This transfer line's performance was measured as 0.1 W/m of total heat load.

4.6 SUPPORTS FOR CRYOGENIC VESSELS AND SIMILAR COLD MASSES

4.6.1 Rod Supports for Large Tanks and Cold Masses

Large storage tanks will use a combination of supports (rods and straps) for vertical and horizontal loading and neck support, which is usually a tube, and sometime compression supports at the base can take advantage of the contact resistance. All structures for inner vessel support are an integral part of the thermal insulation system for a large-scale tank. The support structures in the vacuum space must bear the entire weight of the inner vessel with cryo-liquid and full thermal movement (expansion and contraction between inner and outer vessels [21]. All lines must be of sufficient length to minimize heat leak and provide for adequate flexibility in thermal expansion and contraction. Liquid lines must always include a P-trap to provide a natural gas pocket to minimize heat leak from the environment. Different options for structural supports within the annular space are depicted in Figure 4.1. Supports in tension are generally have several advantages, such as minimum cross-sectional area and maximum heat conduction length, and thus minimize the heat leak to the inner vessel.

The tank supports for a large spherical LH_2 tank are shown in Figure 4.13 [21, 22]. For this tank, there are 40 vertical support rods and 20 horizontal sway rods. The support rods hold the full weight of the inner sphere when filled to capacity and in addition handle the thermal stresses resulting from expansion and contraction between the inner sphere and outer sphere. The sway rods prevent lateral, vertical, and rotary shift of the inner sphere. Both the support and sway rods permit both spheres to expand and contract without damage to either sphere, the support system, the manway, or the piping. The suspension utilizes a minimum path heat principle to prevent the formation of cold spots on the outer sphere and minimize the total heat leak to the inner vessel when operating at cryogenic temperatures. A design for a tank support rod with inner vessel

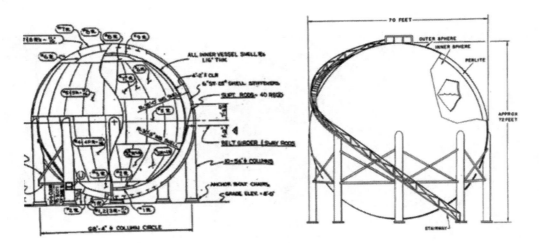

FIGURE 4.13 Tank supports for a large spherical LH_2 tank built in 1965: elevation view (left), top overall elevation (right) [21, 22].

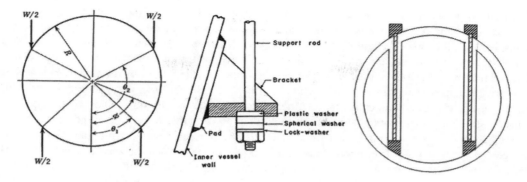

FIGURE 4.14 Tank support rod design: inner vessel loading (left) [21], support rod mounting detail (center), and simple re-entrant tube support to increase the path (right) [2].

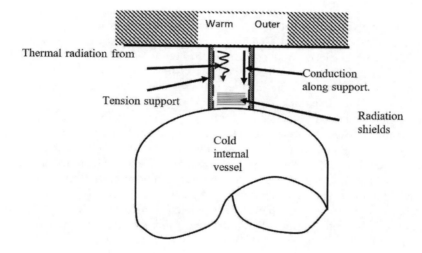

FIGURE 4.15 Top support under tension showing the placement of multilayer insulation radiation shields.

loading diagram, support rod mounting detail, and a way to increase the thermal path are shown in Figure 4.14.

Because of the low weight (density) of liquid hydrogen, the limiting factor of the VJ tank structural design is not the inner vessel but the shell thickness of the outer vessel to withstand the force of full vacuum loading. In a personal correspondence with CB&I in 1990, an estimate for a 5,700-m³ (1.5 M-gallon) sphere was provided with an outer shell diameter of 24.7 m (81 feet) and a thickness of 35.6 mm (1.4 inches). It was further suggested at this time that the upper limit of practical size for LH_2 tanks of the Horton-sphere design is about 7,600 m³ (2.0 M gallons) based on outer shell fabrication.

4.6.2 TUBULAR SUPPORTS FOR MEDIUM AND SMALL VESSELS

Large, medium, and small in terms of cryogenic vessels are really relative concepts. For example, an 800-liter LHe dewar is large for a physics lab currently or even for an industrial application before 1980 but may be thought small or medium for modern SC accelerators or SC tokamak.

The vertical support from a warm outer vessel is illustrated in Figure 4.15. In many applications, where internal vessels are used for pressurization and storage of cryogenic liquids, several storage vessels may be contained inside of a single vacuum chamber. In some of these instances, a warm to cold tension support is used in which the cold inner vessel is suspended from the warm wall of the outer vessel at the top of the tank. This can be done with a suitably sized tube or pipe as opposed to a solid rod of the same cross-section. To minimize the heat load, a long conduction path is provided, limited mostly by the available space. Also, the neck tube is cooled either by the venting cold gas or additional anchors at intermediate temperatures if necessary. The vessel is typically covered with a multilayer insulation blanket to limit the thermal radiation from the warm outer shell to the cold inner vessel, but a significant amount of thermal radiation from the warm shell to the cold vessel is possible inside the support tube. This region must also be fabricated with thermal radiation shields. Large storage tanks will use a combination of support straps or rods for vertical and horizontal loading and neck support, which is usually a tube. In principle, the tubular support can be used to suspend any cold mass, which can be inner vessels of cryogens, instruments being cooled, SC devices, and associated structures.

4.6.3 STACK SUPPORT OF PLATE DISKS

A cold vessel inside a vacuum jacket can be supported by tubes at the top of the vessel or by using compression members on the bottom. Compression supports at the base can take advantage of contact resistance by using stacks of disks or plates at the bottom of the support (warm end), as shown in Figure 4.16A. The plate stack can be made from G-10, with several sheets to provide contact resistance between the plates, causing the thermal resistance of the plate stack to be higher than using a G-10 block of the same thickness as the stack.

The reduction in the heat transfer can be significant, as shown in Figure 4.17. Calculations were performed assuming a stack of four G-10 plates or disks, which are 6.25 mm thick with a cold boundary of 80 K and a temperature rise across the stack of 10 K. For contact conductance over 1,000 W/m²-K, there is no difference between a stack of plates and a solid block of G-10 twenty-five mm thick. Below this value, the amount of heat transfer drops off considerably, demonstrating the benefit of having multiple plates with an imperfect contact.

Figure 4.16B introduces a detachable design of the combination of tubes and decks. When the internal vessel is filled with cryogenic liquid, the upper thermal support contracts on cooling and the gap appears. So, there is no physical thermal contact between the bottom of the cold vessel

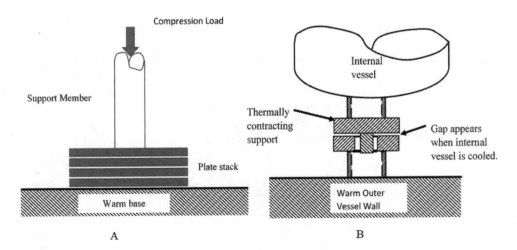

FIGURE 4.16 A: G-10 stack support, and B: concept for a shipping support that contracts due to thermal contraction of the internal vessel when it is cooled.

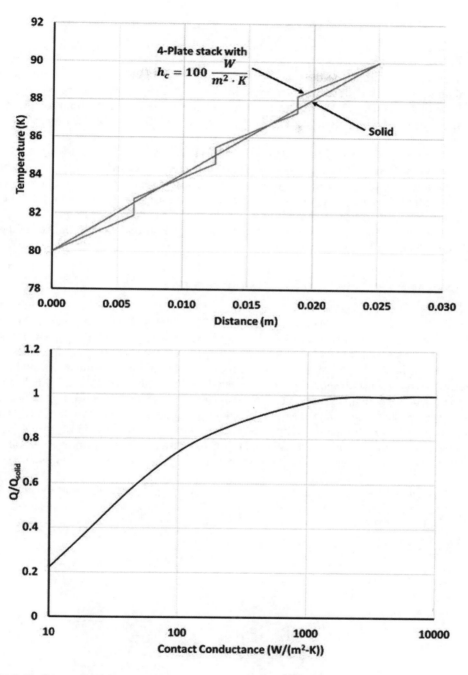

FIGURE 4.17 Upper: Calculated temperature distribution through a four-plate G-10 stack with and without contact conduction effects in the stack. Lower: The calculated ratio of heat transfer with contact conduction to solid conduction in a G-10 stack.

and the warm outer vessel. If the internal vessel is empty (warm), the upper support expands when warming up and supports the bottom of the inner vessel firmly. The structure is useful for vessels in transportation with no liquid.

4.6.4 SUPPORT RINGS FOR CRYOGENIC VESSELS

Light Support Ring. Light support rings, as shown in Figure 4.18A, are designed for a special LH_2 vessel of about 150 liters of LH_2 with a narrow vacuum jacket [23]. The design principle of this G-10 ring-shaped support is similar to the type of supports introduced in Figure 4.7. The inner vessel is an aluminum-lined, carbon fiber-wrapped pressure vessel. The outer jacket measures 120 cm long and 58.5 cm outer diameter (OD). The vessel outer volume is 298 L, and the total weight is about 187 kg. It can be filled with pressurized GH2 or LH_2 at 350 bar and 20 K for a hydrogen-driven car (1,050 km).

Heavy Load Support Ring. The similar but different design illustrated in Figure 4.18(b) is a heavy load support ring, the so-called GFRP support structure that achieves both thermal insulating performance and strength [24]. The double wall, stainless-steel, vacuum thermal insulation and supporting strength of the GFRP support structures achieve both thermal insulating performance and strength of the structure necessary for a 1,000-m^3 LH_2 shipping tank being installed on an overseas ship—the Pilot Ship LH_2 Carrier. The overall targeting thermal insulation performance of the tank with MLI insulation and ring supports is ≤0.01 W/m-K.

4.6.5 SIMILAR SUPPORTS UTILIZED FOR SC COLD MASSES

Superconducting (SC) magnets and superconducting RF cavities can be seen as cold masses of temperature around 2 K–4.5 K, while magnets and the like in high-temperature superconductors are working at higher temperatures. The similar support systems being utilized in cryogenic liquid systems have also been employed for SC cold masses. In practice, the support systems of magnet cold masses make use of high thermal resistive straps, rods, tubes, ring supports, and their combinations.

Support for Accelerator Magnets. As an example, Kaiser introduced HERA SC dipole magnets (about 15 tons and 10 m long) to support the cold mass inside a vacuum vessel using glass fiber straps (bands) as vertical suspensions and glass fiber tie rods (titanium bar used at other labs) as transversal adjustments, which are both intercepted at LN_2 temperature 77 K, illustrated in Figure 4.19, Right [25]. The position of the magnet can be adjusted from outside the cryostat, but only when warm. The radiation shields are also supported by the suspension system between the cold mass and vacuum tank.

The advantage of these support systems is relatively low heat loads because of the space available. A certain disadvantage is the fact that many penetrations through the radiation shield must be closed carefully. A different system is used for SSC and RHIC and LHC magnets, to be discussed in Sections 4.6.6 and 4.6.7.

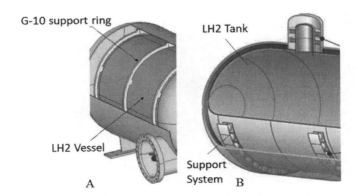

FIGURE 4.18   A: Light ring support for a 150-letter LH_2 vessel, [23]. B: Heavy load support ring for a 1000-m^3 LH_2 tank of Pilot Ship LH_2 Carrier [24].

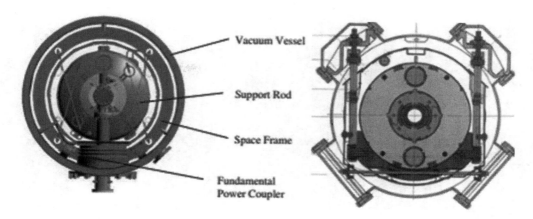

FIGURE 4.19 Left: Nitronic-50 support rod attached to SC cavity assembly to the space frame (CEBAF). Right: HERA dipole SC magnet cold mass supported by glass fiber straps and rods [25]. Source: *Left: Courtesy JLAB [26]. Right: Courtesy BNL*

Support for SRF Cavities. There are tens of SC accelerators, which use several hundred SRF cavities to accelerate charged particle beams. As in the SRF cavity cryomodule of CEBAF, a copper shield 2.37 mm thick operating between 35 K and 50 K is used to intercept thermal radiation. The shield that covers both the cavities and interconnect regions is cooled by supercritical helium gas (4 atm, 35 K) through pipes 22.2 mm in outer diameter by a 1-mm-thick wall. There are Nitronic-50 support rods attaching the helium vessel assembly to the space frame. The space frame serves as an additional thermal barricade to reduce heat flows from the room T vacuum vessel to the SRF cavities' LHe vessel, as shown in Figure 4.19 [26]. The horizontal and vertical helium vessel Nitronic-50 support rods are heat-stationed to the shield segments by copper straps, $1 \times 25 \times 125$ mm long. All wiring and cabling routed to 2 K is heat stationed at 50 K on the nearest shield segment as well. Due to the long length (>600 mm), small cross-sectional area (20 mm²), and difficulty during assembly, the two axial support rods are not heat-stationed.

Support for MRI Magnet. Magnet cryostats for nuclear magnetic resonance imaging (MRI) are being designed for a zero boil-off requirement where the coolant, liquid helium, is re-condensed with the use of a cryocooler [27]. An example of an MRI is provided by Suman et al. In this design, a 1.5-T MRI magnet is inside an annular helium vessel. This is a zero boil-off design, so heat loads must be kept below the refrigeration capacity of the cryocooler. High vacuum reduces any gas conduction. The use of intermediate heat shields and multilayer insulation reduces the radiative heat load. The self-centering suspension system consists of 12 S-glass epoxy support bands. There are four support bands on each end for the vertical direction and four along the side to maintain the horizontal location. The location of the S-glass epoxy support bands on the magnet is shown in Figure 4.20(a). The top and side support assembly designs are shown in Figure 4.20(b) and (c). The supports are heat-stationed as well at an intermediate temperature between 60 K and 80 K. The straps have a cross-section of 190 mm² and are in a racetrack configuration. Each strap has a thermal intercept block made of stainless steel 304L, which is connected to the thermal heat sink by flexible copper strips.

4.7 COMPRESSION AND TENSION POST SUPPORTS

Since the 1980s, based on the continuing international collaboration through the projects of HERA, SSC, LHC, and the International Linear Collider (ILC, TESLA, XFEL), post-type supports between

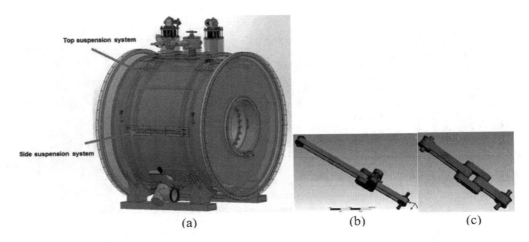

FIGURE 4.20 Indian MRI magnet support for a zero-boil-off MRI cryostat [27]. (a): Overall suspension system, (b): top support assembly, (c): side support assembly.

300 K and 2 K were successfully developed and utilized in large superconductivity projects. The support post minimizes heat leak and can withstand axial and radial loads. For example, LHC is a 27-km-long superconducting accelerator based on the large-scale use of high-field twin-aperture SC magnets (each >10 m long) operated at 1.9 K. The ILC and XFEL also developed a similar post with very efficient thermal performance for SRF cavities (cryostat length >10 m). In the reverse of the LHC arrangement, the support posts are fixed on the top of the vacuum shell. The ILC/XFEL superconducting cavities, placed upside down through the large helium pipe, are suspended to the post in the SRF cavity cryostat. In many applications, there are multiple stages of thermal intercepts available, which are typically around 4 K, 20 K, and 80 K. Post-shaped support geometries are favorable in these applications, which have been based on single-section uniform tubes, multiple-section stepped tubes, and re-entrant tube section designs.

4.7.1 REENTRANT POST SUPPORT

The unique and sophisticated reentrant post design for the suspension system of superconducting supercollider dipole magnets meets rigorous thermal and structural requirements. The design is shown in Figure 4.21. First, it resists structural loads imposed on the cold mass assembly, ensuring stable operation over the course of the magnet's operating life. Second, it serves to insulate the cold mass from heat conducted from the environment. The evolution and selection of the suspension system for SSC magnets has been well documented by Nicol et al. [8–9].

The magnet assembly is supported vertically and laterally at five places along its length. To accommodate axial shrinkage during cooldown, the magnet assembly is free to slide axially at all but the center support. The center serves as the anchor position. The cylindrical post is constructed with fiber-reinforced plastic (FRP, such as G10 and G11) tubing and metallic heat intercepts and end connections. All FRP-to-metal (SS-304) connections are made by mechanical shrink fitting and do not employ adhesives or fasteners. The post can operate in tension, compression, and flexure or in combinations of these loads. In the design, the inner and outer tubes bear compression forces, but the transition tube carries tension force. The thickness of the outer tube (G-11) is 0.109 inches, and the inner tube (carbon FRC) is 0.129. Both tubes are sized based on the ultimate tensile strength. At room temperature, the thermal conductivity of G-11 is four times less than that of the GRP; however, at approximately 40 K and lower, the GRP has lower thermal conductivity. During a model test, the measured heat leaks for a warm boundary at 300 K were 2.103 W for 80 K, 0.320 W for 20

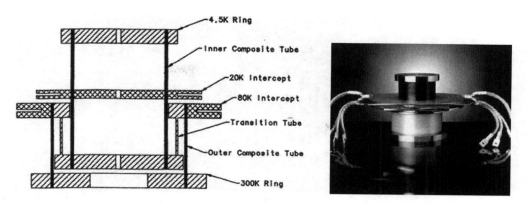

FIGURE 4.21   Sketch and photo of the SSC magnet post [8].

K, and 0.015 W for 4.5 K. The post meets the shipping and handling loads specified of 2.0 g in the vertical direction, 1.0 g in the lateral direction, and 1.5 g in the axial direction.

4.7.2 SINGLE-TUBE COMPRESSION POSTS FOR HEAVY SC MAGNETS

The reentrant post is thermally efficient, mechanically robust, and compact in size, but its assembly is complex and expensive. It requires precise tube wall thicknesses and tolerances to obtain a required shrink connection. The single-section uniform tube posts for heavy load cold mass have been successfully developed and widely utilized by the collective efforts of Fermilab, BNL, CERN, DESY, and others [27–29]. One of the successful developments is the SC magnet support posts of the LHC, as shown in Figure 4.22 [29]. Each LHC dipole magnet (about 30,000 kg and 15 m long, i.e. 2,000 kg/m) has three single-tube post support systems. The LHC support post is made with a glass-fiber reinforced epoxy (GFRE) composite material and has two aluminum heat intercepts positioned in optimal positions. For the LHC support posts, with fixed $T_H = 293$ K and $T_C = 2$ K, the optimal positions of two heat intercepts, at 5 K and 75 K, were chosen to minimize the energy cost function, with $C_1 = 16$ W/W, $C_2 = 220$ W/W, and $C_3 = 990$ W/W as cost factors at 75 K, 5 K, and 1.8 K, respectively. The key dimensions of the post are as follows: GFRE tube OD 240 mm, ID 228 mm, total height 250 mm, height from 293 K to 77 K 102 mm, height from 293 K to 5 K–10 K 193 mm. The calculated heat loads of the support post with and without heat intercepts are listed in Table 4.5.

In practice, the intercept temperatures are also determined by the available cooling resources of the entire cryogenic systems of the LHC, as indicated in Figure 4.22. The heat intercepts are 10-mm-thick aluminum plates close-fitted and glued to the external wall of the composite column. The heat intercepts were shrink-fit onto the composite column during cool-down, ensuring an improved heat exchange due to contact pressure. Active cooling of the bottom intercept at 50 K–65 K is ensured by an all-welded aluminum connection to the thermal shield, while the top intercept is a cast aluminum plate integrating a stainless-steel cryogenic line at 5 K–10 K. The design of the LHC magnet cryostats progressively evolved during the construction and completion of hundreds of magnets [31]. One of the important arguments was whether to keep the 5 K shields with MLI on them or omit the 5 K shields but wrap about 5–10 MLI layers on the 2 K magnet cold mass. Finally, it was decided to remove the 5 K shield. However, the design of the single-tube post for heavy compression loads was not changed. Figure 4.23 shows the arrangement of the supports along a magnet string and inside a magnet cryostat. Figure 4.24 illustrates a single-tube tension post alignment system.

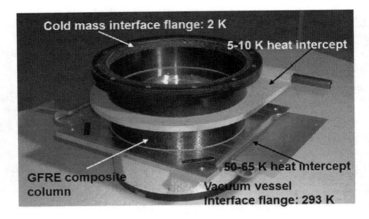

FIGURE 4.22   LHC support post in GFRE with two heat intercepts [30].

TABLE 4.5
Heat Load with and without Heat Intercepts and Total Cost of Heat Extraction per LHC

Number of Heat Intercepts (in Optimal Positions)	$\dot{Q}_{1.8\,K}$ (W)	$\dot{Q}_{5\,K}$ (W)	$\dot{Q}_{75\,K}$ (W)	$\dot{Q}_{elec}$ (W)
No intercept	2.79	–	–	2790
1 (at 75 K)	0.54	–	6.44	638
2 (at 75 K and 5 K)	0.047	0.42	7.1	252

Source: Post [30]

4.7.3 SINGLE-TUBE TENSION POSTS FOR HEAVY SRF CAVITIES

Single-tube posts for heavy tension loads have been successfully utilized in many SRF accelerator projects to support the SRF cavity assembly, such as TESLA, ILC, XFEL, and so on in DESY, INFN, Fermilab, JLAB, KEK, CERN, and others [31–33]. This type of SRF cryomodules originated from the International TESLA Collaboration for the International Linear Collider (ILC). The ILC consists of 16,000 superconducting cavities and stretches more than 20 miles in length. Because of the cost, the ILC has been slowed down and other projects keep the momentum with similar and improved methodologies.

General Arrangement. The ILC cryomodule includes eight-cavity (each 1.3 m) cryomodules with one magnet package and is about 12 m long. The SRF cavities are cooled by 2 K LHe initially with 5 K and 70 K thermal intercepts in the cryostat. The cavity LHe vessels are attached to the 300-mm-diameter helium gas return pipe (HGRP). Three G-10 single-tube posts for one cryomodule through the HGRP hold eight cavity jackets, a quadrupole, cooling channels, and thermal shields from the vacuum vessel. Each post carries the weight load for one post, 750 kg. The center post in the module is fixed to the vacuum vessel, and the posts at both ends have a sliding structure to remove the effect of thermal contraction of the GRP.

The 300-mm-diameter helium gas return pipe serves both as the "backbone" support for the RF cavity string in each cryomodule and as the low-pressure, 2 Kelvin, saturated vapor return line to the cryogenic distribution box. The support posts consist of a fiberglass tube terminated by two shrink-fit stainless-steel flanges. Two additional shrink-fit aluminum flanges are provided to allow

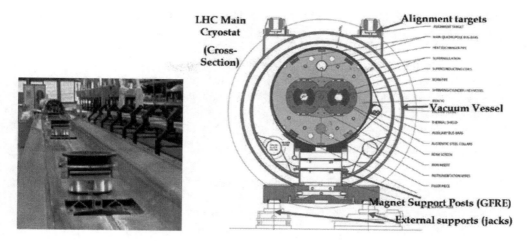

FIGURE 4.23 Right: LHC main cryostat indicates the support post (GFRE) and external support. Left: LHC supporter post prepared for assembling with magnet cold mass [30].

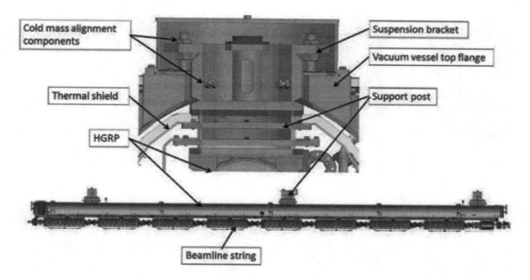

FIGURE 4.24 Single-tube tension post and alignment system. Source: *Courtesy Fermilab [34]*

intermediate thermal intercept connections to the 5 K–8 K and nominally 70 K temperature levels; the exact location of these flanges has been optimized to minimize the heat leakage.

Thermal Design of Support Post. Positions of the thermal intercepts are calculated in order to minimize the heat load to the helium refrigerator. The heat load is not sensitive to the distance L_2 between 5 K and 70 K. The G-10 post heights are as follows: total 140 mm, L_1=27 mm, L_2=37 mm, L_3=10 mm. A G-10 support post and the installation in a cryostat are shown in Figure 4.25.

Outer diameter=300 mm, thickness=2.2 mm, heat load 0.1 W to 2.0 K, 0.65 W to 5 K, 5.9 W to 70 K. Area of 5 K shield plate 30 m², 10 layers MLI on 5 K aluminum plate. Area of 80 K shield plate 35 m², 30 layers MLI on 80 K aluminum plate.

4.8 SUPPORTS FOR LONG COLD MASSES WITH VERY LARGE WARM BORES

4.8.1 SUPPORTS OF FERMILAB COLLIDER DETECTOR FACILITY MAGNET

Huge magnets with thin solenoid magnets and very large warm bore have been widely utilized in many detectors around the world, and two photos of detectors are shown in Figure 4.26. The cold mass must be supported inside the cryostat in a stable and rigid fashion. For example, the solenoid coil for the Collider Detector Facility (CDF) at Fermilab consists of a single-layer helix winding of aluminum-stabilized superconductor located inside a support cylinder and surrounded by radiation shields, superinsulation, and vacuum shell [35]. The solenoid physical characteristics

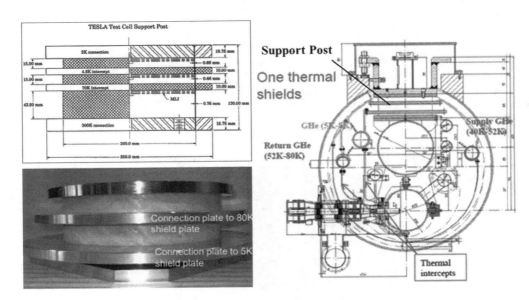

FIGURE 4.25   Left: photo and sketch of G-10 post. Right: The support post in the cryomodule [32].

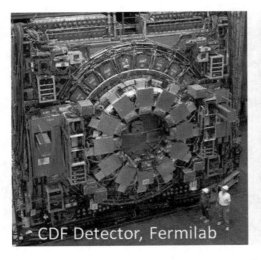
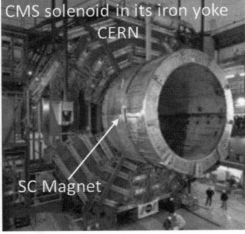

FIGURE 4.26   CDF detector and CMS detector, both with huge, long solenoid SC magnets with large warm bores. Source: *Courtesy Fermilab and CERN [35]*

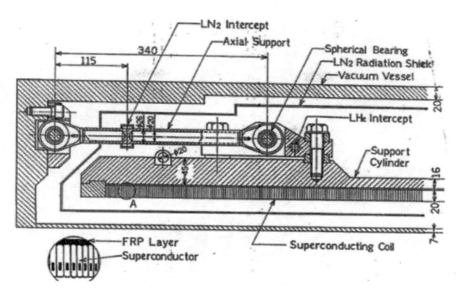

FIGURE 4.27   Fermilab CDF solenoid magnet low thermal conduction support system [35].

are summarized as follows: outer diameter 3.3 m, inner diameter 2.8 m, overall length 5.1 m, total weight 13 tons, cold mass 5,570 kg. The central field is 1.5 T, and the axial magnetic force on coil is about 100 tons. The cold-to-warm support system consists of six axial members, all on one end to provide axial stiffness. Each support element is 26 × 20 mm, made with low-conductivity Inconel 718. Twelve Inconel rods on each end carry the cold mass and provide radial stiffness. The members are thermally intercepted at 77 K and 4.4 K to reduce the heat flux and avoid hot spots. Spherical bearings on both ends of each member eliminate bending stresses due to differential thermal contraction. The coil and support cylinder are thermally screened from 300 K radiation by inner and outer LN_2-cooled shields. Figure 4.27 shows the CDF axial support and the cryostat [35]. Conduction heat leak is 0.25 W for each axial support and 0.31 W for each radial for a total support heat leak of ~9 W.

4.8.2   Supports for CMS and ATLAS Magnets

Support for CMS Detector Magnet. The Compact Muon Solenoid (CMS) of the LHC [36–37] (Figure 4.28, Upper) is a high-field superconducting magnet (4 T): operating current of 19.5 kA in a 5.9-m-diameter and 12.5-m-long warm bore, leading to stored energy of 2.7 GJ. The CMS has higher stored energy per unit mass of coil (12 kJ/kg) than any large superconducting magnet ever built. The cold mass total weight is about 225 tons. It is supported inside the vacuum tank through a system of radial and longitudinal tie-bar supports.

Supports for ATLAS Detector Magnet. The ATLAS magnet with a 2 T axial magnetic field is designed to be extremely lightweight as a single layer coil wound inside a thin AL alloy cylinder. It was developed by KEK in Japan and was cooled to 4 K with a 21,000-A current. The inner diameter is 2.3 m, the length is 5.3 m, and the coil weighs 5.5 tons [38]. Figure 4.28, Lower, shows the structure of the warm-to-cold support in ATLAS, and Figure 4.28(c) is a photo of the conceptual design of the CERN CMS detector cryostat, which is 12.5 m long by 6 m diameter and provides a 4 T magnetic field [38]. The two end toroid magnets were developed by the Rutherford Laboratory. All the performance tests were successful, and ATLAS is now the largest superconducting magnet system in the world.

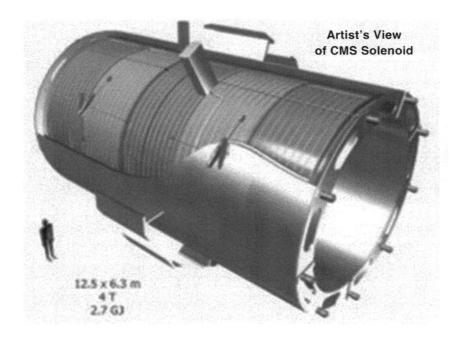

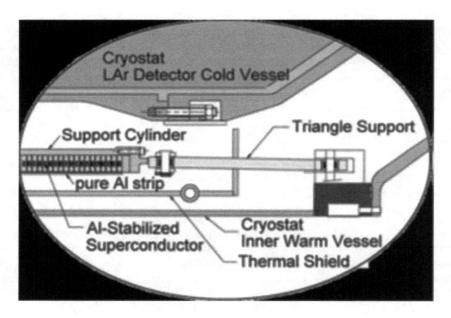

FIGURE 4.28   Upper: CMS solenoid cryostat with supports from CERN [37]. Lower: ATLAS detector support [38].

4.9   CONTACT-FREE SUPPORTS WITH MAGNETIC LEVITATION

Magnetic levitation (maglev) provides a unique contact-free support, which is thermally and mechanically highly efficient and has been exploited in extensive cryogenic engineering domains. Magnetic levitation has been used and is potentially being utilized in the fields of cryogenic transfer

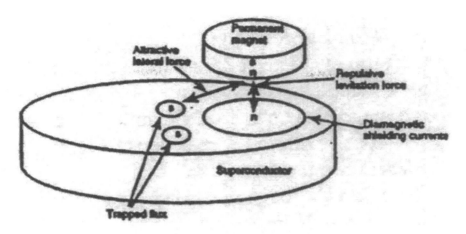

FIGURE 4.29 Levitation and stabilization mechanisms used in the maglev transfer line: (diamagnetic response for levitation and flux pinning for stabilization. Low thermal loss cryogenic transfer line with magnetic suspension [39].

lines and storage vessels, SC bearings, navigation gyro and flywheel energy-storage and high-speed trains and airplane/rocket launches [39]. With HTSs and permanent magnet (PM), the levitation force develops because a shielding current forms immediately beneath the PM. The non-superconductive regions that exist in Type II superconductors, called pinning centers, trap the flux from PM. This flux pinning phenomenon creates horizontal stability (attracts the PM back to its starting position) of the levitation element, as shown in Figure 4.29 [39].

4.9.1 HTS Maglev Support for Cryogenic Transfer Lines and Vessels

Prototype energy-efficient cryogenic transfer lines with magnetic suspension have been built and cryogenically tested by Shu et al. with NASA support [40]. The prototype transfer line exhibits cryogen saving potential of 30–35% in its suspension state as compared to its solid support state. Key technologies developed include novel magnetic levitation using multiple-pole high temperature superconductors, rare earth permanent magnet elements, and a smart cryogenic actuator as the warm support structure. The technology also has applications in zero-boil-off vessels, which have potential in saving launch fuel and/or increasing payloads. This discussion is focused only on the maglev support design.

Three performance indices are emphasized in all maglev configurations: (1) sagging distance, (2) the final levitation gap, and (3) levitation forces. Applied in this development is the field cooling (FC) levitation mechanism. This means that the HFS is initially cooled at a certain cooling height, and due to the gravity of inner line and cryogen, the HTS (Yttrium Barium Copper Oxide or YBCO) will approach the Neodymium PM (NdFeB) to accumulate enough levitation force to balance the loads. In a compact maglev transfer line design, 2–3 mm travel is allowable, and larger sagging of the inner line may compress the superinsulation and cause solid contacts between the inner and outer vessels in some regions. Figure 4.30A–B presents the initial configuration design, in which the levitation force is about 10 kg after improvement. Figure 4.30C illustrates the design employed in the 6-m maglev transfer line, which has a sandwich structure of PM and larger levitation force.

4.9.2 HTS Maglev Support for Bearings and Flywheels

Three types of bearings are used in engineering structures: mechanical bearing, active (electro-) magnetic bearing (AMB), and superconducting magnetic bearing (SMB). Mechanical bearing needs

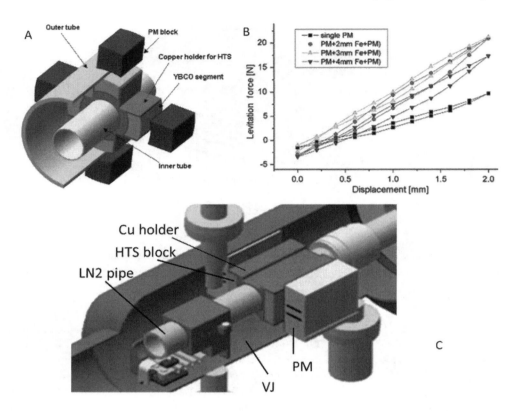

FIGURE 4.30 A–B: Testing magnetic support configuration and support forces. C: Final design sketch with larger forces used in the 6-m prototype transfer line [40].

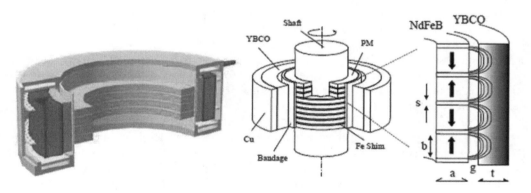

FIGURE 4.31 Maglev support configuration for bearings and flywheels.

lubrication, and the friction energy loss increases as speed and load do. SMB offers a "frictionless" solution with levitated shafts that can spin at very high speeds. With an HTS stator, the coolant can be liquid nitrogen, which is cheap to procure. The use of extremely low-loss high-temperature superconducting bearings is necessary to meet the long idle time requirements of future lunar and Martian missions. High specific energy and high energy storage are demanded, so that rotational speeds of perhaps 150,000–200,000 rpm will be needed.

Figure 4.31 introduces the simplified maglev support configuration for bearings and flywheels, and the sandwich support can hold cold mass above 1 ton. Mukoyama et al. report one of the world's largest class flywheel energy storage systems (FESSs) [41]. The FESS, connected to a 1-MW mega-solar plant, effectively stabilizes the electrical output fluctuation of the photovoltaic (PV) power plant caused by the change in sunshine. The FESS uses a superconducting magnetic bearing to levitate a heavyweight flywheel rotor without mechanical contact. The SMB consists of HTS coils and HTS plates made of YBCO. The HTS plates in the rotor axis receive the levitation forces in the magnetic field generated by the HTS coils in the stator. In the factory test, the SMB was confirmed to levitate a 4,000-kg load, and its performance and basic reliability were verified: The FESS rotor reached a maximum of 2,950 rev/min and was charged-discharged at 300 kW in the PV plant.

REFERENCES

1. Flynn, T, *Cryogenic Engineering*, Marcel Dekker, New York, 2005.
2. Barron, R, *Cryogenic Systems*, Oxford University Press, New York, 1985.
3. Shu QS, Demko J, Fesmire J, Developments in advanced and energy saving thermal isolations for cryogenic applications, *IOP Conf. Series: Materials Science and Engineering,* Vol. 101, p. 012014, 2015.
4. Omet M, Ayvazyan V, Branlard J et al., *Operation of The European XFEL Towards The Maximum Energy*, SRF2019, Dresden, Germany, 2019, mathieu.omet@desy.de
5. Ganni V, Fesmire J, Cryogenics for superconductors, *Adv in Cryo Eng*, Vol. 57A, pp. 15–27, 2012.
6. Serio, L., Challenges for cryogenics at ITER, *AIP Conference Proceedings 1218*, pp. 651–662, 2010.
7. Lebrun, P., Twenty-three kilometres of superfluid helium cryostats for the superconducting magnets of the Large Hadron Collider (LHC), 2016, doi:10.1007/978-3-319-31150.
8. Nicol T, Niemann R, Gonczy J, *Design and Analysis of the SSC Dipole Magnet Suspension System Supercollider I*, Plenum Press, 1989 Name of magazine? [E-E] Nicol T, et al., Design and Analysis of the SSC Dipole Magnet Suspension System 1989 at the 1989 International Industrial Symposium on the Super Collider (IISSC).
9. Nieman RC, Gonczy JD, Nicol et al., Design, construction, and performance of a post type cryogenic support, Fermilab TM-1349, September 1985.
10. Beer F, Johnston Jr., DeWolf J, Mazurek D, *Mechanics of Materials*, Sixth edition, McGraw-Hill, New York, 2012.
11. Shigley J, Mitchell L, *Mechanical Engineering Design*, Fourth edition, McGraw-Hill Book Company, New York, 1983.
12. Byers EF, Snyder RD, *Engineering Mechanics of Deformable Bodies*, Intext Educational Publishers, Scranton, PA, 1975.
13. Young WC, *Roark's Formulas for Stress and Strain*, Sixth Edition, McGraw-Hill, New York, 1989.
14. McAshan M, *Optimization Studies for the Conceptual Design of the SSC Cryogenic System*, Internal Communication, 1986.
15. Rode C, *Tevatron Cryogenic System*, Report at Fermilab, USA.
16. Salerno L, Kittel P, *Thermal Contact Conductance*, NASA Technical Memorandum 110429, Ames Research Center, Moffett Field, CA, 1999.
17. Deng B et al., Simulation and experimental research of heat leakage of cryogenic transfer lines, *IOP Conf. Ser.: Mater. Sci. Eng.*, Vol. 278, p. 012017, 2017.
18. Duda P et al., Entropy analysis of support systems in multi-channel cryogenic lines, CEC 2019, *Materials Science and Engineering,* Vol. 755, p. 012065, 2020.
19. Knoll D, Willen D, Fesmire J, Johnson W, Smith J, Meneghelli B, Demko J, George D, Fowler B, Huber P, Evaluating cryostat performance for naval applications, *AIP Conf. Proc.*, Vol. 1434, pp. 265–272, 2012, doi:10.1063/1.4706929.
20. Nakai H et al., Superfluid helium cryogenic systems for superconducting RF cavities at KEK, *AIP Conference Proceedings*, Vol. 1573, p. 1349, 2014.
21. Fesmire J, Efficient Supports for Cryogenic Tanks, Private communication, 2020.
22. Catalytic Construction Inc., LC-39 drawing package, 1965.
23. Aceves S, Petitpas G, Safe, long range, inexpensive and rapidly refuelable hydrogen vehicles with cryogenic pressure vessels, *International Journal of Hydrogen Energy*, Vol. 38, Issue 5, 2013, E-mail addresses: petitpas1@llnl.gov, guillaume.petitpas@ensmp.fr

24. Nishimura M, Long distance transportation of LH$_2$–hydrogen road, *Japan-Norway Hydrogen Seminar*, 28 February 2017.

25. Kaiser H, *Design of Superconducting Dipole for HERA*, Deutsches Elektronen-Synchrotron DESY, Hamburg, Germany.

26. Biallas G, Brindza P, Rode C, Phillips L, The CEBAF superconducting accelerator cryomodule, *IEEE Trans. Magnetics MAG-23*, Vol. 2, pp. 615–618, March 1987.

27. Suman N, Soumen K, Kumar M, et al., Stress analysis of cryogenic suspension system of superconducting MRI magnet cryostat, *Indian Journal of Cryogenics*, Vol. 43, 2018.

28. Blin M, et al., Design, construction and performance of superconducting magnet support posts for the Large Hadron Collider, *Advances of Cryogenic Engineering*, Vol. 39A, pp 671–677, 1994.

29. Peterson T, *Cryogenic Considerations for Cryomodule Design,* Presentation at SLAC USPAS, 2017.

30. Parma V, *Cryostat Design*, CERN, Geneva, Switzerland.

31. Castoldi M, Pangallo M, Parma V, Vandoni G, Thermal performance of the support system for the LHC SC magnets, *Advances in Cryogenic Engineering*, Vol. 45, 2000.

32. Weisend J, Pagani C, Bandelmann R, Bosotti A, et al., The TESLA test facility (TTF) cryomodule: A summary of work to date, *Advances in Cryogenic Engineering*, Vol. 45, 2000.

33. Pagani C, et al., Design of the thermal shields for the new improved version of the TTF cryostat, *Advances in Cryogenic Engineering*, Vol. 43, 1998.

34. Peterson TG, Weisend II, JG, *TESLA & ILC Cryomodules*, FERMILAB-TM-2620-TD.

35. Fast R, CDF solenoid design note #95, February 22, 1985, Fermilab.

36. Herve A, et al., Status of the CMS magnet MT17, *IEEE Trans on Appl Supercond,* 04, 2002.

37. Védrine P, *Large Superconducting Magnet Systems*, CAS-CERN Accelerator School, Geneva, Switzerland, 2013.

38. Yamamoto A, et al., Design and development of the ATLAS central solenoid magnet, *IEEE Transaction on Applied Superconductivity*, Vol. 9, No. 2, pp. 852–855, June 1999.

39. Shu QS, Cheng G, Yu K, Hull J, Demko J, Colin C, Fesmire J, Low thermal loss cryogenic transfer line with magnetic suspension, *Advances in Cryogenic Engineering*, Vol. 49, 2004.

40. Shu QS, Cheng G, Shusta J, Hull J, Fesmire J, Augustanowicz S, Demko J, Werfel F, Magnetic levitation technology and its applications in exploration projects, *Cryogenics*, Vol. 46, No. 2–3, 2006.

41. Mukoyama S, Development of superconducting magnetic bearing for 300 kW flywheel energy storage system, *IEEE Applied Superconductivity*, Vol. 27, No. 4, 2017.

5 Thermal Anchors and Shields

Jonathan A. Demko, Quan-Sheng Shu, and James E. Fesmire

5.1 INTRODUCTION

To maintain components at cryogenic temperatures, a low-loss connection must be made to the source of the cold. The source of cooling can be a refrigerator or cryocooler, or it can come from a supply of a cryogenic fluid such as helium, hydrogen, or nitrogen. Thermal anchoring is the design feature that makes the thermal connection with minimal temperature difference, that is, a low thermal resistance, between the cold heat sink and the component being kept cold, such as a thermal shield.

Thermal shields are placed between a high temperature boundary and a cold surface in order to limit the flow of heat by thermal radiation. There are two main types, floating shields, which passively reach thermal equilibrium and reflect heat away from the cold space, and actively cooled shields, which must be anchored to a low-temperature thermal sink in order to remove the intercepted heat load reaching the shield to maintain it at the operating temperature.

5.2 THERMAL SHIELDS

Thermal shields are inserted between the warm boundary and the cold space to limit thermal radiation from reaching the cold space. Typical configurations can be flat plates, concentric cylinders, or spheres, as illustrated in Figure 5.1. The warm and cold surfaces may have different emissivities than the shields since these are usually a structural boundary. The shield material is selected for its thermal properties and typically has no structural or mechanical function. Thermal shields are generally made from high thermal conductivity materials such as aluminum or copper so that they are nearly uniform in temperature. In addition, thermal shields are typically highly reflective (low emissivity) so that thermal radiation is mostly reflected from the surface so as not to absorb it at low temperatures only to be removed by a refrigerator. The surface may be treated to lower the emissivity by polishing, electropolishing, or plating with a thin layer of highly reflective metal such as nickel, silver, or gold. When the shields are not connected to a low-temperature heat sink, they can be classified as passively cooled heat shields. This is the principle behind multilayer superinsulation, where the layer temperatures are a result of energy balances for each layer.

5.2.1 PASSIVE THERMAL SHIELDS

A simple approach to calculate the reduction of heat transfer due to the presence of passive thermal shields can be explained using Figure 5.1. Figure 5.1(a) shows a simple planar shield configuration, and Figure 5.1(b) shows a cylindrical or spherical geometry.

The warm surface temperature and emissivity are T_w and ϵ_w, respectively, and the cold surface temperature and emissivity are T_c and ϵ_c, respectively. Let all the shield layers have the same emissivity, ϵ_s; then the heat transfer in watts becomes:

$$Q = \frac{\sigma A\left(T_w^{\,4} - T_c^{\,4}\right)}{1/\epsilon_w + 1/\epsilon_c - 1 + N\left(2/\epsilon_s - 1\right)} \qquad 5.1$$

DOI: 10.1201/9781003098188-5

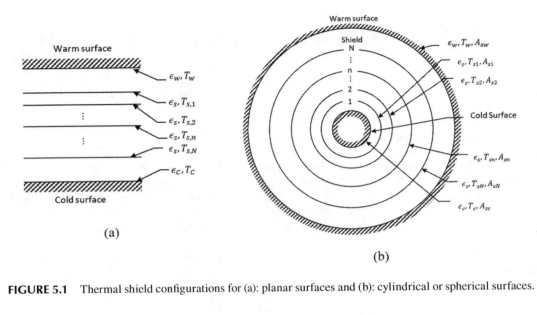

FIGURE 5.1 Thermal shield configurations for (a): planar surfaces and (b): cylindrical or spherical surfaces.

where A is the surface area in square meters. If the walls have the same emissivity as the reflector shield, the expression simplifies to the following:

$$Q = \frac{\sigma A \left(T_w^{\,4} - T_c^{\,4} \right)}{\left(N + 1 \right)\left(2 / \epsilon_s - 1 \right)}$$

5.2

From these expressions, it can be clearly seen that the heat transfer decreases with an increase in the number of shields. For example, for heat transfer from a warm surface at $T_w = 300$ K and a cold surface at $T_c = 4.2$ K , if $\epsilon_w = 0.2$, $\epsilon_c = 0.1$, and $\epsilon_s = 0.05$, the ratio of heat transfer with shields to that without any shield is shown in Figure 5.2. From the figure, it is seen that even a single thermal shield will reduce the heat load to the low temperature to roughly a quarter of the unshielded value.

For cylindrical as well as spherical configurations, the area of the surface varies with the radius. Assuming the shield surfaces have the same emissivity, an expression for the radiation heat transfer is given by Siegel and Howell [1] as:

$$Q = \frac{A_{sc}\sigma \left(T_c^{\,4} - T_w^{\,4} \right)}{1 / \epsilon_c + \left(A_{sw} / A_{sc} \right)\left(1 / \epsilon_w - 1 \right) + \sum_{n=1}^{N} \left(A_{sc} / A_{sn} \right)\left(2 / \epsilon_s - 1 \right)}$$

5.3

They provide details on analyzing the case with variable shield emissivities.

5.2.2 Actively Cooled Thermal Shields

Actively cooled shields require connection to a form of refrigeration or a low-temperature heat sink. If a coolant is circulated to the shield, some form of cooling tube is typically attached to the shield. The shield is typically a thin plate of high thermal conductivity material such as aluminum or copper. In most cases, the shield is wrapped with multilayer superinsulation to intercept thermal

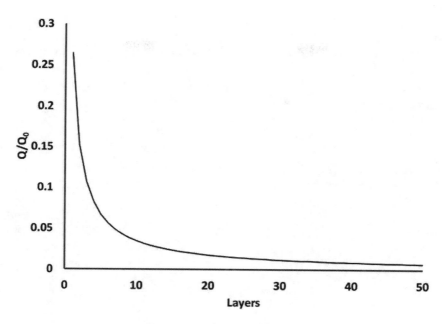

FIGURE 5.2 Ratio of heat transfer between 300 K and 4.2 K with thermal shields to that with no shield as a function of the number of shields.

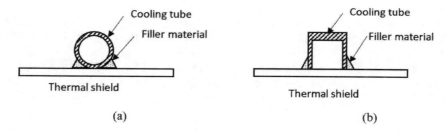

FIGURE 5.3 Simplified sketches of the attachment of a cooling tube to the thermal shield surface. (a): Round tube, and (b): rectangular channel.

radiation from reaching the cold space. The shield can be plated with nickel to reduce the emissivity to intercept a radiant heat load.

Attachment of cooling tubes is frequently specified, as shown in Figure 5.3(a), where a round tube is secured to a surface by welding or soldering. This configuration is often simple to fabricate and is satisfactory for many applications. The disadvantages are, first, that there may be a contact resistance between the tube and the shield surface. Second, there is a thermal resistance of the tube wall, which is usually small. An alternative is to attach a channel, as shown in Figure 5.3(b). This eliminates the cooling tube material thermal resistance and allows the cooling fluid to directly cool the shield surface.

The spacing of cooling tubes has been discussed by McIntosh [2], where the maximum temperature difference on the shield, ΔT, is specified for a line spacing of $2L$. The expression for ΔT is:

$$\Delta T = \frac{q"L^2}{2kt}$$

TABLE 5.1

Optimum Temperatures for Cooled Shields and Related Carnot Refrigerator Power

Number of Shields	$T_1(K)$	$T_2(K)$	$T_3(K)$	$T_4(K)$	Refrigeration Power (W/m²)
Cold end temperature = 4.2 K					
1	97.82	–	–	–	1,293
2	65.78	181.6	–	–	506
3	52.86	137.57	222.65	–	309
4	45.63	114.08	182.02	243.74	222

Number of Shields	$T_1(K)$	$T_2(K)$	$T_3(K)$		Refrigeration Power (W/m²)
Cold end temperature = 1.8 K					
1	81.14	–	–	–	1,632
2	52.75	171.76	–	–	569
3	42.54	127.09	217.42	–	334
4	34.42	104.10	175.28	240.42	236

Source: From Hilal and McIntosh [3]

where the heat flux is $q"$ (W/m²), k is the thermal conductivity at the shield temperature (W/m/K), and t is the thickness of the shield. The spacing can readily be determined from this equation.

Hilal and McIntosh [3] propose a "workable" approach to reducing thermal radiation using refrigerated shields. They provide an approach for minimizing the refrigeration power required by optimizing the operating temperature of the refrigerated shields. The following results in Table 5.1 are provided by [3].

Eyssa and Okasha [4] analyzed the optimization of refrigerated thermal shields that were covered by floating shields or multilayer superinsulation and suggested this can further reduce the Carnot cooling power.

5.3 THERMAL SHIELDS FOR SUPERCONDUCTING MAGNETS AND SUPERCONDUCTING RADIO-FREQUENCY CAVITIES

Thermal shields intercept heat radiated from the surfaces of the vacuum vessel surrounding the magnet or cavity, which are at temperatures higher than the operating temperature of the devices. It is typical to have a thermal shield operating in the 50 K–80 K range, depending on the details of the cryogenic system. There is sometimes a second shield operating in the 5 K–20 K range, again depending on the details of the system and the operating temperature of the SRF devices. The shields are usually segmented to minimize thermal bowing. The material is usually aluminum or copper depending on cost, weight considerations, structural strength, ease of fabrication, attachment needs, and so on.

An illustration of the use of thermal anchors and shields for a superconducting magnet system is discussed in Nilles [5] and provided in Figure 5.4. This figure shows a typical configuration for a superconducting magnet used in a particle accelerator. In Figure 5.4, the thermal shields are anchored to coolant lines by wrapping a portion of the shield around the coolant line. Another application of a

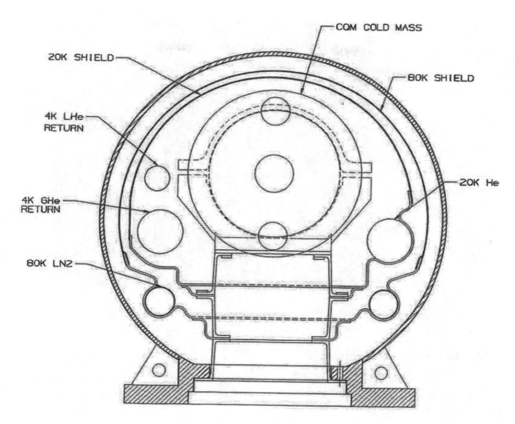

FIGURE 5.4 Typical arrangement of thermal shields and anchors for a particle accelerator superconducting magnet string.

thermal anchor is illustrated in this figure where a portion of the actively cooled shields is extended to the magnet support post, providing a heat station to remove conduction heating from the support.

The magnet cold mass rests on a structural support and is enclosed by a thermal shield cooled by helium gas at 20 K and a second warmer shield cooled by liquid nitrogen to 80 K. The shields are cooled by circulating these fluids through pipes anchored to the shields. The shields are typically made of aluminum in order to provide a high thermal conductivity so that the shields can be at a uniform temperature. The pipes supplying the cooling cryogenic fluid are stainless steel. These are clamped to the C-shaped shield section to minimize the thermal resistance across the joint and provide adequate surface contact to transfer the heat load from the shield to the fluid.

In addition, the support post is thermally anchored to the shields to heat-station the support and remove heat from the support to minimize the thermal loads to the 4 K magnet.

Thermal shields are frequently covered with multilayer insulation (MLI) to reduce the amount of thermal radiation heat intercepted by the shield. The outermost thermal shield is the warmest and intercepts heat from the room-temperature outer vessel. It is usually covered with 30–60 reflector layers since this is where the thermal radiation is the highest. On lower-temperature surfaces, fewer layers are used, for example, ten. Most of the rationale below about 20 K is to reduce heat load in case of loss of vacuum. Material is generally double-aluminized Mylar with fabric, nylon, or spun-bonded material spacers. Some installations use aluminum foil. Typical heat transfer rates are ~1 W/m^2 for 30 layers from 300 K to 70 K and 50 to 100 mW/m^2 for ten layers below 70 K [6]. Figure 5.5 shows the installation of a thermal shield for the International Linear Collider (ILC) and the MLI blanket covering the shield.

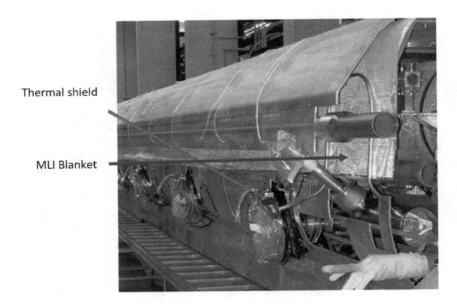

Thermal shield

MLI Blanket

FIGURE 5.5   Thermal shields for ILC cryomodule.

5.4   DEWAR THERMAL SHIELDS

Dewar shields can be built with trace cooling using boil-off of the stored cryogenic liquid (helium, nitrogen, etc.). The attachment of the trace coolant lines to the shield can be accomplished by fusion to the surface by welding, brazing, or soldering, as previously mentioned. For dewars, McIntosh [2] discusses the use of an all stainless-steel trace line system that can be crimped into special 6063 aluminum extrusions. This is shown in Figure 5.6.

5.5   THERMAL SHIELDS IN MAGNETIC FIELDS

In many low-temperature applications, the shield can be subject to a strong time-varying magnetic field. It is necessary to prevent forming a continuous electrical conducting loop that would permit the formation of induced currents in the shield. This can be simply accomplished with a slit longitudinally along the shield, forming an electrical break.

In Figure 5.7, a cryocooled system for measuring the voltage-current characteristics is shown where a nickel-plated copper thermal shield was used to limit thermal radiation to the test block. Measurements were made in magnetic fields up to 6 Tesla of the voltage current (V-I) characteristics of the high-temperature superconductor (HTS), as discussed in Young et al. [7]. Thus, the magnet was ramped slowly to the maximum field and held at different levels to obtain the required data. In this situation, the copper shield was slit to prevent the generation of eddy currents. The shield was wrapped with MLI, but the spacer material separated the thin reflective layers, and a continuous electrical circuit was not made.

5.5.1   THERMAL SHIELDS AND ANCHORS IN VARYING MAGNETIC FIELDS

Kamiya et al. [8] discuss the application of electrically insulated MLI, thermal shields, and anchors as applied to the full superconducting tokamak referred to as the JT-60 Super Advanced (JT-60SA). The superconducting coils of the tokamak are kept at 4.4 K. Thermal shields surround the magnet and are cooled by pressurized helium gas at temperatures between 80 K and 100 K. The shields intercept thermal radiation from the 300 K outer vacuum vessel.

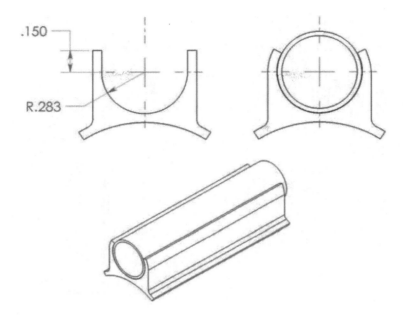

FIGURE 5.6 Stainless steel tube in extrusion.

This tokamak is operated in a pulsed mode, so the shields and the MLI must have electrical breaks to prevent the development of induced eddy currents. These eddy currents can be large enough to produce significant additional joule heating and potentially damage the MLI. The shields and MLI are designed to withstand a maximum of 50 volts. This is accomplished by dividing the shield and MLI into 18 sectors in the toroidal direction and 10 sectors in the poloidal direction. Each individual MLI sector is covered with a polyimide film 25 µm thick. The reflector layers have 3 mm of non-deposited slits at 21-mm intervals to suppress eddy currents within the blankets. Special considerations are detailed regarding the joints between the sectors.

The thermal shield of the JT-60SA consists of several components, including the vacuum vessel thermal shield (VVTS), the port thermal shield (PTS), and the cryostat thermal shield (CTS). The MLI is applied only to the CTS to provide access for maintenance. Figure 5.8 shows an isometric of the CTS and one electrically insulated CTS sector in the toroidal direction, respectively. The structure of the CTS cooling pipes and two SUS316L plates with a thickness of 1 mm on the ambient temperature side (outside) and 3 mm on cryogenic side is shown in Figure 5.8(c). The MLI is located on the higher-temperature side. One MLI sector is designed by closely tracing one CTS sector. The cryostat is assembled around the CTS after completion of the CTS. The MLI needs to withstand the heat from the welding of the cryostat sectors through a heat-resistant insert between the cryostat and MLI. [8]

5.6. THERMAL SHIELDS WITH CRYOCOOLERS

The schematic arrangement of a superconducting magnet system cooled using a cryocooler is illustrated in Figure 5.9 [9]. In this approach, a two-stage cryocooler cools the thermal shield, the current leads, and the magnet. The thermal shield is anchored to the first stage of the cryocooler and intercepts heat flowing from room temperature through the cryostat insulation, the cold mass supports, instrumentation wires, and conventional current leads that carry current into the magnet. The current leads can be the largest load, depending on several factors, including the magnet current and

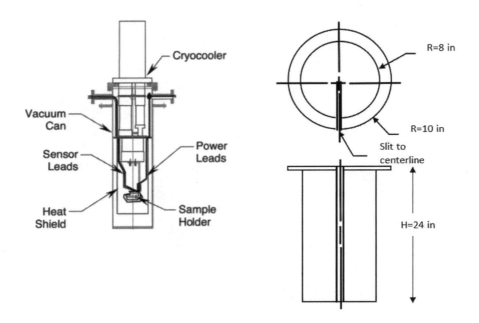

FIGURE 5.7 Measurement apparatus inside the stainless-steel vacuum vessel.

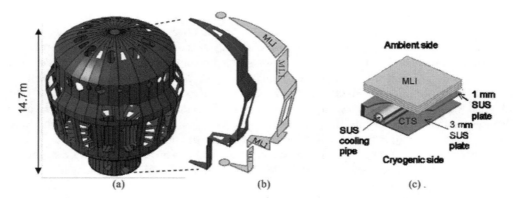

FIGURE 5.8 (a): Isometric view of the cryostat thermal shield of the JT-60SA, (b): electrically insulated MLI sectors, and (c): internal structure [8].

surface area of the cryostat. Systems such as MRI magnets have large warm bores, which results in a relatively large surface area as opposed to a superconducting magnetic energy storage (SMES) magnet, which is contained similarly to the figure.

The second stage of the cryocooler removes the heat flowing into the 4 K region coming from thermal radiation, conduction down the cold mass supports and instrumentation wires, and conduction down high-temperature superconducting leads between the first stage of the cryocooler and the magnet The largest 4 K heat loads are from the HTS current leads and the cold mass supports.

The use of 4 K cryocoolers permits the operation of various magnet systems in a stand-alone cryogen free environment. In order to minimize the size of the cryocooler, for the most energy efficiency, a low heat leak cryostat is required. The incorporation of HTS leads permits the superconducting magnet to be operated with continuous current flow through the leads. Persistent magnets

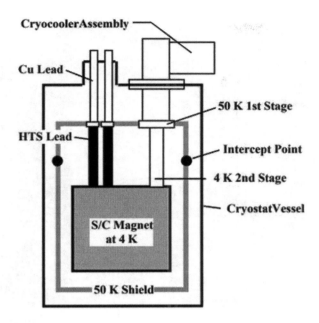

FIGURE 5.9 A schematic diagram of a superconducting magnet, thermal shields, and current leads with a two-stage 4 K cryocooler.

that are cooled using a 4 K cryocooler can be operated with their leads connected. Green points out that there are limits to the use of 4 K cryocoolers to cool large magnets. He suggests that the mass be less than 5 tons or the maximum lead current between 600 and 1,000 A. In general, the heat going down the leads into the magnet limits the temperature of both stages of the cryocooler.

5.7 CRYOGENIC SHIELDS FOR COLD MASSES BELOW 1 K

Experiments which must be conducted at very low temperatures, below 1 K, have especially difficult challenges due to the difficulty in providing refrigeration at these temperatures. A discussion of one such experiment, the Cryogenic Underground Observatory for Rare Events, or CUORE, experiment was provided by Barucci et al. [10]. Figure 5.10 shows the arrangement of the dry dilution refrigerator (DDR) cooling the experiment and the pulse tube (PT) refrigerator used to cool the thermal shields.

The two stages of the PT refrigerator are thermally connected to the 50 K shield by a 400-mm-diameter flange and to the inner vacuum chamber (IVC) by a 340-mm-diameter flange. Thermal gradients in the cold plates and radiation shields must be as small as possible, so relatively thick mechanical parts (flanges, shields) are needed, which also results in long precooling times. The 50 K shield intercepts the thermal radiation coming from the outer room-temperature vessel. It is covered by 25 layers of MLI. The IVC and 50 K thermal shield are made from aluminum alloy Al-5083. The IVC vessel has a diameter of 330 mm, a length of 720 mm, and a thickness of 3 mm. The IVC is maintained at around 3 K. The IVC cold plate has a total mass of about 12 kg, approximately the same as the whole 50 K shield, with the wall thickness reduced to 2 mm.

The 3 K IVC acts as the main radiation shield for the DDR to intercept the thermal radiation from the 50 K shield. There are inner thermal shields thermally connected to the DDR providing thermal shielding to the experimental space. A total heat load of 15 W was measured on the first PT stage at 40 K, while for the second stage, 0.2 W was measured at 3 K. In order to reduce the transmission

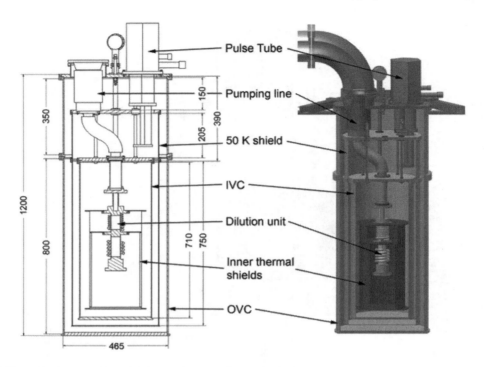

FIGURE 5.10   Layout of the dilution refrigerator for the CUORE experiment.

of vibrations from PT to the dilution unit, several copper braids of about 15 cm are welded to solid copper parts, which were bolted on to the PT cold stages and to the DDR flanges.

The incoming ³He gas temperatures must be well below 5 K–6 K before entering the dilution unit (DU). The pulse tube dilution refrigerator condensation line requires sufficient thermal grounding at the refrigeration stages in order for this to take place. The thermal grounding of the ³He inlet capillary takes place at the first cold stage of the pulse tube refrigerator (PTR) at six points along the regenerator of the second stage and at the 2.5 K cold plate of the PTR cold head. Then the ³He incoming capillary passes through the still pumping line, reaching the main flow impedance above the still. With this configuration, the incoming ³He stream reaches temperatures below 2.8 K after the last thermal grounding at the PT cold plate almost independently of the ³He flow rate.

The lower section of the DU consisting of the still, heat exchangers, and mixing chamber is part of a MNK-CF-500 unit produced by Leiden Cryogenics with about 500 μW at 120 mK with an ³He flow of 10^{-3} mol/s. The DU has two copper thermal shields anchored to the still and to the 50 mK cold plate.

5.8 THERMAL ANCHORS

Several methods have been employed to thermally anchor a system component to the cold sink. These include the use of thermal straps; direct connection to a cold head of a cryocooler or refrigerator; or, as illustrated in Figure 5.3, direct attachment of a coolant tube to a thermal shield.

Anchors are used to remove conduction heat loads from structural components that directly transfer mechanical loads between a warm component and a cold component such as a superconducting magnet support post or current leads

5.8.1 Thermal Anchors for Structural Components

Support structures, such as magnet support posts and dewar necks, have cross-sections determined primarily from structural requirements. To limit the heat conducted from the warm boundary to a cold component, these are heat-stationed, as discussed in the chapter on supports. The thermal link to cold temperatures must be of high thermal conductivity to minimize the temperature rise between the link and the cold sink. Also, the thermal anchor must be able to cool the structural component uniformly at the place of application through the support. One such thermal anchor design is detailed in Nieman et al. [11] for a re-entrant support for the superconducting super collider magnets, as shown in Figure 5.11. This design uses a shrink fit of rings on support tubes to provide the structural and thermal load transfer capacities needed. The concept is illustrated in Figure 5.12 and consists of the support tube, a central disc, and an outer ring. The outer ring has a thermal anchor which attaches to the cold heat sink. The support tube material is usually a glass-reinforced composite such as G-10, G-10CR, G-11, or G-11CR. The disc and ring were made from stainless steel and aluminum. The selection of the materials and their dimensions must satisfy both the structural and thermal requirements.

Nieman et al. [11] provide thermal and structural analysis approaches since design must satisfy both sets of criteria simultaneously. Table 5.2 provides a sample of the shrink fit analysis for the 300 K joint. The design analysis determined that the heat loads to 80 K, 20 K, and 4.5 K are 2.103 W, 0.320 W, and 0.015 W, respectively, which met the required heat load budget.

An example of the thermal management for a superconducting radio frerquency (RF) cavity is discussed in [6]. A support post for superconducting RF cavities is shown in Figure 5.13 [6], showing the helium cooling line attachment to aluminum heat intercept plates. These are attached to the G-10 support post using an adhesive. Aluminum strips provide a place to attach the 50 K to 75 K thermal shield. Flexible copper braids are used to provide a thermal path to cool the shields in this structure, as shown in Figure 5.14.

5.8.2 Thermal Anchors for Cryogenic Sensors and Wires

Cryogenic sensors are also thermally anchored in order to limit heat conduction flow along the sensor leads that come from a warm connection to the low temperature sensor to increase the temperature measurement accuracy at low temperatures.

Instrumentation leads, such as for thermometers, are anchored in several ways. If a thermometer is placed on the outside of a pipe, the leads may be wrapped around the pipe in three or more loops to intercept the heat from the lead wires and provide an accurate temperature reading. This is shown in Figure 5.15 where thin film thermometers were used to measure the temperature at the cold surface of a vacuum insulated line and the temperature at the outside of the multilayer insulation blanket. There are also mounting fixtures for thermometers that take the form of different types of bobbins or a carefully designed probe if inserting into a flow.

Anchors for Thermometers. For example, a stainless-steel sheath is used to surround the probe, as illustrated in Figure 5.16. The leads would be wrapped around the pipe or run against the cold outer wall of the pipe for sufficient distance to limit lead conduction to the sensor.

A fixture for mounting an external thermometer to measure the temperature of helium flow in a tube in the interconnect between dipole magnets is described by Datskov et al. [12] and shown in Figure 5.17. The components of the attachment consist of a thermometer (1) and thermal anchor (2). These are mounted in holes through a stainless-steel plate (3) using thermal-conducting grease (9) to enhance the thermal contact of the sensor. The fixture is clamped on the helium tube (8) with a hose clamp (5) and covered by multilayer insulation (6). During ASST Run #3, this fixture measured the tube temperature to within 0.01 K at 4.4 K from another carbon-glass thermometer mounted nearby in the magnet cold mass in the liquid helium stream. These were also installed on the 20 K and 80 K shields.

An indirect temperature measurement technique, used in the Russian accelerator the NUCLOTRON, is presented by Datskov [12] and was developed to avoid installing thermometers

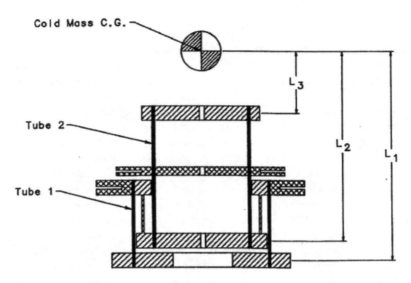

FIGURE 5.11 Re-entrant tube magnet support concept for the SSC from Nieman et al. [11].

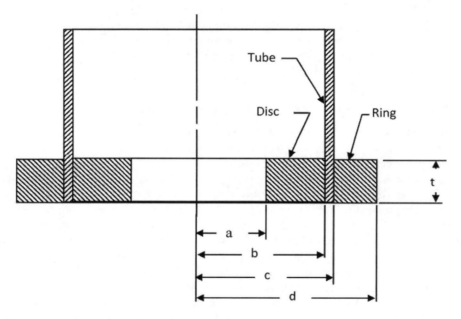

FIGURE 5.12 Shrink fit joint notation from Nieman et al. [11] for a support post thermal anchor.

in helium flow lines, eliminating many vacuum feedthroughs and reducing the possibility of leaks. The mounting fixture for a helium tube is shown in Figure 5.18. Twisted wires (8) from the hermetic connector at an ambient room-temperature sink feed into the screw-thermal anchor (9) on the copper plate (6), which is soldered onto the He tube (7). The screw-thermometer (2) measures the temperature of helium tube (7) with a 1–2% error over 4 K–300 K. The thermometer (3) is a Russian carbon resistor (TVO model) and bifilar winding (4) using a special technique for mounting in the screw (2) covered by a copper cover (1). All surfaces are polished and nickel plated.

TABLE 5.2
Summary of a Shrink Fit Analysis for the 300 K Joint

Shrink Fit Analysis Input	Shrink Fit Analysis Results
Dimension $A = 1.500$ in	Interface required at inner surface $= 0.0190$ in
$B = 3.391$ in	outer surface $= -0.0127$ in
$C = 3.500$ in	
$D = 4.500$ in	Total radial interference required $= 0.0063$ in
$t = 0.75$ in	Contact pressure at inner interface $= 8,695.5$ psi
Material property, $E_1 = 2.8 \times 10^7$ psi	Outer interface $= 8,163.3$ psi
$v_1 = 0.333$	
$E_2 = 2.9 \times 10^8$ psi	Actual force to slip $= 41,695.3$ lb
$v_2 = 0.20$	Resisting moment $= 96,000$ in-lb
$E_3 = 2.8 \times 10^9$ psi	
$v_3 = 0.3$	
Friction coefficient, $\mu = 0.3$	
Force to slip, $F_{slip} = 0.0$ lb	
Resisting moment, $M_{Res} = 96,000$ in-lb	

Source: From Nieman et al. [11]

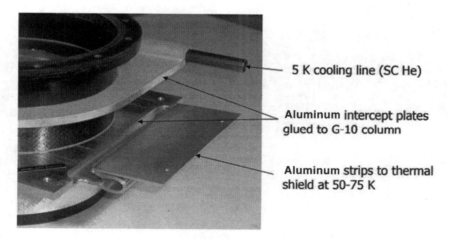

FIGURE 5.13 An example of the thermal anchor cooling structure for the support post [6].

5.8.3 THERMAL ANCHORS FOR RF INSTRUMENTS

Cryogenic anchoring of instrumentation lead wires is another crucial design detail. Batey [13] and Pagano [14] separately reported new developments in advanced anchor design in detail. Figure 5.19

FIGURE 5.14   Cavity interconnection detail with the thermal anchors of copper braids among cryogen tube, shield, and components [6].

                    (a)   (b)

FIGURE 5.15   (a): Photograph of the installation of an RTD on one of the rigid ends of the inner wall of the cryostat. (b): Photograph of the installation of an RTD on the outside of the superinsulation blanket covering the inner wall of the cryostat.

shows how an RF wiring cartridge can be anchored on multiple cold stages [13]. Measuring wires from the ambient environment to a working cold mass causes additional heat leak. It is particularly serious when the cooling capacity is very small at ultralow temperatures and/or cooling power is difficult to supply in aerospace applications. Thermal anchoring of wires to heat sinks in cryogenic equipment is required at each intermediate stage [13, 14]. Special attention is needed if coax cables are used. Thermal anchoring in large-scale applications is quite different compared to ultralow-temperature small apparatus. Effective and precise thermal anchoring of wires is mandatory to measure

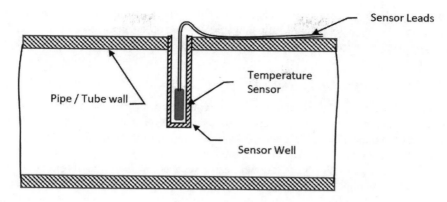

FIGURE 5.16   Temperature probe inserted in a well. Sensor leads are thermally anchored to the surface of the pipe/tube.

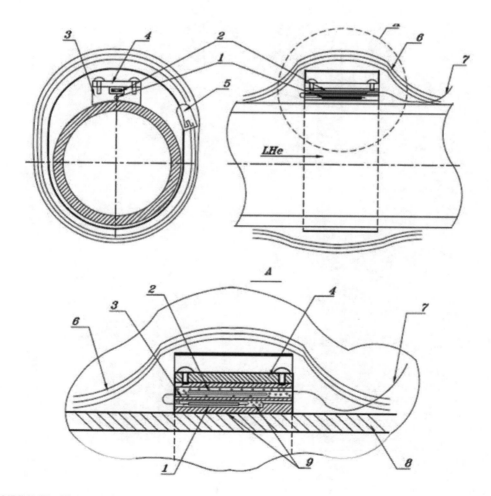

FIGURE 5.17   Thermometer mounting fixture on helium line.

temperature at milli-Kelvin accuracy and to avoid unnecessary cooling power due to additional heat conduction. Besides mechanical clamp and solder, special cryogenic epoxies are used that are electrically conductive, thermally conductive/electrically insulative, and low outgassing approved.

5.8.4 THERMAL ANCHORS FOR CURRENT LEADS AND SUPERCONDUCTOR JOINTS

Current leads bring electric power from room-temperature power supplies to devices at low temperatures such as superconducting magnets. The design and operation of current leads were discussed in another chapter. The interest here is that low-temperature connections must be in thermal equilibrium with the cold connection. This requires understanding of the heat flow and generation at the low-temperature joint. For heat-stationed leads, such as those with an intermediate temperature intercept where a high-temperature superconducting portion is connected to a conventional electrical conductor such as copper, aluminum, or brass, the thermal anchor must provide a stable,

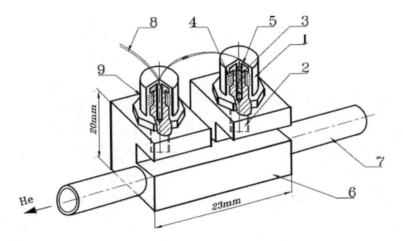

FIGURE 5.18   Thermometer mounting fixture on external surface of helium cooling tube.

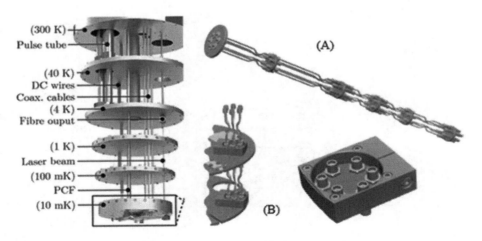

FIGURE 5.19   Framework of a dilution refrigerator: A: RF wiring cartridge with hermetic feedthroughs from 300 K to 10 mK; B: split clamps to thermal anchor the cartridge [13].

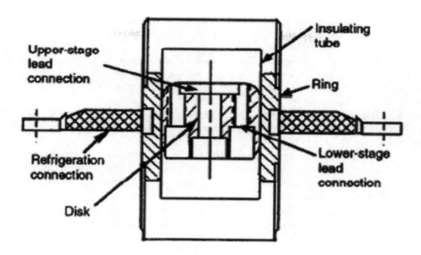

FIGURE 5.20   A low thermal resistance, high electrical isolation heat intercept design by Niemann et al. [15].

sufficiently low temperature at the warm end of the HTS material to prevent the HTS from becoming a normal conductor and burning out. But there is the additional requirement of electrical isolation from the heat sink. For typical superconducting magnets, this electrical isolation is fortunately at low voltages. Recent developments in high-temperature superconducting power introduce additional complications in that the operating voltages will be greater than 10 kV. A properly sized area for removal of the lead heat is needed.

A low thermal resistance, high electrical isolation heat intercept design was discussed by Niemann et al. [15]. The heat transfer along a conventional upper-stage lead is intercepted at the transition to the lower-stage high-temperature superconducting lead. The electrical insulating tube is made from an epoxy/fiberglass material similar to G-10. The wall thickness of 0.43 mm is based on the 8500-V breakdown requirement and fabrication considerations of machinability. The disc and ring are electrolytic-tough-pitch copper. The nominal disc/ring/tube diameter is 0.064 m. The ring is 0.070 m long and 0.0127 m thick. The thermal resistance as a function of the average tube temperature was measured to vary from 0.0068 m^2 K W^{-1} at 54 K to 0.0028 m^2 K W^{-1} at 119 K.

REFERENCES

1. Siegel, R. and Howell, J. R., *Thermal Radiation Heat Transfer*, Second edition, Hemisphere Publishing Corporation, McGraw-Hill, 1981.
2. McIntosh, G., "Cryogenic concepts," *Cold Facts*, Vol. 28, No. 1, p. 13, Winter 2012.
3. Hilal, M. A. and McIntosh, G. E., "Cryogenic design for large superconductive energy storage magnets," *Adv in Cryo Eng*, Vol. 21, pp. 69–77, 1976.
4. Eyssa, Y.M., And Okasha, O., "Thermodynamic optimization of thermal radiation shields for a cryogenic apparatus," *Cryogenics*, Vol. 18, No. 5, pp. 305–307, May 1978.
5. Nilles, M. J., and Lehmann, G. A., "Thermal contact conductance and thermal shield design for superconducting magnet systems," *Adv in Cryo Eng*, Vol. 39, Plenum Press, pp. 397–402, 1994.
6. Pagani, C., *ILC Cryomodule and Cryogenics*, International Accelerator School for Linear Colliders, 19–27 May 2006. Sokendai, Hayama, Japan. Inspirehep.net site.
7. Young, M. A., Demko, J. A., Gouge, M. J., Pace, M. O., Lue, J. W., and Grabovickic, R., "Measurements of the performance of BSCCO HTS tape under magnetic fields with a cryocooled test rig," *IEEE Transactions on Applied Superconductivity*, Vol. 13, No. 2, pp. 2964–2967, June 2003.
8. Kamiya, K., et al., "Electrically insulated MLI and thermal anchor," *AIP Conference Proceedings*, Vol. 1573, p. 455, 2014, https://doi.org/10.1063/1.4860736, (Advances in Cryogenic Engineering).

9. M. A. Green, "The effect of low temperature cryocoolers on the development of low temperature superconducting magnets," *IEEE Transactions on Applied Superconductivity*, Vol. 11, No. 1, pp. 2615–2618, March 2001, doi: 10.1109/77.920404.

10. Barucci, M., Martelli, V., and Ventura, G., "A dry dilution refrigerator for the test of CUORE components," *Journal of Low Temperature Physics*, Vol. 157, No. 5, pp. 541–549, December 2009.

11. Nieman, R. C., Gonczy, J. D., and Nicol, T., "Design, construction, and performance of a post type cryogenic support," *Fermilab TM-1349*, September 1985.

12. Datskov, V. I., Demko, J. A., Weisend, J., and Hentges, M., "Cryogenic thermometry in superconducting accelerators," *Proceedings of the 1995 Particle Accelerator Conference*, Vol. 3, pp. 2034–2036, May 1995.

13. Batey, G., and Chappell, S., "A rapid sample-exchange mechanism for cryogen-free dilution refrigerators compatible with multiple high-freq signal connections," *Cryogenics*, Vol. 60, pp. 24–32, 2014.

14. Pagano, C., "Cryogenic trapped-ion system for large scale quantum simulation," *Quantum Science Technology*, Vol. 4, p. 014004, 2019.

15. Niemann, R. C., Gonczy, J. D., Phelan, P. E., and Nicol, T. H., "Design and performance of low-thermal-resistance, high-electrical-isolation heat intercept connections," *Cryogenics*, Vol. 35, pp. 829–832, 1995.

6 Cryogenic Transfer Pipes and Storage Vessels

Quan-Sheng Shu, James E. Fesmire, and Jonathan A. Demko

6.1 INTRODUCTION

Cryogenic transfer pipes and storage vessels (dewars/tanks) are common crucial devices in almost all cryogenic systems. During transfer and storage, changes in the temperature, vapor quality, and pressure of the cryogenic liquids should be negligibly small. Therefore, the performance of cryogenic pipes and vessels is determined directly based on the thermal design, mechanical configuration, and implementation of new materials of thermal insulation and high thermal resistive supports. Cryogenic pipes and storage vessels have been successfully developed and commercially available for many years. Technologies and design methodologies of cryogenic pipes and vessels have been comprehensively advanced and developed for new platforms [1–3]. This new progress can particularly be seen in, but is not exclusive to, three main aspects:

1. New thermal insulation materials, composite thermal insulation systems, and high thermal restive solid materials/systems have been widely employed in cryogenic pipes and vessels.
2. Capability, mutuality, and configuration are fully improved in successful designs and constructions of both ultra-large cryogenic liquid vessels (about hundreds of m^3 and up) and very sophisticated cryogenic transfer pipes with multiple channels for different cryogens at different T and P values in a common vacuum jacket.
3. Successful implementation and application of design methodologies and construction techniques in cryogenic pipes and vessels for extremely difficult environments, such as under-sea pipes for transferring cryogens, microgravity applications for detective and explorative satellites and space stations, extra-large zero-boil-off cryogenic vessels for launch missions, handling hundreds of tons and thousands of m^3 of LHe for experiments in high-energy physics and nuclear fusion, and overseas transportation and onshore storage of hundreds of km^3 of liquefied natural gas (LNG).

The materials, calculations, and performance of thermal insulation and solid supports have been discussed in previous chapters of the book. In this chapter, the newest developments and successful examples of design methodologies and unique, integrated structures of cryogenic pipes and vessels are systematically introduced and demonstrated. This will contribute to new development of technical challenges and advanced resolutions, while the basic design features of cryogenic pipes and vessels are also briefly introduced. In summary, our interests can be briefly outlined as the following.

Besides regular cryogenic transfer pipes, we will emphasize undersea LNG pipes and large cryogen transfer pipes for LH_2 and LO_2 at space launches. Since Fermilab's LHe transfer pipes successfully deliver a large amount of LHe along the 6.3-km accelerator tunnel to cool the superconducting (SC) magnets, various large superconducting applications such as accelerator (>30 km) and fusion systems have been developed. Extra-large LHe tanks and complex transmitting pipes with multiple channels of different cryogens and temperatures in a common vacuum enclosure at CERN, ITER, KEK, KATRIN, Fermilab, and FRIB are introduced herein.

In space launches and space flying missions, extra-large LH_2 and LO_2 tanks (up to thousands of m^3) and microgravity vessels have been successfully developed. Extra-large cryogen tanks, such as

DOI: 10.1201/9781003098188-6

200–3,000 m³ for LH$_2$ and LO$_2$ are very costly and must last for a long service time. Therefore, the techniques of diagnostics, repair, and modification are crucial to maintain and improve their performance for a long time. First, a project to detect the reason for degradation, locate failure spots, and top-fill perlite for a 3,200-m³ LH$_2$ tank is introduced. Then we discuss the successful results in improving the boil-off rate of a 318-m³ tank by using glass bubbles to replace the existing perlite of the LH$_2$ tank. The current modification of a normal 218-m³ LH$_2$ tank to a zero-boil-off tank by integration of a refrigerator, an immersion heat exchanger of the fish skeleton style, and the tank are demonstrated.

As examples, extra-large tanks of LNG, a cryogen transfer line with maglev suspension, bayonets of cryogen pipes, joint techniques of cryogen pipes in the field, and cryogenic tests for pipes with a boil-off method and different entropy methods are introduced briefly.

While various unique cryogenic vessels and transfer pipes are discussed in detail, the cryogenic thermal insulation materials and high thermal resistive solid structures are only introduced briefly, and focus is given to the merits and structural advantages of insulation materials and structures for different applications.

6.2 BASIC CRYOGENIC TRANSFER PIPES

6.2.1 CRYOGENIC PIPES WITH FOAMS, FIBERS, AND POWDERS

For transmission of LNG and large amounts of LO$_2$, LN$_2$, and LH$_2$, foams, fibers, and powders are traditional thermal insulation materials [4, 5]. The key points for the design of cryogenic pipes are: (1) thermal conductivity of the insulation is one of the most important features in limiting heat gain into the pipe. (2) Water resistance is critical to prevent water and water vapor from infiltrating the insulation system. (3) Ease of fabrication and assembly and compression resistance to keep the designed and specified shapes. (4) Low combustibility. (5) Cost.

Currently, closed-cell cellular glass foams, polyisocyanurate (PIR) like TRYMER, phenolic foams, extruded polystyrene, and so on are widely used in cryogenic pipes for transferring LNG and large amounts of LO$_2$, LN$_2$, and LH$_2$. These materials have relatively low thermal conductivities, low permeability for water/vapor, reasonably high stiffness, and low cost. One of the primary benefits of using a cellular glass and PIR insulation is that it is easy to fabricate both in the fabrication shop and in the field. Well-designed and fabricated cryogenic transfer pipes with cellular glass or PIR usually have no need of a vacuum jacket between 350 K to 77 K, but high thermal contraction is a weakness of foams.

As an example, Figure 6.1 illustrates LNG transfer pipes with closed-cell cellular foam. The straight pipe has double-layer forms (curved sidewall segments). The inner layer has the insulation joints unsealed. The outer layer insulation joints are sealed to the full depth with a non-curing butyl-based sealant. The joints of the outer layer are offset from the joints of the inner layer. The outer layer of insulation is secured with stainless steel bands. A vapor retarder is applied over the insulation. Finally, an optional metal jacket finish is applied. Dickey et al. reported the typical thicknesses for cellular glass insulation on LNG piping ranges from 140–220 mm, depending upon pipe diameter and other design factors. Insulation materials can be foamed in situ and prefabricated in a workshop. In some applications, it can be directly spread on pipe outer surfaces. The thermal conductivity is around 20–40 mW/m-K [5].

In jacketed cryogenic transfer pipes, powders, fibers, and foams function to block radiation and reduce convection heat transmission. Insulation spaces may also be filled with gas of lower boil-off T than the cryogen being transmitted or evacuated to a soft vacuum level. Baking and replacing gas several times is a routine procedure before sealing the insulation jacket.

6.2.2 CRYOGENIC PIPES WITH AEROGELS AND AEROGEL LAYERED COMPOSITES

Aerofoam, Aeroplastic, and Aerofiber are newly developed aerogel composites for use in thermal insulation systems and are discussed in detail in Chapter 2. The composites and aerogels are now

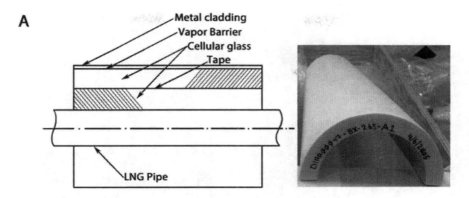

FIGURE 6.1 Simplified cross-section of LNG straight pipe with closed-cell cellular foam insulations.

commercially available and implemented by industry. They provide superior energy efficiencies and enable new design approaches for more cost-effective cryogenic systems [6, 7]. Lightweight blankets optimized for the high-vacuum environment of space or vacuum-jacketed equipment are now under development.

6.2.2.1 LH$_2$ Transfer Pipes for Space Launch Facilities

Fesmire et al. have developed various types of cryogels for cryogenic applications, particularly in the area of LH$_2$ transfer and distribution. The benefits of cryogels include a decreased insulation thickness requirement, a simplified insulation system design (i.e. no contraction joints), and fast installation, even during operating conditions. Cryogels are also useful for insulating difficult areas such as between flanged sections of vacuum-jacketed piping.

LCI and LCX are two types of newly developed advanced aerogel composite structures for temperatures ranging from –269°C to +650°C. Layered composite insulation (LCI) systems can provide the ultimate thermal performance for soft vacuum (SV) environments or degraded vacuum. Layered composite extreme (LCX) systems provide excellent, long-life thermal performance for non-vacuum (NV).

A durable, lightweight, hydrophobic composite insulation system is designed to reduce heat leak in complex piping or tank systems that are difficult or practically impossible to insulate by conventional means. Thermal insulation systems for non-vacuum applications including a multilayer composite have been patented (US 9,617,069 B2). LCX is composed of different layered materials and has been manufactured as commercial and custom-designed products for externalapplications. For example, silica aerogels with fiber matrix reinforcement are super-hydrophobic and mechanically durable. The k-value of aerogel insulated pipe is smaller than that of cellular foam insulation if the insulation thickness is the same. LCX's structure, assembly with an LH$_2$ transfer pipe, and finally use in a launch facility are shown in Figure 6.2.

6.2.2.2 Subsea-Buried LNG Pipeline Technology

Subsea-buried cryogenic pipelines can be used to transport LNG from/to an offshore terminal over up to about 32 km, thereby eliminating the need and cost for a connecting trestle [8]. Subsea cryogenic pipeline designs usually employ vacuum systems for insulation and Invar pipe materials to reduce differential stress in the pipelines. The thermal performance of vacuums is better than insulation with ambient pressure. However, the high cost and reliability of maintaining vacuum through the life of pipe operation are serious concerns. If there is a leak and the vacuum is lost, then all thermal performance is lost.

Many types of subsea pipeline configurations are now being considered. Phalen et al. discussed thermal nano-porous foam insulation being used in the annular space between the inner and outer

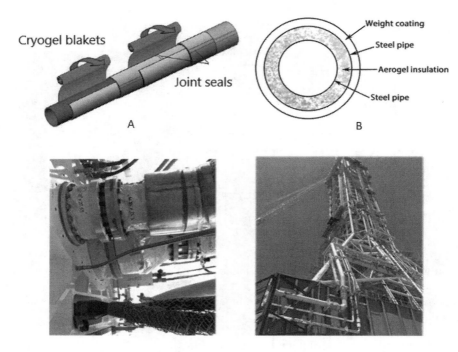

FIGURE 6.2   A: Aerogel-insulated pipe. B: Simplified sketch of underwater LNG pipeline with aerogel insulation. Lower: LH$_2$ pipe with LCX insulation assembled to a flange and LH$_2$ distribution pipelines at a launch facility [7]. Source: *NASA*

pipes in ambient pressure. The type of insulation foam must also have strong water resistance capability. Metallic bulkheads are located at both ends of the pipe. The thermal conductivity of nanogel insulation for cryogenic applications is about 11 mW/m-K [8]. The inner pipe is 9% nickel steel, while the jacket pipe can be carbon steel. The thermal insulation is approximately 50 to 75 mm thick, in blanket or granular form installed within the annular space without vacuum and under ambient pressure. For the 24" outer diameter (OD) inner pipe of a subsea-buried cryogenic LNG pipeline, two design options are briefly introduced (with different insulation thickness): (1) when nanogel blankets (one of aerogel type) are used, the heat leak is around 0.13 W/m^{2}K. (2) When polyurethane foam is used, around 0.13 W/m^{2}K is also reached.

6.2.3   CRYOGEN PIPES WITH VACUUM JACKETED + MULTILAYER INSULATION

Vacuum-jacketed (VJ) pipe plus multilayer insulation (MLI), as shown in Figure 6.3A, is widely utilized to reduce the heat leak up to 100 times less than in foam/aerogel insulated piping [1, 9]. The stainless steel inner and outer pipes can be rigid piping with expansion joints and/or flexible hose assemblies. Solid conduction is greatly reduced by low thermal conductivity radial supports designed for minimizing cross-section and small contact areas. MLI is wrapped on the inner pipe to reduce the radiation heat from the warm outer pipe. The vacuum is maintained for a long life by getting materials to minimize the gaseous conduction heat transfer. The loss of vacuum in MLI-insulated lines produces a rapid, sharp increase in the heat load [10], making vacuum maintenance crucial for long-term installations. Transfer lines for small and medium amounts of LH$_2$, LN$_2$, and LO$_2$ are generally preferred with VJ+MLI insulation, while a great amount of liquid cryogen is handled by the foam/aerogel-insulated pipes. Regarding LHe, VJ+MLI is normally the only choice.

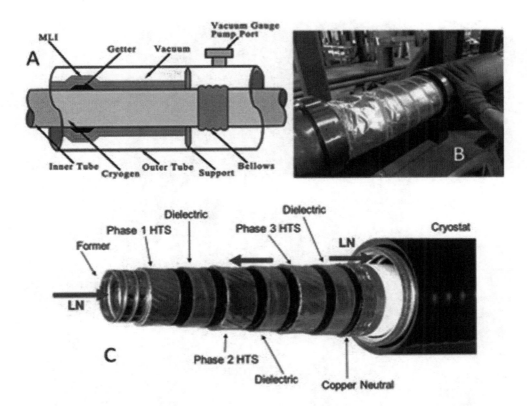

FIGURE 6.3   A: Cross-section of a simplified cryogenic pipe with VJ+MLI. B: LH$_2$ transfer pipe with VJ+MLI close to the outer pipe [11] C. Triaxial HTS high current cable [12]. Source: *Southwire*

In most cases, stainless steel (304, 316, etc.) is used for inner and outer pipes, while carbon steel can be used for outer pipes for budget reasons. MLI has been discussed in detail in Chapter 2. In general, MLI is a flexible blanket of double-aluminized Mylar spaced by polyester fabric-type layers (or polyester net-type layers) between each Mylar layer.

6.2.3.1   LN$_2$ and LH$_2$ Transfer Pipes with VJ+MLI

To support implementation of high-temperature superconducting (HTS) systems, kilometer-long flexible cryogenic LN$_2$ pipelines have been developed, tested, and operated with the collaboration of private industries and national labs. Some cryogenic pipes have concentric multipipe channels, housing the HTS cable and the LN$_2$, which provide flows through the pipe to cool the cable as an integral thermal envelope. In these applications, the thermal insulation channel is very tight in radio. Therefore, the spacer rod pairs (or wires with small beads) are chosen as thermal resistive supports [13]. A demonstration photo of the HTS cable by Southwire is presented in Figure 6.3C [12]. Figure 6.3B shows a part of the LH$_2$ transfer pipe with VJ+MLI, which has larger VJ space than in HTS cable, being assembled at the NASA Kennedy Space Center [11].

6.2.3.2   LHe Transfer Pipes with VJ+MLI

Liquid helium (LHe) is expensive and easily vaporized due to having the lowest boiling temperature (4.2 K) and very small latent heat (20.6 J/g). Significant design features of LHe transfer pipes can be summarized briefly using the world's first long LHe transfer lines of 6.3-km circumference for Tevatron at Fermilab [14]. Figure 6.4 shows the original Tevatron accelerator in the tunnel and

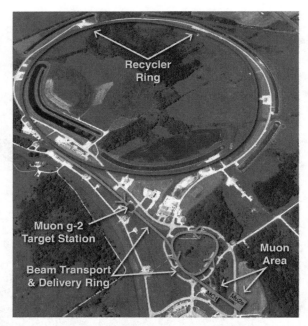

FIGURE 6.4   Upper: Landscape of the Muon project partly modified from Tevatron. Lower: Fermilab Tevatron superconducting accelerator with LHe and LN$_2$ transfer pipes in the 6.3-km tunnel and above ground. Source: *Fermilab*

the landscape of the Muon project, partly modified from Tevatron. The design of cryogenic liquid transfer pipes in the original Tevatron can be briefly introduced as follows:

1. Use of an annular screen (at ~77 K or intermedium T) to intercept a large fraction of heat load from ambience (as shown in Figure 4.7). (For Linac return pipes, LHe is protected with a 50 K GHe screen, not LN$_2$.)

2. Uniquely designed low heat-conduction spacers (fiberglass/epoxy composite) with minimal cross-section area (possibly small thickness), which extended the conductive heat transmission path, and high-impedance thermal contacts to both inner and outer tubes.

3. An optimized number of MLI layers that were carefully wrapped on the outer surfaces of the LHe pipe and the LN_2 screen pipe to reduce radiation heat load (keep proper layer density). Here, there were 60 layers on the LN_2 pipe and 15 layers on the LHe pipe.

4. In some cases, the annular screen is cooled by return low-T GHe from the cold mass instead of LN_2, depending on what cryogens are available and convenient.

The Tevatron transfer line consists of twenty-six 250-meter sections interconnected with vacuum-jacketed U-tubes, forming a loop. The pipe is made out of commercial 304 stainless steel with G-10 supports. LHe: flow rate 5,000 L/hr; input 3.5 atm, 4.6 K; output 2.8 atm, 5.5 K; length 6.5 km; heat leak 240 W. LN_2: flow rate 3,000 L/hr; input 3.0 atm, 85 K; output 2.0 atm, 90 K; length 6.5 km; heat leak 3,600 W.

6.2.4 CRYOGENIC TRANSFER PIPES WITH MAGLEV SUSPENSION

Shu et al. reported an energy-saving cryogenic transfer line with magnetic suspension replacing traditional solid support, which was developed and prototyped by AMAC Inc. [15, 16]. The prototype transfer line was tested at CTL, Kennedy Space Center. Fesmire reported that the test data exhibit cryogen saving potential of 30–35% in its magnetic suspension state as compared to its solid support state [17]. Key technologies include novel magnetic levitation using multiple-pole YBCO high-temperature superconductor, rare earth permanent-magnet (PM) elements, and a smart cryogenic actuator as the alternative support structure when the pipe is warm. Maglev support is introduced in Chapter 3. These techniques have vast applications in extremely low thermal leak cryogenic storage/delivery containers, superconducting magnetic bearings, smart thermal switches, and so on.

In the prototype of the Maglev transfer line, thermal conduction is eliminated because of the non-contact support provided by Maglev units, while high vacuum and 30-layer MLI wrapped on the inner pipe minimize the convection and radiation heat from the outer pipe at ambience. In the full-scale prototype, an initial minimum clearance of approximately 7 mm is maintained between the outmost MLI layer and the inner wall of the outer pipe. This is to ensure that when the inner line sags for 1–2 mm, the MLI will not touch the outer pipe. The inner tube is made of 304 ss of 6 m length. There are two bellows at both ends of the inner tube to cancel the thermal stress. Three YBCO blocks with copper thermal enhancements are soldered on the inner pipe. Facing the HTS magnets, three permanent magnet elements and a smart cryogenic actuator are mounted on the inner surface of the outer pipe. Section #1 includes the maglev suspension assembly and most of the instruments for temperatures and the inner pipe's central location and actuator, as shown in Figure 6.5 (Lower). The ruler, mirror, and light tube enable people to watch inside the vacuum jacket through the inspection port.

Prototype Cryogenic Test. The pipeline was cooled down and allowed to cold-soak for over 24 hours before starting the testing. The cold and warm boundary temperatures were approximately 77 K and 290 K. The boil-off flow rate showed the warm support device on for the start of the test and then off for the remainder of the stabilized period, as shown in Figure 6.5 (Upper). The boil-off flow rate was recorded, and k_{oafi} profiles were calculated. The average thermal performance was calculated to be 2.9 mW/m-K at an approximately 80% full condition. Preliminary results show that heat leak reduction of approximately one-third to one-half is achievable through such transfer lines with a magnetic suspension system. The heat leak rates of the prototype with many penetrations and observation instruments inside the vacuum jacket are similar to those of high-performance pipelines with multilayer insulation.

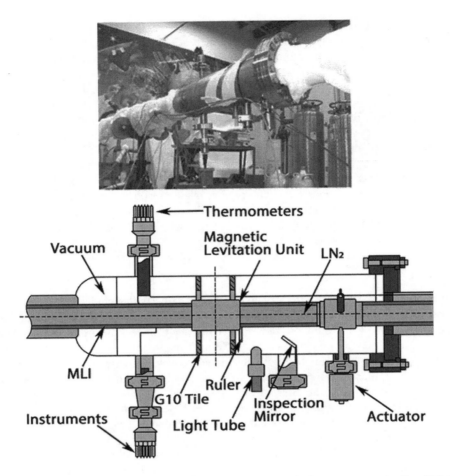

FIGURE 6.5 Upper—Photo of boil-off test showing the section # 1 of prototype maglev line [17]. Lower—Sketch of a section of the maglev transfer line with viewports for levitation and mechanical movement and monitors of T and vacuum level [16].

6.3 COMPLEX PIPELINES WITH MULTIPLE CHANNELS AND CRYOGENS

Complex cryogen pipe systems have been designed to meet the special requirements in cryogen transmittal for extra-large and/or extra-complicated cryogenic applications, such as the International Thermonuclear Experimental Reactor (ITER), Large Hadron Collider (LHC), Linear Particle Accelerator (LINAC), and so on. The technical features of the systems can be briefly pointed out: very large size (several km or longer), very large cold mass (hundred-thousand kg), multiple cryogens with different Ts, and multiple transmitting directions. Therefore, cryogenic thermal management is a crucial task in their design, development, construction, and operation. Complex cryogenic pipe systems that combine multiple transfer lines into one large common vacuum pipe have been developed for comprehensive tasks. Designs of this type have significant advantages of thermal management with lower cost, less space, and energy savings. For example, the Karlsruhe Tritium Neutrino (KATRIN) experiment high-performance multi-cryogen transfer line has as many as nine independent channels carrying different cryogens with different T, P, and mass flow rates thermally guarded by the LN_2-cooled shields inside one common vacuum jacket [18]. A 3D model of isometric views of the Fermilab PIP2IT multi-channel transfer line with fixed supports at the $90°$ turn is illustrated in Figure 6.6 [19].

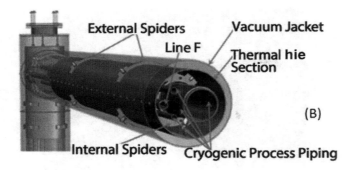

FIGURE 6.6 CAD model of part of the Fermilab PIP2IT transfer line [19].

6.3.1 ITER CRYOGENIC PIPELINE SYSTEM

The ITER project in the south of France is demonstrating the feasibility of nuclear fusion energy. The ITER cryogenic system [20, 21] provides cooling at three main nominal temperature levels, 4 K, 50 K, and 80 K. Complex cryogenic transfer lines of about 5 km connect various cryogenic equipment, valve boxes, cryo-distribution systems, superconducting magnets, and cryo-pumps. ITER cryolines circulate about 25 tons of LHe from the cryoplant through the ITER magnet system to cool 10,000 tons of superconducting coils down to 4 K. The cryoline system forms a structured network of multiple- (two to eight) and single-process pipes localized inside the tokamak machine building on a dedicated plant bridge inside the cryoplant, as shown in Figure 6.7 (Lower).

Thermal Requirements. The total maximum budgeted heat leak of the cryolines at 4.5 K is about 3% of the average cooling capacity of the ITER cryoplant (2.3 kW at the 4.5 K level). The budgeted heat leak of the lines is about 150 W at 50 K and 9.7 kW at 80 K. To limit heat leak by radiation from ambient T, thermal shields are actively cooled by the supply process pipe at about 80 K. Both the shield and the process pipes are wrapped with MLI.

Design Considerations. The most demanding scenarios (mass flow rate, P, T combination) have been chosen for the design. The facts to be considered are: (1) acceptable pressure drop limits of all the lines, (2) heat load. (3) cross-section center of gravity, spacing between pipes for possible use of orbital welding machine, and (4) orientation of thermal shield for assembly and so on. The maximum expected outer jacket diameter of the largest line is 1,016 mm.

Key Points of Cold Test. As an example, in the PTCL-X cryoline system (Figure 6.7, Lower), the T on process pipes at inlets and outlets along with helium P and flow rate were measured during steady state to calculate heat load using the enthalpy difference method. The estimated overall error for the Cernox sensors was less than 0.05 K per sensor, or 0.1 K in ΔT. The helium flow rate was measured at the exit side using a thermal mass flowmeter (2% error) after heating the helium gas to room T. The T of the internal fixed supports, internal sliding supports, and MLI was measured and recorded during the cold test, as these T are important to decide whether steady-state conditions are achieved. The inner and outer spacers are separated by a thermal shield. There were three MLI blankets on the thermal shield (each with ten layers), and T sensors were mounted at a few locations.

6.3.2 LHC CRYOGENIC PIPELINE SYSTEM

The LHC at CERN is a 27-km-long superconducting collider, which is cooled at 1.9 K by superfluid LHe from a huge refrigeration system. At about 2.17 K, LHe undergoes a second phase change and becomes a superfluid. Superfluid LHe has a high thermal conductivity and becomes the best choice to cool and stabilize large superconducting systems. There are eight cryogenic plants, and each

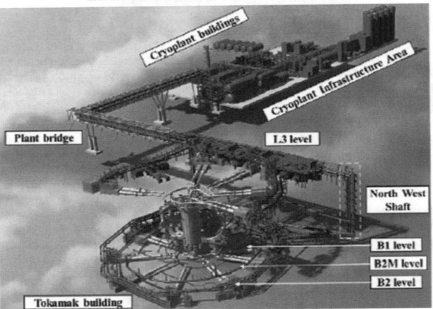

FIGURE 6.7 Upper: Cross-section photo of the prototype cryoline—group X (PTCL-X) [20]. Lower: Schematic layout of the ITER cryoline system.

distributes and recovers kilowatts of refrigeration across 3.3 km at 1.8 K with a temperature change of less than 0.1 K [22].

As there is a high thermodynamic cost of refrigeration at 1.8 K, the thermal design of the LHC cryogenic components is to intercept the largest fraction of heat loads at higher temperatures: (1) 50 K to 75 K thermal shield, a major heat intercept from the ambient environment, and (2) 4.6 K to 20 K for low temperature heat interception.

The QRL sector is a representative pipeline sector, which is a continuous cryostat of about 3.2 km length without any header sub-sectorization but divided into nine vacuum sub-sectors. The QRL, following the LHC lattice, is a repetitive pattern of pipe modules (about 100 m) and service modules (about 6.6 m). Five pipes (inner diameters 80 to 269 mm) are housed in the QRL (outer diameter

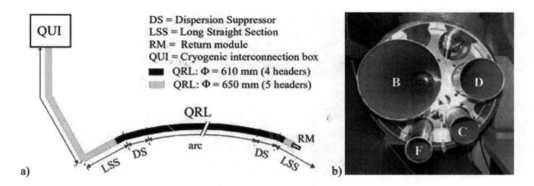

FIGURE 6.8   (a): QRL sector layout of a typical feeding point, (b): cross-section [22].

650 mm), as shown in Figure 6.8(a). A service module consists of the subcooling heat exchanger, control valves, and the monitoring instruments.

6.3.3   ANOTHER EXAMPLE: COMPLEX MULTICHANNEL PIPES

Design and development of complex multichannel pipeline systems have been continuously carried on with the efforts of industries and national labs such as KEK (High Energy Accelerator Research Organization, Japan.), DESY (Deutsches Elektronen-Synchrotron Laboratory), and Jefferson Lab. All these systems are different in detail but are similar in design methodology. They basically consist of two main components: distribution boxes and multichannel pipes with a common vacuum jacket [23, 24]. The cryogenic distribution pipelines at the MSU-FRIB (Michigan State University operates the Facility for Rare Isotope Beam) accelerator are a cost-effective, efficient, and reliable system. As shown in Figure 6.9, three distribution boxes plus cryogenic pipes handle cryogens between the refrigerator room and accelerator tunnel.

6.4   CONNECTIONS (BAYONETS) FOR CRYOGENIC PIPING

6.4.1   TRADITIONAL BAYONETS

There are various designs and products of cryogenic pipe connections (bayonets) to meet different applications, which are based on principles similar to Herrick Johnston's design, introduced in R. Scott's book *Cryogenic Engineering* [25]. One vacuum-insulated male pipe slides easily into another vacuum-insulated female pipe, and then two flanges on both pipes seal together tightly. The male and female design provides easy field assembly and a long heat path from the outer jacket to the inner pipe. Bayonets of this type are commercially available worldwide, with an example shown in Figure 6.10.

Many improvements focus on how to protect cryogen flow to the warm sealed end, such as using a Teflon O-ring and so on. If the cold end is higher than the warm end, it is necessary to make the arrangement, and if the cold end is lower than the warm end, the cold end may not need additional seals, since the vaporized cryogen, as a dead gas, can plug the liquid cryogen flow to the warm end. McIntosh introduced two designs of orientation-independent cryogenic bayonets in detail [26]. This seeks to eliminate free convention by narrowing the annular gas space to approximately 0.02 mm. These bayonets work well but tend to have higher conduction heat leak, as thicker isolator tubes are required to support the machining operation. Wrapping the male bayonet with a flexible Teflon or Dacron cord is also effective in blocking free convention, especially at lower pressure.

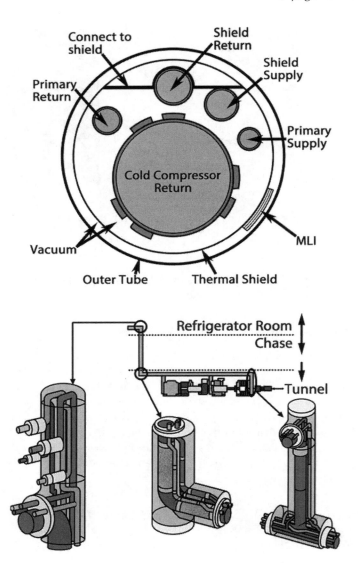

FIGURE 6.9 Upper—A cross-sectional view of the MSU-FRIB transfer pipe. Lower—Stack at the top of the shift, and expansion cans connecting transfer pipes to the accelerator at the bottom [23].

6.4.2 LH₂ Bayonets for Field Joint Connections

Traditionally a field joint bayonet is insulated with multilayer insulation, and a vacuum is pulled on the can to minimize heat leak through the bare section and prevent frost from forming on the pipe section. Vacuum-jacketed LH$_2$ lines for the mobile launch platform were about a combined 650 m, with 60+ pipe sections and field joint cans. Historically, Kennedy Space Center has drilled a hole in the long sections to create a common vacuum with the field joint can to minimize maintenance on the vacuum-jacketed piping. However, this effort looked at ways to use a passive system that didn't require a vacuum but may cryopump to create its own vacuum. Various forms of aerogel, multilayer insulation, and combinations thereof were tested to determine the best method of insulating the field joint while minimizing maintenance and thermal losses. A 98 × 152-mm vacuum-jacketed pipeline test apparatus with a 200-mm field joint can was developed to test the various insulation systems [27, 28]. The tests showed that using a combination of aerogel blankets and aerogel beads with a

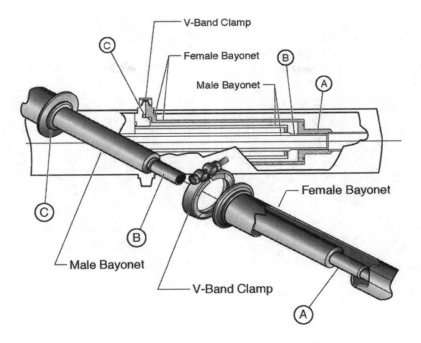

FIGURE 6.10 Connection (bayonet) for cryogenic transfer pipe. Source: *Chart LLC*

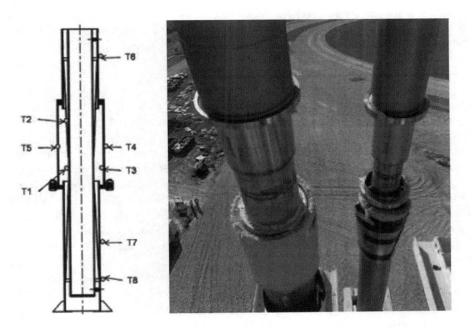

FIGURE 6.11 LH_2 pipe joints with bayonets used in the field for very long LH_2 transfer pipes. Left: cross-section of the joint bayonet. Right: photo of the field arrangement [27].

residual gas of CO_2 appears to be the best-performing non-vacuum system. Due to eliminating the high vacuum requirements, monthly pressure readings don't have to be taken. Figure 6.11 shows a test sketch and photo of the practical assembly in the field. Inside the vacuum can, T1 measured the temperature at the weld joint, T2 measured the temperature halfway up the cone connecting the

FIGURE 6.12 Design of cryogenic interconnection between two SRF cavity modules with multi-cryogen tubes [29].

inner and outer pipes, and T3 measured the warm boundary temperature of the insulation. Outside of the test article, T4 and T5 measured the can exterior temperature, T6 and T8 measured the cone junction temperatures on the outer pipe weld, and T7 measured the skin temperature of one of the simulated pipes. Liquid nitrogen was used to chill the system and provide a constant cold boundary temperature for the entire system.

6.4.3 INTERCONNECTIONS FOR CRYOGENIC MULTI-CHANNEL PIPING

In cases of cryogenic transfer pipes with complex multiple channels and multi-cryogens, they must be designed to be composed of a defined number of modules, constructed in a manufacturing workshop, and then assembled in the application facilities away from the workshop. At its final location, the cryomodules are connected to each other with so-called cryogenic interconnections. Figure 6.12 shows a design example of a cryogenic interconnection [29]. For some large applications, the multi-channel pipes inside the common vacuum enclosure with many valves and control elements (so-called valve boxes) are very complex. MLI blankets were penetrated by these devices and elements, and the thermal performance of MLI is greatly degraded by these penetrations. Therefore, systematical research was conducted by Shu et al. to understand the mechanism and search for approaches to recover the MLI performance [30, 31]. Carlo Pagani introduced external envelope sections of two adjacent superconducting radio-frequency (SRF) cavity cryo-modules ending with a welding ring that the interconnection sleeve is welded to. The sleeve is made of a pipe section slightly larger in size than the cryo-module external envelope. Such a sleeve can be moved aside on one of the modules in order to get access to the internal parts of the interconnection. Some of the interconnection sleeves are equipped with bellows, allowing for required axial or lateral displacements. In this case, the internal process lines should also be equipped with axial or lateral compensators.

6.5 THERMAL TESTS OF CRYOGENIC TRANSFER PIPING

6.5.1 BOIL-OFF TEST (STATIC) METHOD

Thermal performance measurement of piping systems under actual field conditions is important for many applications, particularly for space launch development and large superconducting applications. Fesmire reported a new 18-meter-long test apparatus for evaluation of cryogenic pipelines in which three different pipelines, rigid or flexible, can be tested simultaneously under different temperature, vacuum pressure, and flow conditions at the Cryogenics Test Laboratory (CTL) of the NASA Kennedy Space Center [32]. The method of testing for the apparatus can be static (boiloff) or dynamic (enthalpy difference). Critical factors in heat leak measurements include eliminating heat transfer at end connections and obtaining proper liquid saturation condition. The static method of liquid nitrogen evaporation has been demonstrated, but the apparatus can be adapted for dynamic testing with cryogens, chilled water, or other working fluids. This technology is suited for the development of an industry-standard test apparatus and method. Another small-scale boil-off pipe testing system is also reported [33].

The cryogenic pipeline test apparatus includes two cold boxes, one upstream and one downstream, between which the test articles are connected. Figure 6.13 gives a simplified schematic of the apparatus configured for the static method of testing. LN_2 is delivered to the test apparatus from a 23-m^3 storage tank. The cryogen is supplied to the individual test pipelines (through a heat exchanger coil), the upstream cold box, and the downstream cold box.

The cold boxes are generally oriented so that the downstream end of the pipeline is slightly elevated. The cold boxes are mobile such that the apparatus can test a variety of different test articles. The data monitoring and recording system is based on Field Point hardware and Lab View software. The cold boxes are filled with the test fluid for thermal conditioning. The pipeline test fluid is routed through a heat exchanger coil inside the upstream cold box to provide single-phase flow through the pipeline. The ends of each pipeline are connected to the cold boxes such that end effects are

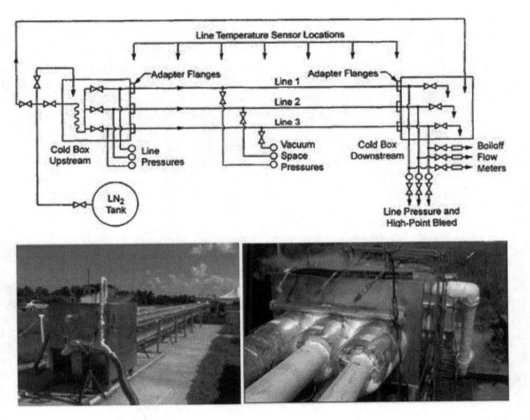

FIGURE 6.13 Simplified diagnostic chat of the cryogenic pipeline test apparatus. Thermal guard boxes at the ends of a pipe under test are used to make the fluid connections and for temperature control [32].

minimized to an inconsequential level. Flow control valves, mass flow meters, and thermometers are used to assure that a wide range of different test articles can be tested.

The system requires 24 to 48 hours of thermal conditioning (cold soak) by maintaining a small replenishing flow for liquid nitrogen through the line. When all temperature sensors are verified to be steady and all heat transfer effects are stabilized, the test can begin. The heat leakage rate (watts) is computed from the boil-off mass flow rate and the latent heat of the cryogen. The overall thermal performance factor, K_{oafi} (mW/m-K), is calculated by Equation 6.2.

$$Q = V \rho h_{fg} \qquad\qquad \text{W} \qquad\qquad (6.1)$$

$$K_{oafi} = \frac{V \rho hfg \ln\left(\dfrac{Do}{Di}\right)}{2\pi L \Delta T} \qquad\qquad \text{mW/m-K} \qquad\qquad (6.2)$$

where V is volumetric flow rate standard (cm^3), ρ is gas density (g/cm^3), h_{fg} is heat of vaporization (J/g), D_o and D_i are outer and inner diameters (m), L is effective heat transfer length (m), and ΔT is temperature difference between the warm and cold boundaries.

The apparatus has been successfully implemented for flexible HTS power cables with Oak Ridge National Lab, for microsphere insulation piping systems with Technology Applications Inc., and for a cryogenic transfer line utilizing magnetic suspension with AMAC International. Thereafter, the apparatus was used many times for different applications. The test data are in good agreement with other sample tests, such as 21 mW/m-K for polyurethane foam and 0.7 mW/m-K for vacuum + MLI at 78 K–310 K. The method may also be extended to LH$_2$ and LNG.

6.5.2 Enthalpy Difference (Dynamic) Method

The enthalpy difference method (cryogen flow-through method) to evaluate the thermal performance of cryogenic transfer pipes is a very useful technique for testing a long transfer pipe, particularly with relatively small cross-section areas, that has a certain temperature difference at both ends to be effectively measured. Demko et al. reported the testing of a vacuum-insulated flexible line with flowing liquid nitrogen at different operational statuses [10, 11]. Long-length vacuum-insulated lines are used to carry flowing liquid nitrogen in several HTS cable projects. A serious, but rare, failure scenario is the abrupt or catastrophic loss of the thermal insulating vacuum, producing a rapid increase in heat transfer to the liquid nitrogen stream. In this experimental investigation, a vacuum + MLI-insulated 88.9 × 141.3-mm nominal pipe size (NPS) flexible cryostat is subjected to an abrupt loss of vacuum in order to measure the thermal response of a flowing liquid nitrogen stream and the temperature response of the cryostat. The measured outlet stream temperature has a slight peak shortly after the loss of vacuum incident and decreases as the cryostat warms up. The heat loads measured before and after the vacuum loss event are reported.

The circulating liquid nitrogen stream was reduced in temperature by vacuum pumping on the subcooler to obtain colder temperatures. This condition simulates a HTS cable installation. The vacuum pump system used on this run has a pumping capacity to produce temperatures of the circulating stream around 73 K. The thermometers for the nitrogen stream inlet and outlet temperatures as well as the temperatures along the MLI are located as shown in Figure 6.14A. Temperatures of the LN$_2$ inlet and outlet and the subcooler bath pressure are measured during a run with a vacuum-pumped subcooler bath as shown in Figure 6.14(B and C). The heat loads, Q, are determined from Equation 6.3.

$$Q = \rho \left(dV / dt \right) C_p \left(T_{out} - T_{in} \right) \qquad\qquad \text{W} \qquad\qquad (6.3)$$

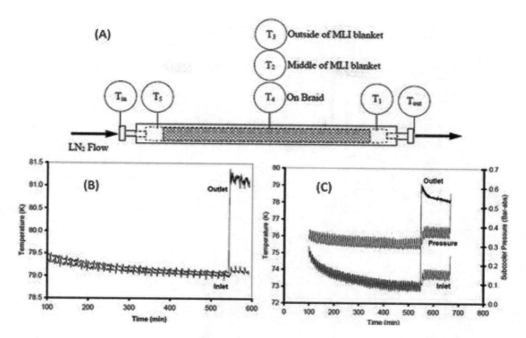

FIGURE 6.14 A: Schematics of instrumented cryogen pipe (flange connections were insulated with poly-urethane foam). B: LN_2 inlet and outlet T with atmospheric subcooler. C: LN_2 inlet and outlet T and subcooler bath P measured with a vacuum-pumped subcooler bath [10].

where dV/dt is volumetric flow rate at the supply pressure and temperature, $(T_{out} - T_{in})$ is the T rise across the section of flexible cryostat, ρ is density, and C_p is specific heat. The volume flow rate and temperatures of LN_2 are measured and averaged for the appropriate times to obtain the heat load before the vacuum break. The scatter in the data is mainly from the level of precision in the T measurements used to resolve the small temperature differences. The platinum sensors used have an accuracy of +/−0.020 K, which is good enough for temperature differences of the levels with insulating vacuum failure but will have a high uncertainty in the case of MLI with good vacuum. The heat loads prior to losing vacuum are about 19.2 W, including the end effects. The average peak heat load after the loss of insulating vacuum is about 644 W.

6.6 REGULAR CRYOGENIC STORAGE VESSELS

6.6.1 Storage Vessels Insulated by MLI

6.6.1.1 Techniques to Minimize Cryogen Boil-Off

Currently small (tens of L) and medium (up to thousands of L) sizes of cryogenic storage vessels (LN_2, LO_2, LH_2, and LHe) have usually employed MLI as thermal insulation instead of powder insulation. A basic design is shown in Figure 6.15. Normally the inner shell is made up of stainless steel, and the outer shell is made up of carbon steel or stainless steel. In some special applications, the fill neck and inner and outer shells can be made of fiberglass or aluminum due to customer needs.

First, the inner container is completely encircled by multilayer thermal insulation at a high vacuum level inside the outer container. The assembly consists of 60–100 layers of MLI, which effectively reduce the heat radiation from the outer container at ambient T [34–37].

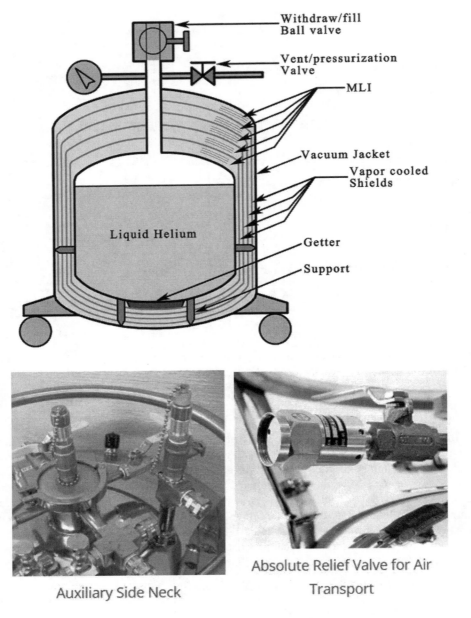

Withdraw/fill
Ball valve

Vent/pressurization
Valve

MLI

Vacuum Jacket

Vapor cooled
Shields

Liquid Helium

Getter

Support

Auxiliary Side Neck

Absolute Relief Valve for Air
Transport

FIGURE 6.15   Upper: Cross-sections of a LHe dewar. Lower: Instruments and control devices integrated on the dewar top. Source: *Cryofab*

Second, it utilizes vapor-cooled radiation shields at intermediate T, which is more thermally efficient due to the second law of thermodynamics. Traditional approaches are to use LN_2 cooled shields or separate cryogen vapor-cooled pipe shields, which are inserted between the MIL or insulation powder assembly completely surrounding the LH_2 or LHe inner container. Recently for small and medium-sized cryogenic vessels, radiation shields have been constructed by thermally anchoring foil shields to the neck tube at different elevations. The considerable cooling capacity of vented gas is utilized to remove heat from the shields though conductive fins on the neck tube.

Third, the neck tube also functions as hanging support for the inner container. The neck tube is constructed with material of high mechanical strength and low thermal conduction, such as stainless steel and fiberglass. The neck tube is also designed to be long with a thinner wall to reduce the heat flow. To fix the inner cryogen container and minimize heat leak, high thermal resistive supports are mounted on the side and bottom of the inner container.

While the cryogenic vessel is evacuated, the vacuum jacket is backed, and the high-quality getters are thermally anchored to the bottom of the outer surface of the inner container to maintain high vacuum.

6.6.1.2 Integration of Regular Cryogenic Vessels

Various devices are employed to monitor and operate the cryogenic vessel, avoiding any instability and misoperation, as shown in Figure 6.15 (Lower). The devices normally consist of an evaporator, vacuum bursting disk, vent/pressurization valve, pressure gauge, inner container bursting disk, liquid fill/withdraw valve, cryogenic liquid level gauge, and pressure-building coil (optional). All devices and parameters (such as liquid level, vacuum status, pressure in inner container) are carefully monitored and regulated manually or by computers.

6.6.2 Storage Vessels Insulated by Powder Material

As shown in Figure 6.16, perlite powder (or silica aerogel or glass bubbles) is filled in the space between the LN_2 shields and outer container at a moderate vacuum. The unfilled space at high vacuum between the shield and LH_2/LHe container provides high thermal insulation for the inner container. Although it is an early design of cryogenic vessels, the insulation principles and structural methodology have been extensively referenced and used for many years [38]. The LN_2 shield completely surrounds the liquid hydrogen container and intercepts all ambient temperature radiation from the outer shell. Thermal radiation to the liquid hydrogen container is then reduced by a factor of at least 200 and becomes the same order of magnitude as the residual gas conduction and solid conduction in the design.

Stainless steel 304 and 312 with good strength and low thermal conductivity are common materials for cryogenic vessels. Aluminum probably maintains its reflectivity better than copper. It is possible to use stainless steel covered with aluminum foil to obtain a high reflectivity.

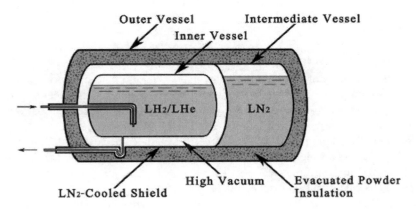

FIGURE 6.16 A simplified sketch of a storage vessel with evacuated powder insulation and LN_2 cooled shield.

Since the solubility of hydrogen in aluminum is much less than that in stainless steel [8, 25], aluminum should be a very good material with respect to hydrogen diffusion. For LN_2-cooled radiation shields, low weight and high thermal conductivity are important, and aluminum is chosen. Instead of being cooled by LN_2, the shields can also be cooled by H_2/LHe vapor before venting. Therefore, the shield LN_2 container is omitted and thinner support tubes of stainless steel can be used to further reduce the heat leak. The daily evaporation for the type of LH_2 vessel (about 1,000 L) can reach about 0.25%–1%. As an additional development, spray-on foam insulation has been successfully applied to a cryogenic tank of the launch vehicle by KSC, NASA [39].

6.6.3 OTHER INTERESTING TOPICS

6.6.3.1 Zero Boil-Off Vessels

Zero boil-off (ZBO) vessels have no boil-off or ignorable boil-off of the cryogen. The cryogen in ZBO vessels is usually cooled below normal boil-off temperature, or the vaporized gas of cryogen is re-condensed to the liquid state. ZBO vessels combine a cryocooler with a long operational lifetime (pulse tube, G-M cooler, or J-T cooler) with a thermally well-insulated cryostat. In missions where cryogenic liquids are required but refilling is difficult, impossible, or very expensive, the ZBO concept is the best approach. ZBO vessels have been successfully pursued and developed for MRI systems of superconducting magnets, instruments in space exploration, and densified LH_2 application in recent years. ZBO techniques include three main methods [40–42]: (1) direct contact of the cryocooler cold tip with the vessel wall, (2) employment of a circulation loop to distribute the cooling over large surface areas or multiple vessels, and (3) the employment of various J-T approaches.

Recently, ZBO vessels of liquid methane, LN_2, LH_2, and LHe have been achieved with cryocoolers. Interesting examples are briefly listed as follows: a commercial Gifford-McMahon cryocooler (Cryomech AL325) was integrated into a 150-liter LH_2 vessel using a heat pipe to transfer refrigeration to copper cooling straps located in the liquid region [42]. With the cryocooler operating at full power, the test bed had the ability to store LH_2 at temperatures close to 15 K. A single-stage pulse tube cryocooler was integrated with a cryogenic methane tank. The flow loop utilizes steady unidirectional (DC) flow to provide cooling at temperatures near 110 K at the remote cooling location and was tested on a 635-liter cryogenic methane tank. A re-condensing cryostat with a two-stage cryocooler is used to intercept heat loads to the radiation shield and helium reservoir. The first stage of the cryocooler is connected to a radiation shield that will cool to approximately 50 K, while the second stage is linked directly to the gas charge above the helium reservoir via an optimized heat exchanger.

6.6.3.2 Qualification Test of Regular Cryogenic Vessels

Although qualification tests of cryogenic vessels include vibration, acceleration, and thermal performance, the most important qualification test is the thermal leak test in the operation environment at manufacturers and customers. Generally, there are two types of heat leak tests, the weight balance test and cryogen boil-off test.

Weight Balance Test. This is a simple and practical method for vessels that are not too large. A test vessel is placed on a weight scale platform and filled and refilled with the cryogen for about one day until thermal equilibrium is reached. The weight changes are recorded for about a day, and the weight change is calculated as the percentage boil-off rate.

Boil-Off Test (High Accuracy). After thermal equilibrium is reached, a mass flowmeter is hooked to the vessel's venting port through a heat exchanger (cooled by water at ambient T) to record the boil-off rate at ambient T and P by a thermometer and a barometer. The boil-off data are used to convert to the standard mass flowrate at 273 K and 1 atmosphere. The percentage boil-off rate is also calculated using the full cryogen capacity of the vessel. The cryogen level in the vessel and the T of the outer jacket are also measured optionally for further thermal analyses of the vessel. More detailed information about the boil-off test can be found in Chapter 8.

6.7 EXTRA-LARGE TANKS FOR LO$_2$, LN$_2$, AND LH$_2$

6.7.1 EXTRA-LARGE TANKS WITH PERLITE, GLASS BUBBLES, AEROGEL

Extra-large tanks of LH$_2$, LN$_2$, and LO$_2$ are used for space launch facilities, centers of cryogen distribution, commercial cryopreservation, and large applications of superconducting facilities since the design methodologies of these tanks have common features in thermal management. Due to high energy per unit and environment friendliness, it is a suitable way to site the tanks in space application centers. Key elements of cryogenic infrastructures that support launch vehicle operations and propulsion testing are large double-walled cryogenic storage tanks of LH$_2$ and LO$_2$ in fields around the world. Perlites, glass bubbles, aerogel, and aerogel composites are popular thermal insulation materials used due to considerations of performance against cost in extra-large cryogen tanks. For example, space shuttle operation accepts a loss rate (boil-off) of approximately 1.6% of LH$_2$ by weight per hour, whereas for long flight-duration aircraft applications, an acceptable rate of boil-off of LH$_2$ would be on the order of 0.1% by weight per hour.

Perlite powder has historically been the insulation materials of choice for these applications, but new bulk-fill insulation materials, including glass bubbles and aerogel beads, have been shown to provide improved thermal and mechanical performance [43–45]. Research was conducted on thermal performance to identify operational considerations and risks associated with using these new materials in extra-large cryogenic tanks. The program for optimization of insulation choices was divided into three main areas: material testing (thermal conductivity and physical characterization), tank demonstration testing (liquid nitrogen and liquid hydrogen), and system studies (thermal modeling, granular physics, and insulation changeout) [46]. Investigations show that more energy-efficient insulation solutions are possible for extra-large cryogenic tanks worldwide. The cost of loss of propellants due to boil-off in large cryogenic storage tanks is on the order of millions of dollars per year.

A numerical model of the full-scale tank was run using perlite and glass bubbles. The boil-off rate using perlite insulation is in agreement with field data [47]. When using glass bubbles instead of perlite as insulation, the numerical model predicts (1) a 28% reduction of boil-off rate in the demonstration tank using LN$_2$, (2) a 38% reduction in boil-off rate in the demonstration tank using LH$_2$, and (3) a 30% reduction in boil-off in the LC-39 LH$_2$ tank at KSC. Recently, tests also showed that aerogel beads and aerogel composites have unique thermal performance, better than perlite and glass bubbles, particularly at the soft vacuum level.

For extra-large tanks with double-walled vacuum jackets, the outer tank wall is sized mechanically by the pressure differential between atmospheric pressure and vacuum (a biaxial compressive loading condition), causing the controlling failure analysis to be stability driven. Consequently, specific compressive stiffness is more important than specific strength. Spherical geography would be a preferable concept based on preliminary material and structural configuration trade studies [48] because it offers: (1) increased safety factors, (2) reduced strain allowable to prevent leakage is imposed, (3) multiple failure criteria, (4) metallic inner tank liner materials can be introduced, and (5) additional materials are utilized for the sizing of both inner and outer tanks. Also, one clear option to lighten the design is to stiffen the outer tank to prevent buckling. Grid-stiffened structural concepts have been considered for outer tank sizing. The inner container with cryogen

TABLE 6.1
Basic Performance Data

	Baseline Perlite	Glass Bubbles	
Normal evaporation (NER)	0.18%/day	0.10%/day	44% reduction
Boil-off rate	386 L/day	216 L/day	
Vacuum pressure	4.5 Pa	1.3 Pa	
	(34 milli-torr)	(10 milli-torr)	

in an extra-large vessel is very large and heavy. Therefore, the support between the inner container and outer vacuum shell must be mechanically strong and thermally highly resistive in order to stabilize the structure and minimize the heat leak. Many structures with high-strength metallic tubes/rods and fiber graphite-reinforced polymeric composites have been successfully developed and implemented.

The world's largest liquid hydrogen storage tanks were constructed in the mid-1960s at the NASA Kennedy Space Center. These two vacuum-jacketed, perlite powder-insulated tanks, still in service today, have 3,200 m^3 of useable capacity (one for LH$_2$ and another for LO$_2$). In 2018, construction began on an additional storage tank at Launch Complex 39B. This new tank will give an additional storage capacity of 4,700 m^3 for a total on-site storage capacity of roughly 8,000 m^3, as shown in Figure 6.17.

6.7.2 Extra-Large Tanks with Multilayer Insulation

With significant improvements to MLI in cost, work handling, and performance, it has also been employed for extra-large tanks. Shown in Figure 6.18, the 140-m^3 LH$_2$ tank with MLI insulation was designed and constructed originally by MVE Inc in 1994 for the Titan IV program at LC-41, NASA. The LH$_2$ tank was then relocated to its new place at the Hydrogen Technology Demonstration Facility (HTDF) at NASA Kennedy Space Center (KSC). The new mission is centered on the modification and repurposing of an existing LH$_2$ storage tank. The original tank has the following specifications: 140 m^3 total volume, 21.3 m overall length (12.2 m length between saddle supports), 2.90 m inner vessel diameter, 20,000 kg inner vessel, 0.584-m-diameter manway, 655 kPa design pressure, and vacuum jacketing with 80 layers MLI of foil/paper.

6.7.3 Extra-Large Cryogenic Movable Tanks

Extra-large cryogenic movable tanks used for transporting cryogen over hundreds of km must successfully undertake serious vibration, acceleration/deceleration, and environmental changes on the road. For safety consideration, movable cryogenic tanks should employ cutting-edge thermal insulation design, which has better thermal insulation performance and reliability than traditional design in regular tanks. It must automatically monitor the cryogen pressure to prevent overpressure. The independent vacuum pumping and vacuum helium mass spectrometer is equipped for the long-term vacuum life of the movable storage tank. The static evaporation rate is designed to be superior to industry standard requirements. The piping systems attached to the tanks should be compact, centralized, and easy to operate. Additionally, a lateral support and lifting lugs of the tank are constructed for safe lifting and transportation.

Even larger quantities of hydrogen can be transported in a truck with a special tank for LH$_2$. Here the costs of liquefaction must be taken into consideration, as liquefaction currently accounts for approximately 30% of the energy stored in the hydrogen. However, as the energy density of

(a)

(b)

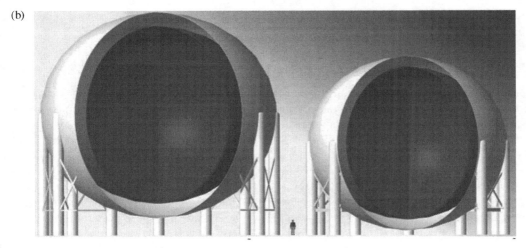

FIGURE 6.17 Lower: Scale comparison between the new 4,700-m^3 LH$_2$ tank and the Apollo-era 3,218-m^3 LH$_2$ spherical storage tank. Upper: Overall elevation view of the new 4,700-m^3 LH$_2$ tank (being constructed). Source: *NASA*

FIGURE 6.18   LH$_2$ storage tank of 125 m^3 with 80 layers MLI at the HTDF NASA KSC. Source: *NASA*

LH$_2$ is considerably higher than that of GH$_2$, these expenses balance out when transporting larger quantities over increasingly long distances, because fewer journeys are necessary to transport the same quantity of energy. Various movable cryogenic tanks are available commercially from many companies around the world. As an example, a Linde movable LH$_2$ tank has a capacity of about 4,000 kg (56 m^3) and ullage 5% at pressure 12 bar.

6.8   DIAGNOSES AND MODIFICATION OF EXTRA-LARGE TANKS IN THE FIELD

Since the design and construction of advanced extra-large cryogenic tanks are projects with very high costs that take long periods of time, this type of tank is usually used for many decades. If the tank performance is degraded and structures are damaged, diagnosis, repair, and modification must be undertaken to recover and even improve its performance instead of abandoning it. Such a project to recover the thermal performance of an extra-large tank is highly technical and very comprehensive [44, 49, 50].

6.8.1   DIAGNOSIS, REFILL, AND RETURN TO SERVICE OF A POORLY PERFORMING LH$_2$ TANK

Krenn reported on the diagnosis, refurbishment and modification of the extra-large LH$_2$ tank [50]. There are two 3,218-m^3 spherical cryogenic storage tanks used to support the launch program at NASA. In 2001, one tank's loss was approximately double that of another and well above the maximum allowed by the specification. In the 1990s, a "cold spot" appeared on the outer sphere that resulted in poor paint bonding and mold formation. Thermography was used to characterize the area, and the boil-off rate was continually evaluated. Borescope examinations revealed a large perlite void in the region where the cold spot was apparent.

If that had been the case, the warming process would have caused the annular vacuum to decay very rapidly. Consequently, the inner sphere temperature and annular space vacuum levels were monitored closely until the tank warmed beyond 77.6 K. Interestingly, the vacuum level only increased to approximately 8 Pa (60 m-torr) during that time. When warmed to 246.5 K, the vacuum level had only risen to approximately 20 Pa (150 m-torr). The vacuum was then broken with a low-flow GN$_2$ purge to the annulus over 24 days, and the sphere had warmed to ambient T and was ready to be investigated. The 30.5-cm annular relief port was opened, and the condition of the perlite was observed and photographed. The insulation was approximately 44.5 cm below the top of the outer sphere and had a downward slope in the direction of the cold spot, as shown in Figure 6.19. Perlite was then trucked in and offloaded into the annular void region until full. After top-off was

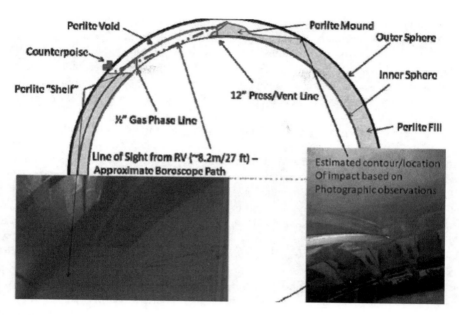

FIGURE 6.19 Estimated contour of perlite based on borescope result [49].

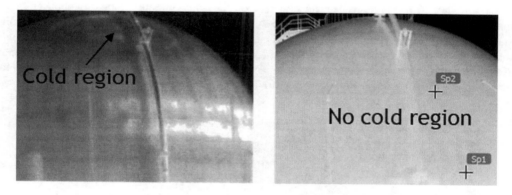

FIGURE 6.20 Thermal images of the LH$_2$ outer vessel walls before and after refilling insulation [50].

completed, an estimated 226.5 m^3 (14%) of perlite had been blown into the annulus. This volume was within the original estimated void range of 198.2–424.8 m^3. The LH$_2$ tank now returned to service in a space launch facility. Thermal imagery of the LH$_2$ tank storage in 2009 showed a clear indication of a cold region due to a perlite void in the annular space, and thermal imagery in 2018 showed no indication of a cold region after refilling insulation, as illustrated in Figure 6.20.

6.8.2 IMPROVEMENT AND MODIFICATION OF ULTRA-LARGE LH$_2$ TANK IN THE FIELD

Sass reports that the six-year field demonstration of the 318-m^3 LH$_2$ tank indicated a boil-off decrease from 386 to 201 L/day when perlite powder was replaced by glass bubbles [44].

A full-scale field application of glass bubble insulation to replace perlite powder was demonstrated in a the spherical LH$_2$ storage tank. This work is the evolution of extensive materials testing, laboratory-scale testing, and system studies leading to the use of glass bubble insulation as a cost-efficient and high-performance alternative in cryogenic storage tanks of any size. The tank utilized is part of a rocket propulsion test complex at the NASA Stennis Space Center. The original perlite that was removed from the annulus showed no signs of deterioration or compaction. Test results showed a significant reduction in LH_2 boil-off when compared to recent baseline data prior to removal of the perlite insulation. The data also validated the previous laboratory-scale testing (1,000 L) and full-scale numerical modeling (3,200 m³) of boil-off in spherical cryogenic storage tanks.

NASA's current interest in high-performance hydrogen storage tank insulation stems from the scope of the Space Launch System (SLS) program. Significantly greater hydrogen storage (8,000 m³)) is needed at the launch pad to reliably launch two successful rockets within mission timelines. Given the relatively infrequent missions (one or two per year), high-performance insulation will minimize life-cycle costs by reducing stand-by propellant boil-off losses.

The Cryogenic Tank. Specifically, the tank is a double-walled tank with an evacuated annulus and internal rod supports near the equator. The diameter of the outer sphere is 9.3 m, and the diameter of the inner sphere is 7.3 m, resulting in an insulation thickness of about 0.90 m. The volume of the annulus is about 200 m³. The tank has been in operation at its current location, shown in Figure 6.21. It has a history of several thermal cycles, good vacuum retention, and normal propellant boil-off performance. The annulus pressure during its final months of operation with perlite averaged 4.5 Pa (0.034 torr). The normal evaporation rate (NER) was 0.18% per day, or 386 L per day [7]. These marks serve as the baseline data for comparing performance of the original perlite insulation to the glass bubble insulation. To determine NER, the liquid level was measured by a differential pressure transducer and examined over many months with the tank vented to atmosphere.

Perlite Removal. The annulus is evacuated via a circular piping manifold near the bottom of the annulus connected to a port on the bottom of the tank, which is further connected to a permanently installed vacuum pump. The vacuum manifold has sixteen 1.8-m-long filter elements threaded into it that consist of slotted 89-mm-diameter pipe covered with filter blanket material. The filter element was connected to a portable vacuum pump with an inline filter. The vacuum pump was operated for approximately 15 hours under high flow conditions, and no bubble material was observed to have

FIGURE 6.21 The 218-m³ LH_2 tank shown with its annulus being filled with glass bubble insulation from an over-the-road pressure differential trailer [44]. Source: *NASA*

passed through the tank filter element. Based on the results of this test, the original vacuum manifold was accepted for use with glass bubbles without modifications. The filter element was then sent back to SSC and reinstalled on the tank vacuum manifold.

Glass Bubble Installation. The overall philosophy employed for the installation of the glass bubbles was to use processes that would be directly applicable to tanks of any size and any location. The main facility requirement for the installation process is physical access for an over-the-road trailer. The portable dust collector requires electricity that can be provided by a portable generator and 790 kPa pneumatic supply that can be provided by a portable gas bottle. A standard glass bubble product was used: 3M glass bubbles Type K1. Using normal industrial handling processes, the glass bubbles were installed through ports on the top of the tank while relying upon gravity and the very fluid-like behavior of glass bubbles to completely fill the annulus.

Slightly more than four trailer loads would be needed to fill the annulus, so six boxes of glass bubbles were on site to top off the annulus. To fill from the boxes, a partial vacuum (approximately 70 kPa) was pulled on the annulus, and the fill hose was inserted into the box. A butterfly valve on the tank port was opened to initiate flow. Several cycles were needed to offload three boxes to finish filling the annulus. The same vacuum cycle process could be utilized using bulk trailers to perform final topping above the level of the fill and vent ports. Slightly more than 15,000 kg of glass bubbles were installed in the annulus, resulting in a bulk density of approximately 75 kg/m^3.

Evacuation of Tank Annulus. After the glass bubbles were installed, the temporary openings to the annulus were welded closed in preparation for evacuation. The vacuum pump down operation lasted about two months. The initial pump down was very rapid, taking about 12 hours to reach 13 kPa from ambient pressure. Subsequently, significant amounts of moisture were being removed from the annulus and collecting in the pump oil, and the pump oil was regularly changed. The collection of moisture began to significantly decrease below 130 Pa (1.0 torr). The moisture removal operations could have been accelerated by use of a cold trap upstream of the vacuum pump but would have required a piping modification. At the end of two months (42 days of pumping operations), the vacuum levels were 13 Pa at the bottom and 40 Pa at the top of the tank. The preferable target for the warm vacuum pressure was 6.7 Pa (0.050 torr), but because of an imminent price increase for liquid hydrogen, the decision was made to go forward with liquid hydrogen loading.

Conclusion. Glass bubble insulation for cryogenic tanks has now progressed from the laboratory to a full-scale field application on a 218-m^3 LH$_2$ storage tank. The thermal, mechanical, and economic indicators all show glass bubbles are an excellent high-performance insulation choice for future large storage tanks and when replacing perlite insulation in existing tanks. The logistical aspects of installing large quantities of glass bubbles and the subsequent evacuation were straightforward to execute. The boil-off rate was reduced by 46% compared to perlite. This field application builds confidence that glass bubble insulation is ready to be adopted in spherical tanks for storing LH$_2$ or any other cryogen.

6.9   ZERO BOIL-OFF ULTRA-LARGE LH$_2$ TANKS

To pursue zero boil-off in extra-large LH$_2$ storage tanks is a crucial task not only to reduce the cost of LH$_2$ boil-off loss but also to ensure the immediate availability of LH$_2$ for launch preparation [2, 51, 52]. Zero boil-off is defined here as the ability to store LH$_2$ without venting and loss for indefinite periods of time. This can be done in continuous processes where the state of the hydrogen is essentially constant or in batch processes where the hydrogen can undergo periods of pressurization followed by depressurization. Finally, the LH$_2$ must be returned to its original state. The research and development of integrated cryocoolers for small-scale liquid hydrogen tanks have been successfully advanced over many years [53]. The demonstration of such an extra-large system at full scale, representative of future space launch vehicle facilities, has been recently undertaken.

6.9.1 INTEGRATED REFRIGERATOR AND STORAGE ZERO BOIL-OFF METHODOLOGY

The core methodology of zero boil-off at NASA is an integrated refrigerator and storage (IRAS) tank to a complete system so called the Ground Operations Demonstration Unit (GODU-LH$_2$). The key challenge of the IRAS is to design and build a large, efficient, and sophisticated heat exchange system within an existing tank. The work reported by Fesmire et al. is to develop and demonstrate efficient methods for launch pad storage and vehicle propellant loading systems for space launches [51, 52].

Focusing on the storage tank, an integrated heat exchanger system, as shown in Figure 6.22, has been designed for incorporation with the existing storage tank and a reverse Brayton cycle helium refrigerator of capacity 850 watts at 20 K. The storage tank is a 125,000-L-capacity horizontal cylindrical tank, with vacuum jacket and multilayer insulation (80 layers) and a small 0.6-meter diameter manway opening.

The heat exchanger (HX) will be used in conducting the following operations: (1) cooling of LH$_2$ for propellant preservation (zero boil-off), (2) cooling of LH$_2$ for propellant densification (increased mass on board a spacecraft or for the ability to accept a tanker trailer without venting), (3) liquefaction of gaseous hydrogen inside vessel as part of propellant production, and (4) cooling of LH$_2$ for propellant transfer advantages (enthalpy margin). The HX system is to be connected to the helium refrigerator. The system loss is about 50 W for a net refrigeration of 800 W. The total allowable pressure drop across the refrigerator load is 20.7 kPa. The transfer lines between the refrigerator and tank manway have a calculated pressure drop of 5 kPa, which means the manway and HX lines can have a pressure drop of no more than 15.7 kPa, including any margin.

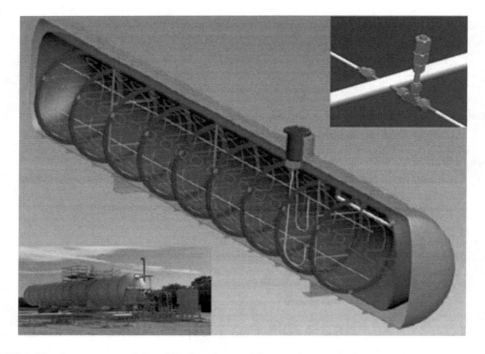

FIGURE 6.22 Arrangement of the stiffening rings and heat exchanger cooling tubes inside the vessel and GODU-LH$_2$ test site showing IRAS tank (refrigeration system is on the opposite side). Cooling tube manifold and helium temperature sensor detail on the upper corner. Source: *[51] Fesmire-NASA*

6.9.2 Advantages and Challenges

In the novel IRAS concept, it cools the LH_2 directly, minimizing the thermal resistance between the refrigerant and LH_2 while using only free convection. Cooling the liquid directly allows for control of the bulk temperature of the LH_2 as opposed to pressure control of the ullage. This also enables conditioning or densification of the propellant and can be used to store refrigeration energy in the liquid to allow greater response during transient operations. Using free convection requires no more active components than forced convection methods and doesn't add heat from a pumping system.

There were five interrelated design challenges: (1) fit all parts and materials through existing manway, (2) have the design be assembled within the tank in a modular manner without welding, (3) recertify the tank per pressure vessel code to a new lower operating temperature (from 20 K to 15 K), (4) provide negative pressure contingency for the densification function, and (5) provide even distribution of heat rejection for the large 8:1 length to diameter ratio, for different modes of operation, and for different fill levels.

6.9.3 Design and Construction of Heat Exchanger

SINDA/FLUINT thermal and fluid modeling and analysis were performed iteratively in concert with the heat exchanger design. The design drivers were needed for high surface area and low pressure drop and the spreading out of the cold sink over the long horizontal tank. A number of different configurations were modeled, but a hoop-and-barrel (or whale-skeleton) approach was the final choice, as shown in Figure 6.22. This gives the most heat rejection, the best cooling distribution, and the lowest pressure drop across the heat exchanger.

Modularity was achieved by using common sections and high-reliability fittings (Swagelok VCR type with silver-plated nickel gaskets and retainers). This approach allowed all parts to fit through the manway and be assembled within the tank. Maintaining the existing vessel integrity meant no welding was allowed on the inner vessel, including the manway joint. In case of loss of vacuum in the annular space coinciding with a propellant operation below ambient pressure in the vessel, a series of nine stiffening rings were designed to provide negative pressure contingency in accordance with the ASME Boiler and Pressure Vessel Code. The rings were divided into three 120-degree segments for fitting through the manway and then bolted together inside the tank. The C channel was made from the same material (304 SST) as the inner tank to match the coefficient of thermal expansion. The rings were designed to accommodate an internal pressure of 0 kPa (absolute) and an external (vacuum jacket) pressure of 101 kPa.

Several key features of HX are listed: (1) distributed "whale-skeleton" approach with heat interception at the tank wall, (2) balanced flow distribution and inline temperature measurements, and (3) placement of heat exchanger at a happy medium to cover different operational modes. The heat exchanger design includes supply and return manifolds (total of 37 m of 25-mm OD tubing) and 40 cooling lobes (total of 244 m of 6.4-mm OD tubing). There are also 25-mm-diameter by 3-m-long flexible lines that connect to the manway feedthrough. The tank is instrumented with 24 silicon diode temperature sensors. Twenty of these sensors are used to measure bulk liquid, while four sensors are positioned inline to measure the helium inlet/outlet temperatures in two of the cooling lobes.

6.9.4 Integration, Test, and Conclusion

A simplified functional diagram of the system is given in Figure 6.23 and depicts how the primary subsystems are integrated. Central to the design is the 125-m^3 IRAS tank, the IRAS heat exchanger is supplied with approximately 22 g/s of cold gaseous helium from the Linde LR1620 refrigerator. The refrigerator has capacities of 880 W and 390 W at 20 K, with and without LN_2 precooling, respectively.

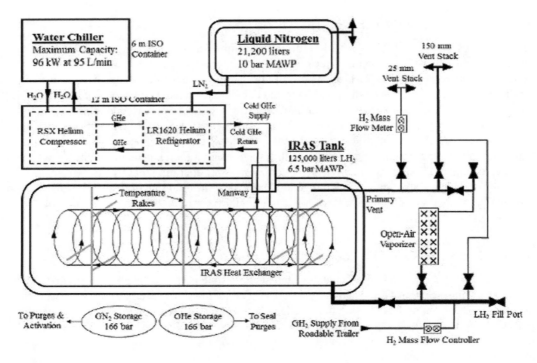

FIGURE 6.23 Simplified functional diagram of zero boil-off LH_2 storage tank testing (GODU-LH_2) [52].

In the testing, ZBO was achieved in three ways: *Temperature control*—to control the helium supply temperature and helium flow rate (the cooling capacity is constant), the LH_2 will eventually reach an equilibrium state in the tank. *Pressure control*—Another continuous ZBO process, the refrigerator output heater is controlled using the IRAS tank pressure. If the tank pressure increases above the set point, the heater power is reduced and more refrigeration capacity is delivered to the tank. *Refrigerator duty cycling*—Turning on and off the refrigerator and running at full power when it is on will create pressure and temperature cycles in the LH_2, but over time, the LH_2 returns to its initial state and there is no hydrogen vented.

Conclusion. Zero boil-off operations were conducted on large quantities of LH_2 for a total period of over 13 months. Three different ZBO methods were tested. Using the tank pressure as the control point demonstrated the most precise control over the state of the fluid, while temperature control of the refrigerant required longer time periods to stabilize and there was less control of the final conditions. The final method of duty cycling was the minimum-energy solution but did not provide continuous control over the fluid state.

6.10 EXTRA-LARGE LHe STORAGE TANKS

LHe has the lowest boil-off temperature, 4.2 K, and the smallest latent heat among all the cryogenic coolants. Therefore, LHe vessels require the best thermal insulation to prevent heat leak from the ambient environment. All extra-large LHe tanks employ MLI as thermal insulation materials. In large engineering applications of MLI systems, there are eventually many overlaps, seams, and penetrations, which can cause serious degradation of MLI performance and need to be carefully considered and designed [30–31]. Moreover, there are usually two actively cooled thermal radiation shields in an MLI system: one cooled by LN_2 and another by returning cold gaseous He. As its latent heat is so small, significant heat leak due to vacuum failure in a large LHe tank will cause a serious safety risk of pressure sharply rising. The design and development of extra-large LHe tanks is a great challenge.

6.10.1   CERN's Ultra-Large LHe Storage Tanks

Benda et al. report the cryogenic system of the Large Hadron Collider under operation at CERN has a total helium inventory of 140 t [54–56]. Up to 50 t can be stored in gas storage tanks. The remaining inventory will be stored in a liquid helium storage system consisting of six 15-t liquid helium tanks in four locations. The 27-km-long superconducting collider at 1.9 K is cooled by a huge refrigeration system. Helium undergoes a second phase change at about 2.17 K and becomes a superfluid. Due to the high thermal conductivity of superfluid helium, it becomes the best choice of coolant for stabilization of large superconducting systems.

In order to store the about-30 t helium inventory in one cryogenic system at the LHC, the heat loss system was erected close to the 18 kW refrigerator. This system consists of two horizontal 15-t liquid helium tanks, a combined liquid helium line (supply liquid helium/return cold helium gas) including a distribution box, an 11-m³ liquid nitrogen tank, liquid nitrogen distribution lines, and low- and high-pressure lines for gaseous helium and related instrumentation, as shown in Figure 6.24. The LHe tank uses MLI insulation and two thermal shields. The first shield is kept at 80 K by liquid nitrogen, and the second one at about 20 K by cold helium vapor.

FIGURE 6.24   Upper: View of the 30-t liquid helium storage system at LHC Point 18. Lower: Side view of large LHe tank with gauges and valves for safety and controls [54]. Source: *CERN*

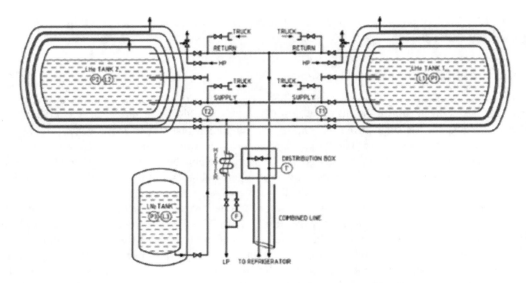

FIGURE 6.25   Simplified process lines of the storage system [54].

The liquid nitrogen tank of 11 m³ is a standard vertical tank with a design P of 18 bar and a working P of 1.25 bar. The tank is equipped with a pressure sensor and a level measurement, and a semi-rigid line supplies liquid nitrogen from the nitrogen tank to the nitrogen shield of the helium tanks.

There are combined cryogenic transfer lines, shown in Figure 6.25, one from the refrigerator to the helium tanks with saturated LHe and another from the helium tanks to the refrigerator with gaseous helium at 5 K. The line consists of four concentric corrugated pipes of external diameter 150 mm. The length of the line is about 100 m. The internal pipe for transferring LHe is supported in the inter circular space (annular space) between the second and third pipes with a cold gaseous return path. A distribution box splits helium between tanks and allows the independent cool-down of the line with a bypass cryogenic valve.

6.10.2   ITER's Ultra-Large LHe Storage Tanks

ITER's helium plant is estimated to handle approximately 24 T of He, which corresponds to the gas needed to fill up 14 million He balloons. The plant will consist of seven warm GHe tanks (7 × 360 m³), one LHe storage tank with a capacity of 175 m³ at 4.5 K (able to store up to 85% of the liquid gas needed in the plant), and two quench tanks [57]. The LHe tank, called the "jewel" of the ITER Cryoplant, is a double-walled vacuum-jacketed vessel with MLI insulation and intermediate temperature shields between MLI layers to minimize thermal loss and remain at 4 K. It has an impressive size of 25.5 m long and 3.8 m in diameter and was manufactured by CryoAB (Sweden) with a capacity of 20 T of LHe. The volume of its inner tank is 190 m³ and measures 23 m in length by 3.5 m in diameter, and it will be assembled subsequently inside a bigger tank. The inner tank is made of stainless steel, and the thermal shield is aluminum. When the tank is filled with liquid helium, its weight will reach 88 tons. Considerable time and effort was required for technicians and engineers to inspect the inner tank with 500 m of linear welds, which join the different sectors of the tank.

6.11   LARGE LNG STORAGE AND SHIPPING TANKS

Liquefied natural gas is a great green energy source. The LNG supply market doubled in the 2010s, and it is anticipated to grow continuously in the next decade. LNG has been stored in onshore

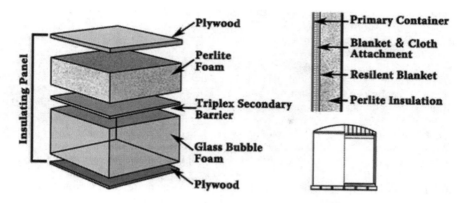

FIGURE 6.26　Simplified thermal insulation sketch for floors and walls of LNG storage tanks.

storage tanks and floating carriers and transported by LNG carriers and pipes [5, 58–60]. The capacities of LNG storage tanks (medium–large) are from 10–260 km³. The design, development, and construction face tremendous challenges in the mechanical structure, thermal insulation, fast filling and pull out, risks of safety and costs.

Design consideration. LNG containment tanks are usually double-walled construction, generally in the neighborhood of 60 m high and 80 m in diameter (wide) or 43 m high and 40 m in diameter (narrow). An LNG tank module on a carrier normally has less capacity than an on-shore tank. The sidewalls of LNG containment tanks are usually insulated with a combination of loose-fill perlite and fiberglass blankets. The tank roof typically has a floating component insulated with closed-cell foam. The tank base is usually insulated with multiple layers of high-density cellular glass foams. The details of the base and sidewall construction vary with the overall tank design. Two simplified sketches are shown in Figure 6.26. The designs of LNG tanks combine different designs codes to trade study in details. There are two criteria for selecting the thickness structures. The first is to limit heat gain to between 8 and 12 Btu/hr ft² (22 and 32 kcal/hr m²). The second is to limit condensation. Value-engineering and cost-cutting measures are also implemented. For LNG storage in excess of 10,000 m³, 9% Ni steel is the choice.

Cellular Glass Insulation for LNG Systems. Cellular glass foam is of sodium silicate glass, and the cell foam walls are impermeable from water-vapor intrusion into the system. Cellular glass insulation has a high compressive strength. Cellular glass insulation is available in a number of higher-density and high-load bearing grades. Cellular glass insulation is 100% glass, does not burn, and increases the fire resistance of the overall insulation system. Cellular glass insulation is primarily used in LNG tank bases and can also be found on the outer sidewalls of double-walled tanks as corner protection.

Sample of tank parameters. For a ground containment storage tank: inner tank: 9% nickel steel; outer tank: pre-stressed concrete; roof: concrete dome with suspended ceiling deck; secondary barrier: 9% nickel steel; corner protection system up to 5 m high from the tank bottom, then polyurethane foam coating; design pressure 29 kPa; diameter of inner tank 84.0 m; design liquid level 36.22 m; maximum operating level 35.92 m.

REFERENCES

1. Shu QS, Demko J, Fesmire J, 2015, Developments in advanced and energy saving thermal isolations for cryogenic applications, *CEC 2015, IOP Conf Ser: Mater Sci Eng*, 101, 012014.
2. Ganni V, Fesmire J, 2012, Cryogenics for superconductors, *Adv Cryo Eng*, 57A, 15–27.

3. Fesmire E, Tomsikb T, Bonnera T, et al., 2014, Integrated heat exchanger design for a cryogenic storage tank, *AIP Conference Proceedings*, 1573, 1365.
4. Ammar A, 2018, Efficient design of a large storage tank for liquefied natural gas, *Ammar J Univ Babylon Eng Sci*, 26(6), 362.
5. Dickey A, 2015, Insulation outlook, cellular glass on LNG systems, *Insulation Outlook*, August 1, 2015.
6. Coffman B, Fesmire J, White S, Gould G, Augustynowicz S, 2010, Aerogel blanket insulation materials for cryogenic applications, *Adv Cry Eng*, 55.
7. Fesmire J, 2018, Aerogel based insulation materials for cryogenic applications, *ICEC-ICMC2018*, Oxford, UK, September 3–7, 2018.
8. Phalen T, Prescott C, et al. 2007, Update on subsea LNG pipeline technology, *Offshore Technology Conference*, Houston, TX, April 30–May 1, 2007.
9. Fesmire JE, Augustynowicz SB, et al., 2008, Robust multilayer insulation for cryogenic systems, *Adv Cry Eng*, 53, 1359.
10. Demko J, Duckworth, Roden R, et al., 2008, Testing of a vacuum insulated flexible line with flowing liquid nitrogen during the loss of insulating vacuum, *AIP Conference Proceedings*, 985, 160.
11. Fesmire J, Johnson WL, 2018, Cylindrical cryogenic calorimeter testing of six types of multilayer insulation systems, *Cryogenics*, 89, 58–75.
12. Southwire cable in Figure 5.3C. Demko.
13. Knoll D, Willen D, Fesmire JF, Johnson W, et al., 2012, Evaluating cryostat performance for naval applications, *Adv Cry Eng*, 1063.
14. The Fermilab Tevatron Cryogenic Cooling System, *An International Historic Mechanical Engineering Landmark*, September 27, 1993.
15. Shu QS, Cheng G, Susta J, Hull J, Fesmire J, Augustynowicz S, Demko J, Werfel F, 2006, Magnetic levitation technology and its applications in exploration projects, *Cryogenics*, 46(2), 105.
16. Shu QS, Cheng G, Susta J, Demko JE, Fesmire J, et al., 2005, A six-meter long prototype of the mag-lev cryogen transfer line, *IEEE*, 1051–8223.
17. Fesmirea J, Augustynowiczb S, Nagyb Z, Sojournerb S, Shu QS, 2006, Testing of prototype magnetic suspension cryogenic transfer line, *Adv Cry Eng*, 51.
18. Grohmann S, Gil W, Neumann H, Weiss C, 2010, Commissioning of the cryogenic transfer line for the KATRIN experiment, *AIP Conference Proceedings*, 1218, 1095.
19. Rane T, Chakravarty A, Klebaner A, 2018, Parameters study for use of stainless steel as a material for thermal shield in PIP2IT transferline at Fermilab, *Adv Cry Eng*, 63.
20. Badgujar S, Bonneton M, Chalifour M, Forgeas A, Serio L, et al., 2014 Progress and present status of ITER cryoline system, *AIP Conference Proceedings*, 1573, 848.
21. Chang H, Vaghela H, Patel P, Rizzato A, Cursan M, 2017, Status of the ITER cryodistribution, *2018, IOP Conf Ser: Mater Sci Eng*, 278, 012018.
22. Riddone G, Trant R, 2002, The compound cryogenic distribution line for the LHC: Status and prospects, CERN-LHC-Project-Report-612.- Geneva: CERN, 15 November 2002.
23. Gannia V, Dixona K, Laverdurea N, Knudsena P, et al., FRIB cryogenic distribution system, *AIP Conference Proceedings*, 1573, 880.
24. Hosoyama K, et al., 2000, Development of a high performance transfer line system, *Adv Cryo Eng*, 45, 1395–1402.
25. Scott R, 1959, *Cryogenic Engineering*, Van Nostrand Publisher, Princeton.
26. Mcintosh G, 2000, Convention design of cryogenic piping and components, *Adv Cryo Eng*, 45.
27. Johnson W, Fesmire J, Meneghelli B, 2012, Cryopumping field joint can testing, *Adv Cryo Eng, 57, AIP Conference Proceedings*, 1434, 1527.
28. Fesmire J, Augustynowiczb S, Demko J, 2001, Overall thermal performance of flexible piping under simulated bending conditions, *Adv Cryo Eng, 47*.
29. Pagani C, 2017, International collaboration view from INFN, *Univ. of Milano & INFN-LASA*, 12–14 December.
30. Shu QS, Fast R, Hart H, 1988, Crack covering patch technique to reduce the heat flux from 77 K To 4.2 K through multilayer insulation, *Adv Cryo Eng*, 33.
31. Shu QS, Fast R, et al., 1988, Theory and technique for reducing the effect of cracks in multilayer insulation from room temperature to 77 K, *Adv Cryo Eng*, 33.
32. Fesmire J, Augustynowicz S, Nagy Z, 2003, Thermal performance testing of cryogenic piping systems, *Adv Cryo Eng*, 49.
33. Deng D, Xie X, Pan W, 2017, Simulation and experimental research of heat leakage of cryogenic transfer lines, *Adv Cryo Eng*, 63.

34. Johnson W, Kelly A, Jumper K, 2012, Two dimensional heat transfer around penetrations in multilayer insulation (Final Report), NASA/TP-2012–216315.
35. Johnson W, Kelly A, Fesmire J, 2014, Thermal degradation of multilayer insulation due to the presence of penetrations, *AIP Conference Proceedings*, 1573, 471.
36. Shu QS, Fast R, Hart H, 1986, Heat flux from 277 to 77 K through a few layers of multilayer insulation, *Cryogenics*, 26, 671.
37. Shu QS, Fast R, Hart H, 1986, An experimental study of heat transfer in multilayer insulation systems from room temperature to 77 k, *Adv Cryo Eng*, 31, 455–463.
38. Birmingham B, Brown E, Class C, Schmidt A, 1957, Vessels for the storage and transport of liquid hydrogen, *Journal of Research of the National Bureau of Standards*, 58(5), May.
39. Fesmire J, Coffman B, Meneghelli B, Heckle K, 2012, Spray-on foam insulations for launch vehicle cryogenic tanks, *Cryogenics*, 52(4–6), 251–261.
40. Notardonato W, Baik J, McIntosh G, Operational testing of densified hydrogen using G-M refrigeration, *Adv Cryo Eng*, 49.
41. Plachta D, Christie R, Carlberg E, Feller J, 2008, Cryogenic propellant boil-off reduction system, *AIP Conference Proceedings*, 985, 1457.
42. Frank D, Roth E, Olson J, Evtimov B, et al., *Development of a Cryocooler to Provide Zero Boil-Off of a Cryogenic Propellant Tank*, Lockheed Martin Advanced Technology Center Palo Alto, CA.
43. Fesmire J, 2018. Aerogel-based insulation materials for cryogenic applications, *International Cryogenic Engineering Conference*, IOP Conf Ser 502, Oxford University, United Kingdom.
44. Sass J, Barrett T, Baumgartner R, Lott J, Fesmire J, 2010, Glass bubbles insulation for liquid hydrogen storage tanks, *Adv Cryo Eng*, 55, 20130011367.
45. Fesmire J, et al., 2006, Aerogel insulation systems for space launch applications. *Cryogenics*, 46(2–3), 111–117.
46. Fesmire J, Sass Nagy Z, Sojourner S, Morris D, Augustynowicz S, 2008, Cost-efficient storage of cryogens, *AIP Conference Proceedings*, 985, 1383.
47. Majumdar A, Steadman T, Maroney J, Sass J, Fesmire J, 2008, Numerical modeling of propellant boil-off in a cryogenic storage tank, *AIP Conference Proceedings*, 985, 1507.
48. Arnold S, et al., 2007, Cryogenic hydrogen tank preliminary design trade studies, *48th AIAA/ASME/ASCE/AHS/ASC Structures, Structural Dynamics, and Materials Con*, 23–26 April (NASA/TM—2007-214846).
49. Krenn A, 2012, Diagnosis of a poorly performing liquid hydrogen bulk storage sphere, *AIP Conference Proceedings*, 1434, 376.
50. Krenn A, Desenberg D, 2019, Return to service of a liquid hydrogen storage sphere, *Adv in Cryo Eng*, 65.
51. Fesmirea J, Tomsikb, Bonnera T, Oliveiraa J, Conyersc H, Johnsona W, 2014, Integrated heat exchanger design for a cryogenic storage, tank, *AIP Conference Proceedings*, 1573, 1365.
52. Swanger A, Jumper K, Fesmire J, Notardonato W, 2015, Modification of a liquid hydrogen tank for integrated refrigeration and storage, *CEC 2015 IOP Conf. Series: 101*, 012080.
53. Notardonato W, Swanger A, Fesmire J, Jumper K, Johnson W, 2017, Zero boil-off methods for large-scale liquid hydrogen tanks using integrated refrigeration and storage, *Adv in Cryo Eng*, 63.
54. Benda V, Bel J, Fathallah, Goiffon M, 2011, The liquid helium storage system for the Large Hadron Collider, *IOP Conf. Ser.: Mater. Sci. Eng.*, 278 012012, 2018.
55. Benda V, Risk analysis of LHC liquid helium storage system, CERN-Report EDMS905921.
56. Serio L, Technical specification for the supply of liquid helium storage tanks for LHC, CERN-Technical specification EDMS 892819.
57. Badgujara S, Bonnetona M, et al., 2014, Progress and present status of ITER cryoline system, *AIP Conference Proceedings*, 1573, 848.
58. Dickey A, Oslica S, 2008, Insulation & liquefied natural gas in production and storage, *Insulation Outlook*, November 1.
59. Dickey A, 2015, Cellular glass on LNG systems, *Insulation Outlook*, August 1, (Pittsburgh Corning and a NIA-certified insulation energy appraiser).
60. Yang TM, 2006, Development of the world's largest above-ground full containment LNG storage tank, *23rd World Gas Conference*, Amsterdam.

7 Vacuum Techniques

James E. Fesmire, Quan-Sheng Shu, and Jonathan A. Demko

Vacuum techniques are a mainstay of cryogenic heat management and cryogenic engineering. The technology of vacuum is the central enabling feature of cryogenic engineering on Earth. If, for example, the ambient environment is 90 K, like Titan, the icy moon of Jupiter, then cryogens like LNG, LO_2, and LN_2 are akin to "room temperature" commodities. As we understand that the temperature difference is the driver for heat flow, we know that vacuum is the necessary second line of defense against reducing the flow of heat. Good design is always the first line of defense in reducing heat conduction to a manageable level, but it is the vacuum enclosure or vacuum jacket that commands the overall design, look, and operational characteristics of the cryogenic system.

Vacuum processes are needed in many different applications. A few of these are as follows:

- Thermal insulation (that is, isolation from the "hot" ambient environment)
- Control and consistency of the operation/process
- Molecular-level (or smaller) processing such as sputter coating
- Simulation of end-use environments
- Scientific research
- Mass spectrometers and leak detectors

Vacuum is necessary for the thermal isolation of cryostats, tanks, and piping to enable basic functionality of the system and control of the cryogenic fluid. Vacuum is essential for conducting many scientific measurements. Vacuum comes into play for a range of experimental physical and chemical processes. The word "vacuum" means the absence of matter, but the reality is always an approximation. Thus, the level of vacuum must be clearly defined and understood for each specific application.

The design, construction, and operation of vacuum systems requires specialized knowledge and use of materials. Materials include metals, polymers, glass, ceramics, composites, and others. Materials also includes the aspects of finishes, treatments, cleaning, and associated techniques. Materials also means the proper specification and use of seals, gaskets, adhesives, compounds, as well as the key working elements of flanges, valves, pumps, instrumentation, feed-throughs, and other components. This complex mix of materials and practical applications, and specialized techniques and design approaches is well addressed by Ekin [1].

Vacuum is not nothing, and nothing is impossible to get. Vacuum is always about a range of pressure or a "level of vacuum." Knowing the proper target level or threshold of vacuum for a given application is the starting point. The main steps in producing a vacuum, following successful leak detection and validation of the system, are purging, baking (heating), pumping (evacuation), and retention. And through it all goes vacuum measurement and monitoring, in accordance with the changes in the internal temperatures, unless the system is "factory sealed" and working as designed.

7.1 DEFINITION OF VACUUM

For practical purposes in cryogenic engineering, the terms vacuum, vacuum pressure, and vacuum level are often synonymous in general discussion. But we understand that the range of vacuum can be, and often is, quite important to specify. The key dividing line in vacuum is the transition from the range of continuum mass flow to the range of molecular mass flow. The continuum flow range has to do with the bulk gaseous (or fluid) flow of a substance, while the molecular flow has

DOI: 10.1201/9781003098188-7

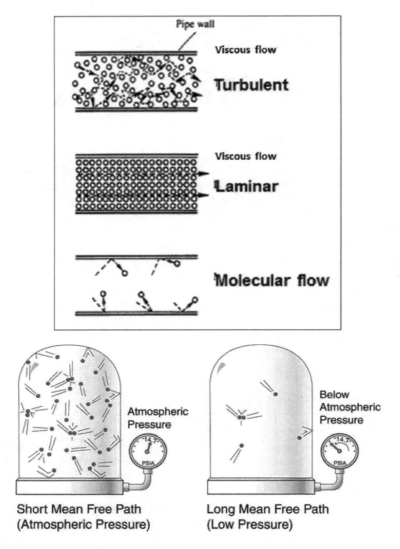

FIGURE 7.1 Physical characteristics of the three basic ranges of pressure: turbulent, laminar, and molecular.

to do with the movement of individual molecules. The bulk flow acts as an organized whole, while the individual molecules act more independently. Figure 7.1 depicts the physical characteristics of the three basic ranges of pressure: turbulent, laminar, and molecular. Turbulent flow happens at ambient pressure to some reduced pressure. Convection heat transfer is dominant in this range. Laminar flow, or continuum flow, is similar but is dominant in the soft vacuum range.

When the mean free path of a molecule becomes larger than its container, there is a sharp transition to molecular flow, which starts the onset of the so-called vacuum range. For most systems, this transition occurs at about 50 millitorr. The ideal gas law equation is arranged in Figure 7.2 to define the mean free path of a given molecule and container. Examples of the mean free path of a typical system are given in Figure 7.3. The sharp transition from continuum flow to free molecular flow is clearly indicated to occur between the pressure levels of 1 torr and 1 millitorr. Systems of different thermal insulation materials are known to be optimum for a particular vacuum level: aerogel composites for soft vacuum, bulk fill glass bubbles or perlite powder for moderate vacuum, and multilayer insulation (MLI) for high vacuum.

$$n_V = \frac{nN_A}{V} = \frac{nN_A}{\frac{nRT}{P}} = \frac{N_A P}{RT}$$

Mean free path

$$\lambda = \frac{RT}{\sqrt{2}\pi d^2 N_A P}$$

FIGURE 7.2 Defining the mean free path of a system.

Pressure (torr)	Mean Free Path (inches)
760	2.5×10^{-6}
100	1.9×10^{-5}
1.0	1.9×10^{-3}
0.001 (1×10^{-3})	1.9
0.0001 (1×10^{-4})	19
0.00001 (1×10^{-5})	190 (16 feet)
0.000001 (1×10^{-6})	1,900 (160 ft or 53 yds)

FIGURE 7.3 Examples of the mean free path of a system.

7.2 VACUUM SYSTEM BASICS

Vacuum comes in two categories: static and dynamic. Static vacuum, or sealed-off vacuum, is used almost exclusively in modern vacuum-jacketed piping or tanks. Dynamic vacuum, or continuous vacuum pumping, is often used in large process systems or experimental devices where specific requirements for control are needed. In either case, there are two main parameters of operation in work: (1) the time required for pumping to a specified vacuum level (threshold value of the desired vacuum-pressure) and (2) the features of the system to enable long-term vacuum maintenance. Leakage is never zero, and outgassing from materials is never zero. Whether that leakage matters is entirely dependent on the system process requirements and conditions.

To accelerate the time required for pumping, there are four main factors: (1) temperature, (2) moisture content, (3) type of vacuum pump(s), and (4) conductance of the evacuation piping. There are three vacuum level designations corresponding to the bulk temperature of the total system. These are the hot vacuum pressure (HVP) at the prescribed bakeout temperature, as applicable; the warm vacuum pressure (WVP) at ambient conditions; and the cold vacuum pressure (CVP) at the cryogenic operating condition. These three vacuum levels can be dramatically different for the same system, and so the distinction becomes particularly important when defining test or operational requirements and procedures.

Bakeout of the system to 100°C or more, for example, will greatly reduce the pumping time compared to 20°C, as the thermophysical effects at the surfaces within the vacuum space are energetically increased by the temperature to the third power, or about 100 times increase in this case. The moisture content, and that of other volatile materials, can be driven off by bakeout, by purging cycles with dry gaseous nitrogen, or a combination of the two, to better dry out the system. Evacuation to a few torr might be accomplished in a few minutes, while evacuation to a few millitorr could require a few weeks for the same system. Thus, any amount of heat and purging will go a long way toward dramatically speeding up the process. It is worth noting that the large surface areas common to thermal insulation materials used within the vacuum space greatly promote the need for heating and purging to the maximum practical extent.

Of course, the selection of vacuum pumps is another important factor in the pumping speed and overall effectiveness. For achieving high vacuum levels, at least two vacuum pumps are needed:

roughing stage (diaphragm or mechanical, for example) and high vacuum (turbo-molecular or dif-fusion, for example) stage. Some pumps such as scroll or mechanical types are all that is needed to moderate vacuum levels (from about 10 to 100 millitorr, for example). For maintaining very high vacuum levels (such as below 10^{-6} torr), ion pumps can be used.

The fourth factor in vacuum pumping speed is the conductance of the vacuum piping system.

The inlet diameter to the first stage (higher vacuum) pump is of the utmost importance. Filters are often required to protect the pump, but any filter will impede the molecules coming through and increase the overall pumping time. Likewise, bellows and flexible hoses will impede the evacuation process but are a practical necessity in most systems and operational facilities. The diameter of the piping can be progressively smaller at the inlet of the roughing pump. The exhaust from the roughing pump is preferably vented external to any enclosed space to reduce noise and minimize any contamination of the room air.

For static vacuum, the maintaining of the required vacuum level is accomplished by getters and adsorbents. Adsorbents such as molecular sieve materials are placed on the colder surfaces to enhance the adsorption and capture of gases and condensable materials within the vacuum space. Getters such as palladium oxide or silver zeolite are used to take away the hydrogen gas by chemical conversion processes. The system seals and weld integrity, along with operational duty cycles (thermal cycles) of the system, all play into the long-term vacuum stability.

7.3 LEVELS OF VACUUM

A range of vacuum levels is given in Table 7.1, spanning 15 decades of pressure, from 10^{-12} mbar to 10^{+3} mbar. Even in the most common fabrication and construction of industrial cryogenic vacuum-jacketed piping and tanks, the relevant range extends at least six decades, from 10^{-3} mbar to 10^{+3} mbar.

For the vacuum application of isolation, or thermal insulation, the vacuum levels of interest cover a wide range depending on the type of insulation materials used and the system heat leak requirements. Multilayer insulation systems require vacuum levels below 10^{-5} millitorr for optimum performance. Evacuated powder-type insulation systems require vacuum levels below 10^{-3} millitorr for optimum performance. The difference of two orders of magnitude is no small difference. Does it really matter to have the optimum thermal performance from the thermal insulation system? Well, that entirely depends on the system design and the system operational requirements. That is, what is the threshold (maximum) heat leak requirement?

TABLE 7.1

Range of Vacuum Levels for Different Units of Pressure

Vacuum Level	Description	psi	torr	millitorr (micron)	mbar	Pa
Ambient	No vacuum (NV)	14.696	760	760,000	1,019	101,325
Rough	Coarse	0.193	~10	10,000	13.3	1,333
Soft	Soft vacuum (SV)	0.019	~1	1,000	1.33	133.3
Moderate	Moderate vacuum (MV), VIP	1.9×10^{-3}	~0.1	100	0.133	13.33
Moderate	Moderate vacuum (MV), bulk-fill	1.9×10^{-4}	~0.01	10	0.0133	1.333
Moderate	MV, bulk-fill, some MLI	1.9×10^{-6}	$<10^{-4}$	0.1	1.3×10^{-4}	1.3×10^{-2}
High	High vacuum (HV), MLI	1.9×10^{-7}	$<10^{-5}$	0.01	1.3×10^{-5}	1.3×10^{-3}
High	HV, empty space	1.9×10^{-8}	$<10^{-6}$	0.001	1.3×10^{-6}	1.3×10^{-4}
Ultra	Ultra-high vacuum (UHV)	1.9×10^{-11}	$<10^{-9}$	0.000001	1.3×10^{-9}	1.3×10^{-7}
Extreme	Extreme high vacuum (XHV)	1.9×10^{-14}	10^{-12}	0.000000001	1.3×10^{-12}	1.3×10^{-10}

With a total of 104 kilometers of piping under vacuum, the vacuum system of the Large Hadron Collider (LHC) is among the largest in the world. The insulating vacuum, at approximately 10^{-6} mbar, is made up of an impressive 50 km of piping, with a combined volume of 15,000 cubic meters, more than enough to fill the nave of a cathedral. Building this vacuum system required more than 250,000 welded joints and 18,000 vacuum seals. The remaining 54 km of pipes under vacuum are the beam pipes, through which the LHC's two beams travel. The pressure in these pipes is on the order of 10^{-10} to 10^{-11} mbar, a vacuum almost as rarefied as that found on the surface of the Moon. The LHC's vacuum systems are fitted with a total of 170 Bayard-Alpert ionization gauges and 1,084 Pirani and Penning gauges to monitor the vacuum pressure.

7.4   VACUUM PUMPING

Conductance is the key to vacuum pumping in the high vacuum range. A high value of conductance means the best rate of pumping in the high vacuum range. The transition point to free molecular flow is again crucial to understand. Conductance does not matter for continuum flow. The important factors for continuum flow, such as smooth bends in the piping, do not matter for molecular flow. A small diameter with smooth transitions can be optimum for evacuation down to soft vacuum. But a large diameter and sharp-edged transitions are the keys to high conductance and optimal evacuation in the high vacuum range. Some example conductance values for different pipe diameters are given in Figure 7.4. Doubling the diameter increases the conductance by eight times. The example

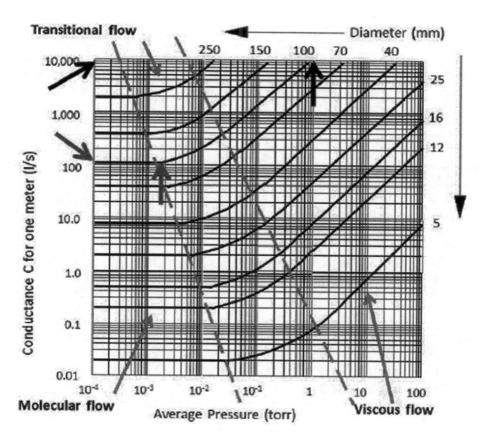

FIGURE 7.4   Conductance vs pressure for different pipe diameters.

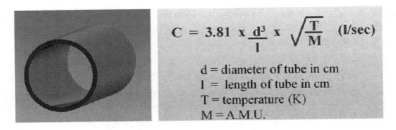

$$C = 3.81 \times \frac{d^3}{l} \times \sqrt{\frac{T}{M}} \quad (l/sec)$$

d = diameter of tube in cm
l = length of tube in cm
T = temperature (K)
M = A.M.U.

FIGURE 7.5   Conductance example for a long round pipe or tube.

of Figure 7.5 for a long round pipe shows that the system conductance is inversely proportional to its length.

The in-between region (moderate vacuum) is indeed in between and must be identified if in fact this range is the target range for system operation. However, it is more common that a high vacuum range is the target. In these cases, the high-vacuum pumping is always the long pole, and the system should be designed for optimal evacuation for the high-vacuum target specified. Knowing the target vacuum range, and the rationale or basis for that target, is the starting place for vacuum system design.

Valuable resources on vacuum technology are found on the website of Normandale Community College (Bloomington, MN). These resources, written by Phil Danielson, include a compilation of 52 articles on vacuum technology (*A Journal of Practical and Useful Vacuum Technology*—https://www.normandale.edu/departments/stem-and-education/vacuum-and-thin-film-technology/vacuum-technology-resources--the-danielson-collection/articles) as well as detailed reviews of 10 technical books on the subject [2] (https://www.normandale.edu/departments/stem-and-education/vacuum-and-thin-film-technology/vacuum-technology-resources--the-danielson-collection/book-reviews).

7.5 VACUUM EQUIPMENT

The construction of vacuum systems is typically done using stainless steel, but aluminum alloys can also be used if conditions are not too stringent. Welding on the inside surface of the chambers is preferred in order to avoid traps for contaminants or the increased potential for virtual leaks. Helium mass spectrometer leak detection to levels below 1×10^{-9} standard cubic centimeter per second (sccs) is the typical minimum standard for most equipment.

The types of flanges and seals include KF (with Viton O-ring material), ISO (with Viton O-ring material), or Conflat (with copper gasket and knife-edge seal). Swagelok VCR fittings have also been successfully used in high vacuum. Seals for moderate vacuum levels may include pipe threads with Teflon tape or grease, Swagelok fittings, or 37-degree flare fittings (KC type with Teflon nose seal).

Vacuum pumping (evacuation) systems generally include at least two stages of pumping: roughing, or backing, and vacuum pumping, or high vacuum. Roughing pumps are used for vacuum levels down to about 0.1 torr. Turbo pumps then extend the evacuation to vacuum levels well below 1×10^{-3} torr and into the very high vacuum range if needed. The necessary components of fittings, valves, seals, feedthroughs, tubing, flexible hose, and so forth are standardized and readily available from many manufacturers.

Vacuum breakers or vacuum relief devices are also required for cryogenic vacuum jacketed equipment and cryostat chambers. These devices can be as simple as "lift plates" or "pop-off plugs." These devices are sized according to the worst-case venting requirement such as from an internal

FIGURE 7.6   Vacuum relief device (25-mm lift plate) on a cryostat vacuum chamber. Source: *J. Fesmire*

leak and subsequent loss of vacuum. Vacuum breakers are not to be confused with "relief valves" required for process pressure systems, as the operational principle and potential effect are distinctly different. An extreme test (LN_2 dump test) of a 25-mm-diameter vacuum breaker is shown in Figure 7.6. The device, with its 5-psi rating (for full flow open), was shown to limit the peak pressure in the chamber to only 1.6 psig (84 torr).

Leak Checking and Troubleshooting

Leak checking and troubleshooting of vacuum systems include both positive pressure and negative (vacuum) pressure tests. Leak-tightness starts with cleanliness and orderliness in all phases of the work. From start to finish, through all phases of the job, the cleaning specifications and treatment of all parts and materials are important. The vacuum system must be considered in an end-to-end (all parts of the system) approach to avoid the one weak link which will defeat the desired vacuum result.

The pressure decay test is to lock up the system with pressure inside and monitor decrease over time (use pressure gage or transducer). This method works for gross or small leaks. The bubble soap test of a pressurized system works for leaks as small as approximately 1×10^{-3} sccs, or "bubble tight." The helium mass spectrometer test, which can be an art form, is for leaks from 1×10^{-4} sccs to 1×10^{-10} sccs and below. There are two main methods of helium mass spectrometer leak testing: direct and indirect (or sniffer). The direct technique, for a quantitative result, is to pull a vacuum on the chamber and then put a very small dose of helium on the outside. In this method, taping or bagging of all suspect locations is done, followed by systematic leak checks starting with the least likely location. The sniffer probe technique includes a wand connected to mass spec machine and the chamber filled with helium. The sniffer method provides a qualitative result for the isolation of a particular leak point.

7.6　VACUUM MEASUREMENT

The measurement and monitoring of vacuum levels can be an essential part of cryogenic system operation and maintenance. Vacuum transducers include at least four main types. The simplest and lowest-cost type is the Pirani or thermocouple gage (with heated filament). For example, the DV-6M by Teledyne Hastings provides a measurement range from about 1×10^{-3} to near 1,000 millitorr. Another type, the Convectron by Granville-Phillips, extends the operation for full-range vacuum of 760 torr down to 1×10^{-4} torr. For vacuum levels below 1×10^{-3} torr and down to 1×10^{-9} torr, an ion gage (or Bayard Alpert style) is used. The MKS Micro Ion is one example. For highest accuracy from ambient pressure to 1×10^{-5} torr, capacitance manometers are used. However, as capacitance manometers rely on a precise strain gauge, their ranges are limited to about three decades of pressure. The MKS Baratron is one example.

7.7　TEMPERATURE MEASUREMENT AND VACUUM

Cryogenic instrumentation is a unique field of measurement requiring novel techniques in part because of the complication of the vacuum shell, chamber, or jacket around the cold mass or cold process element. That is, getting the temperatures affixed to where they are needed and then the lead wires routed through the inside cold boundary, through the vacuum space, and through the vacuum shell to the air side is no simple matter. Factors involved include the attachment method, the mounting technique and design, the exposure (or environment), the temperature and pressure cycling, the feedthrough design, the wiring, and connectors, and many more. In summary, the sensor must function within the intended environment, the sensor must be placed to sense that which is intended to be sensed, and the sensor lead wires must be conveyed to the ambient side environment of the system.

FIGURE 7.7 Thermocouple temperature sensors being prepared for cold mass testing at 20 K. Source: *J. Fesmire*

Temperature sensors include thermocouples (from 600 K to 70 K), RTD or platinum resistance thermometers (from 14 K to 400 K), capacitance (from 1.4 K to 290 K), and silicon diode (from 1.4 K to 500 K). Silicon diodes are usually the most accurate, depending on the calibration applied, and are usually required below 20 K. Thermocouples using Type-E wire have been successfully used down to about 20 K, with crude accuracy but functionality for monitoring the process. Figure 7.7 shows the installation of Type-E thermocouples onto a cold mass for testing at 20 K using liquid helium.

7.8 LARGE-SCALE VACUUM SYSTEMS FOR CRYOGENIC APPLICATIONS

The LHC, one of the largest vacuum systems in the world, is a unique three-in-one vacuum system. The three separate vacuum systems are listed as follows: one for the beam pipes, one for insulating the cryogenically cooled magnets, and one for insulating the helium distribution line. To avoid colliding with gas molecules inside the accelerator, the beams of particles in the LHC must travel in a vacuum as empty as outer space. In the cryomagnets and the helium distribution line, the vacuum serves a different purpose. Here, it acts as a thermal isolator, to minimize the heat leak from the surrounding room-temperature environment into the cold mass, which is maintained at 1.9 K (−271.3°C) by superfluid liquid helium that circulates through the entire 27-km circuit.

Ultra-high vacuum is needed for the pipes in which particle beams travel. The LHC includes 48 km of arc sections, kept at 1.9 K, and 6 km of straight sections, kept at room temperature, where beam-control systems and the insertion regions for the experiments are located. The volume to be pumped in the LHC beam pipes is 150 m^3, which is equivalent to an entire a 60-m^2 apartment. One portion of the LHC circuit is shown in Figure 7.8.

In the arcs, the ultra-high vacuum is maintained by cryogenic-vacuum pumping of 9,000 m^3 of gas. This volume is equivalent to the central nave of a large cathedral. As the beam pipes are cooled to extremely low temperatures, the gases condense and adhere to the walls of the beam pipe by adsorption. Almost two weeks of pumping are required to bring the pressures down below the target of 10^{-10} mbar (or 10^{-13} atmospheres).

FIGURE 7.8   Large Hadron Collider at CERN.

Two important design features maintain the ultra-high vacuum in the room-temperature sections. First, these sections make widespread use of a non-evaporable "getter coating" developed by CERN that absorbs residual molecules when heated. The coating consists of a thin liner of titanium-zirconium-vanadium alloy deposited inside the beam pipes. It acts as a distributed pumping system, effective for removing all gases except methane and the noble gases. These residual gases are removed by a total of 780 ion pumps in connected about circumference of the LHC.

Cryopumping of gas on the cold surfaces provides the necessary low gas densities, but it must be ensured that the vapor pressures of cryo-adsorbed molecules, of which H_2 and He are the most critical species, remain within acceptable limits.

The vacuum system for the LHC will be at cryogenic temperatures (between 1.9 K and 20 K) and will be exposed to synchrotron radiation emitted by the protons. A stringent limitation on the vacuum is given by the energy deposition in the superconducting coils of the magnets due to nuclear scattering of the protons on residual gas molecules because this may provoke a quench. This effect imposes an upper limit to a local region of increased gas density (e.g. a leak), while considerations of beam lifetime (100 hours) will determine more stringent requirements on the average gas density. The proton beam creates ions from the residual gas that may strike the vacuum chamber with sufficient energy to lead to a pressure "run-away" when the net ion induced desorption yield exceeds a stable limit. Synchrotron radiation-induced gas desorption, well known from electron rings, also affects the dynamic vacuum in the cold LHC by the gradual accumulation of easily re-desorbable, condensed gas and by the steeply rising H_2 vapor pressure as the coverage exceeds a monolayer. These dynamic pressure effects will be limited to an acceptable level by installing a perforated "beam screen," as indicated before, which shields the cryopumped gas molecules at 1.9 K from synchrotron radiation and which also absorbs the synchrotron radiation power at a higher and, therefore, thermodynamically more efficient temperature.

7.9   THERMAL ISOLATION AND VACUUM

Thermal isolation is achieved in the LHC in a similar way to vacuum-jacketed cryogenic piping and tanks: with the combination of vacuum and radiation shields (or MLI). The installation of MLI blankets on a cryostat section is shown in Figure 7.9. The fastener details, an important detail for both assembly and performance, are shown in Figure 7.10. Further performance levels are achieved by the incorporation of cooled shields.

FIGURE 7.9   Prefabricated MLI blankets upon installation.

FIGURE 7.10   Fasteners (Velcro type) for attaching MLI blankets.

Vacuum always works in concert with the insulation materials, the arrangement of those insulation materials, and the operational parameters of the system, including vacuum level and thermal profiles. In addition to insulation materials for thermal isolation, thermal shields can also be incorporated as an integral part of the vacuum system (cryostat) design. Thermal shields can be conduction cooled or vapor cooled.

7.10   VACUUM AND THERMAL SHIELDS

Vacuum chambers for cryogenic equipment, or cryostats, can be designed to house cooled thermal shields. The thermal shields come in two basic categories: conduction cooled and vapor cooled. For all designs, he accounting for heat leaks through supports and connections is crucial so that the benefit of the shield is not undone by the additional heat load required by the shield. The LHC cryostat gives extraordinary examples of each type, as shown in Figures 7.11–12.

Conduction-cooled thermal shields are often of a cylindrical configuration and are cooled either by cooled coils affixed to a conductive metal surface (copper or aluminum) or by attachment to a cold surface (cryocooler cold head) at one end by a conductive link (copper or carbon fiber strap, for example) [3]. Calculating the temperature profile along the shield is centered on the one-dimensional heat conduction. If the surface is well insulated with MLI and under high vacuum, then the

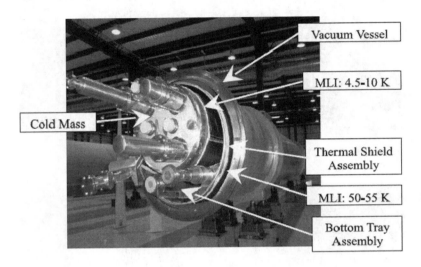

FIGURE 7.11   Cryomagnet dipole assembly of LHC.

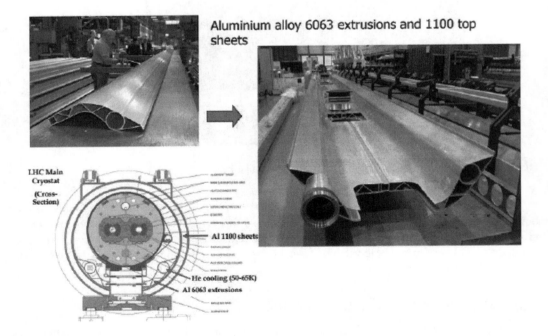

FIGURE 7.12   Cross-section and prefabricated section of the LHC main cryostat.

FIGURE 7.13   Aluminum LHe dewar with vapor-cooled thermal shield. Source: *Meyer Tool & Mfg.*

heat flux from a 300 K surface is typically 1 W/m². With a uniform heat flux on the surface, the temperature will reach a maximum value halfway between the cooling coils.

In the design of a vapor-cooled shield, the cold vapor evaporating from the liquid volume is used to cool the thermal shields and thereby intercept some of the incoming heat load on the cryogenic liquid. An example is shown in Figure 7.13. While vapor-cooled thermal shields are usually associated with liquid helium dewars, the same principle can be applied to any cryogen. Liquid nitrogen dewars with carefully designed vapor-cooled thermal shields have been used to achieve very long holding times in instruments for spacecraft and balloon-borne applications. Shields may be constructed by diverting all or part of the cryostat boil-off through cooling coils attached to thermal shields. In sub-Kelvin cryostats, for example, vapor from an evaporation stage may be drawn off to cool a 1 K heat shield while vapor from a separator vessel cools an outer 4 K heat shield.

Boiling a gram of liquid helium at 4.5 K absorbs 21 J of heat. Warming a gram of helium vapor from 4.5 K to 300 K requires 1,550 J of heat. Effectively using the enthalpy of the cold helium vapor allows liquid helium vessels to achieve very low boil-off rates. Traditionally, many laboratory liquid helium cryostats and magnet dewars have employed liquid nitrogen-cooled thermal shields to intercept heat from surrounding 300 K surfaces. The heat flux from a 300 K surface to an 80 K surface insulated with 60-layer MLI blanket and high vacuum level is typically about 1 W/m². If the 80 K surface surrounds a 4 K surface, which is also well insulated, a heat flux on the inner 4 K surface of about 0.1 W/m² is expected. The liquid nitrogen-cooled thermal shield intercepts 90% of the heat flux at a much lower cost than would be achieved if the 4 K surface were not shielded.

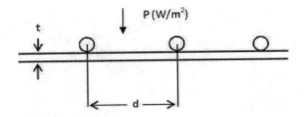

FIGURE 7.14   Heat shield with cooling coils

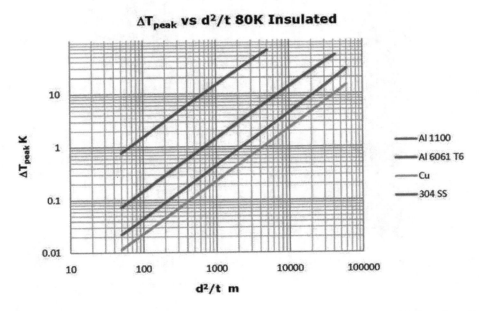

FIGURE 7.15   Peak temperature rises for various shield materials for an insulated 80 K surface. Source: *Meyer Tool & Mfg.*

A well-designed helium vapor-cooled shield for an LHe dewar should be able to perform almost as well as an LN$_2$-cooled shield. Many helium dewars with vapor-cooled leads do have holding times that approach those of nitrogen dewars [4]. A heat shield with cooling coils is shown schematically in Figure 7.14. The maximum temperature difference to be obtained is approximated by Equation 7.1. The estimated peak temperature rises for various shield materials in given for an insulated 80 K surface in Figure 7.15 and for an uninsulated 80 K surface in Figure 7.16.

$$\Delta T_{MAX} = \frac{1}{8k}\left(\frac{d^2}{t}\right)P \tag{7.1}$$

7.11   VACUUM CHAMBERS FOR TESTING

Vacuum chambers are a necessity for testing materials, components, and systems under relevant-use environments. These include laboratory-scale thermal vacuum chambers and large-scale vacuum chambers. Cryogenics plays a role for generating the high vacuum condition or for providing cooling for below-ambient temperature conditions. Large-scale vacuum chambers are enabled by large

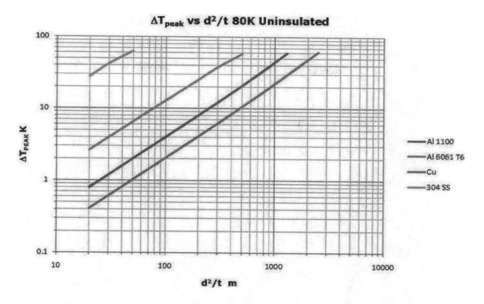

FIGURE 7.16   Peak temperature rises for various shield materials for an uninsulated 80 K surface. Source: *Meyer Tool & Mfg.*

FIGURE 7.17   James Webb Space Telescope being prepared for vacuum chamber test. Source: *NASA*

FIGURE 7.18 Vacuum chamber at NASA Goddard Space Flight Center for testing of elements of the James Webb Space Telescope and other spacecraft. One of the cryopumps is shown in the middle right. Source: *J. Fesmire*

FIGURE 7.19 One of several thermal vacuum chambers at NASA Goddard Space Flight Center for testing of materials, systems, and components under different environments. Source: *J. Fesmire*

FIGURE 7.20 Cold box vacuum chamber for a hydrogen liquefier at NIST Center for Neutron Research in Gaithersburg, Maryland, USA. Source: *J. Fesmire*

cryopumping units in combination with roughing pumps to achieve high vacuum levels of 10^{-6} torr and below. Examples of different vacuum chambers are given in Figures 7.17–19.

Another type of vacuum chamber is the cold box for a cryogenic process system. One example of the cold box vacuum chamber for a hydrogen liquefaction system in given in Figure 7.20.

Whether for test chambers, cold boxes, cryostats, cryogenic piping, or cryogenic tanks, vacuum technology plays an integral role in the design, construction, and operation of those systems. Vacuum and insulation go together to make cryogenic heat management a possibility. Vacuum and cryogenics are in many ways two sides of one coin.

REFERENCES

1. Ekin, J, *Experimental Techniques for Low-Temperature Measurements*, Oxford University Press, Oxford (2006).
2. Danielson, P, A Journal of Practical and Useful Vacuum Technology, Normandale College, https://www.normandale.edu/departments/stem-and-education/vacuum-and-thin-film-technology/vacuum-technology-resources--the-danielson-collection
3. Meyer Tool & Mfg., Custom Vacuum Chamber Articles, www.mtm-inc.com/custom-vacuum-chambers.html
4. Meyer Tool & Mfg., Custom Cryogenic Technology Articles, www.mtm-inc.com/custom-cryogenic-technology.html

8 Cryogenic Calorimeters for Testing of Thermal Insulation Materials and Systems

Quan-Sheng Shu, James E. Fesmire, and Jonathan A. Demko

8.1 INTRODUCTION

The advent and continued development of advanced cryogenic thermal insulation materials and novel insulation systems along with growing industrial and scientific needs in cryogenic applications have brought significant challenges to technical societies. One of the critical issues is that accurate thermal performance information, including effective thermal conductivity data, is needed under relevant end-use conditions. Cryogenic thermal insulations have been successfully developed and utilized in many scientific and engineering projects around the world since the 1960s [1, 2]. However, there are still less reliable standards for the performance data, working procedures, design handbooks, and constructive guidance, which push many engineers and scientists to develop special calorimeters in testing and obtaining particular thermal information for applications [3–5]. The dilemmas are typically for multilayer insulation (MLI), since a 50% aviation of MLI performance is not unusual in some large and complicated applications. Beside MLI, the performance of other cryogenic insulation materials also highly depends on the process and working conditions in different applications.

It is crucial to develop customers' own data and procedures before design and construction of large projects. In order to accomplish such tasks, it is better to have their own test apparatuses, so-called calorimeters, to test the performance of thermal insulation in various shapes and conditions.

Therefore, testing calorimeters that are able to test the performance of thermal insulation in systems similar to the real cryogenic system for scaled-up real applications are crucial. First, design principles and structures of several boil-off meter calorimeters (BOMCs) with LHe and LN$_2$ as cryogens are introduced that can test cylindrical and plant specimens with boundary temperatures of 77 K–300 K, 77 K–4 K, and many special intermediate temperatures. Second, thermal conductive meter calorimeters (TCMCs) with multi-boundary temperatures for various specimen shapes are discussed. How to employ cryogens or cryocoolers as cooling sources for specimens and guarding shelters in TCMCs is deliberated. Then, two specially designed multipurpose calorimeters (SDMCs) for MLI with penetrations, cracks, and holes are introduced. Then, a 1,000-liter (L) spherically shaped calorimetric (SSC) tank is discussed for obtaining thermal insulation data more directly for application of extra-large cryogenic tanks. Finally, successful demonstrations of cryogenic heat management with the assistance of calorimeters in the structural heat intercept, insulation, and vibration evaluation rig (SHIIVER) in large space applications are presented. The design methodologies, unique structures, and merits in comparison of various cryogenic calorimeters are our focus in the detailed discussions. Some definitions of common symbols are given in Table 8.1.

8.2 CYLINDRICAL BOIL-OFF CALORIMETER

Cryogenic boil-off calorimetry is developed for determining the effective thermal conductivity k_e and heat flux (q) of a test specimen at the fixed conditions (boundary temperatures, cold vacuum

DOI: 10.1201/9781003098188- 8

TABLE 8.1

Definitions of Some Common Symbols

Symbol	Description	Unit
V_{STP}	Volumetric flow rate (boil-off) at STP	m³/s
ρ_{STP}	Density of GN$_2$ (boil-off) 0.0012502	Kg/m³
h_{fe}	Heat of vaporization	J/g
do and di	Outer and inner diameters of insulation specimen	m
x	Thickness of insulation specimen	M or mm
L_e	Length, effective heat transfer	m
A_e	Area, effective heat transfer area	m²
ΔT	Temperature difference (WBT-CBT)	K
k_e	Effective thermal conductivity	mW/m-K
WBT	Warm boundary T	K
CWT	Cold boundary T	K
VP	Vacuum pressure	torr, mtorr
Q	Heat flow rate	W
q	Heat flux	W/m²
stp	Standard T and P (0°C and 760 torr)	

pressure, and residual gas composition). The vaporization heat of cryogenic liquid (LN$_2$, LHe, or LH$_2$) serves as the energy meter. The energy meter provides a direct measure of the heat flow rate, and the heat flow rate is directly proportional to the boil-off flow rate. The insulation specimen to be evaluated is cylindrically shaped between a cylindrical cryogen container and temperature controlled cylindrical shields (or another container with higher boil-off cryogen in it). The heat through the insulation specimen is only taken into account in the test cryogen container, but the thermal effect at both ends is canceled by the carefully designed end of the cryogen container or other thermal structure.

Although both cylindrical and flat-plate cryostats have been standardized for laboratory operation [6, 7], cylindrical configurations are better at minimizing (or even eliminating) unwanted lateral heat transfer or "end effects." Cylinders also align better with most applications, including tanks and piping. The vertically oriented cold mass assemblies of the cryostats can cause some convection problems when tests are conducted at ambient pressure, but otherwise, these assemblies provide a stable platform for testing over a wide range of heat flows.

The heat flow rate, Q, through the insulation test specimen and into the side wall of the test chamber of the cold mass assembly is directly proportional to the cryogen boil-off flow rate, V, as given by Equation 8.1:

$$Q = V_{stp}\rho_{stp}h_{fe}(\rho_f / \rho_{fe}) \tag{8.1}$$

where *stp* is standard T and P, and (ρ_f/ρ_{fe}) is the density correction, if any, between the liquid and the saturated liquid conditions. The value of k_e is determined from Fourier's law of heat conduction through a cylindrical wall, as given by Equation 8.2.

$$K_e = (Q/\Delta T)(A_e / x) = [Q\ln(d_o / d_i)]/2\pi L_e \Delta T \tag{8.2}$$

$$Ae = (2\pi x L_e) \ / \ \ln(d_o \ / \ d_i) \tag{8.3}$$

The heat flux (q) is calculated by dividing the total heat transfer rate by the effective heat transfer area, as given by Equation 8.4.

$$q = Q / Ae \tag{8.4}$$

8.2.1 300 K–77 K Cylindrical Boil-Off Calorimeters

Cryogenic insulation systems used in the temperature region of 300 K–77 K encompass a wide range of material combinations. An insulation test specimen is a system composed of one or more materials (homogeneous or nonhomogeneous, with or without inclusion of a gas) whose thermal transmission properties are measured through its thickness under sub-ambient temperatures. Several cryostat instruments have been developed and standardized for laboratory testing of thermal insulation systems in a cylindrical configuration separately by J. Fesmire [6], K. Kamiya [8], Van Sciver [9], and others.

8.2.1.1 Cryostat CS-100

Cryostat CS-100 is a fully thermally guarded design, which has tested several hundred specimens over the years by Fesmire et al. at NASA KSC, following the guidelines of ASTM C1774, Annex A1 [10]. It is a primary (absolute)-type instrument of thermal transmission measurement. Absolute instruments produce the data by which other instruments, such as the comparative (secondary), can be calibrated. This apparatus is guarded on top and bottom to eliminate end effect for absolute thermal performance measurement.

Test Apparatus Design and Setup. The basic schematic and a photo of the overall arrangement are shown in Figure 8.1. It is composed of a cold mass assembly, including the top and bottom guard chambers; a middle test chamber is suspended from a domed lid atop the vacuum canister, including the mechanical lift mechanism.

Each of the three chambers is filled and vented separately through a single feedthrough for easy operation and minimum overall heat leakage. All instrument wires pass other special feedthroughs and are connected from the lid. A novel thermal break design is used between the liquid chambers (to preclude solid-conduction heat transfer from one liquid volume to another [11]). Fesmire and Dokos introduced the technique in U.S. Patent No. 8,628,238 B2 [12]. Such isolation is critical for achieving very low heat measurements because even small temperature variations between the liquids in the chambers can produce dramatic errors in results. The liquid within each individual chamber is allowed to stabilize in its natural stratified state.

Cryostat-100 includes an external heating system for bakeout and high-heat load tests, as well as an internal heater system for fine control of the warm boundary temperature (WBT). Three funnel filling tubes (7.93-mm OD) interface with the three LN_2 feedthroughs (12.7-mm OD). The filling tubes are removed when not being used. Connected to the top ports of the LN_2 feedthroughs are the plastic tubing assemblies that route the boil-off flow from all three liquid chambers to their respective mass flow meters. The vacuum pumping system includes a directly connected turbopump and a separately plumbed mechanical pump. In addition, a gaseous nitrogen (GN_2) supply system provides purging and residual gas pressure control to vacuum levels as low at 5×10^{-5} torr. All instruments are connected to a customized LabVIEW data acquisition system for data recording and monitoring. The overall assembly of Cryostat-100 is shown in Figure 8.1 (right). The core of the cryogenic calorimeter is hung on the support and wrapped with MLI layers. The outer container stands at the

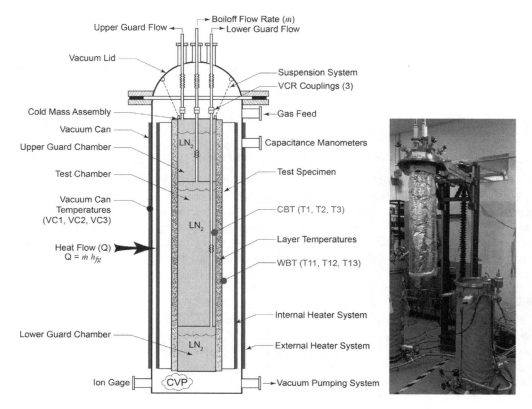

FIGURE 8.1   Left: Cryostat CS-100: basic schematic. Right: Overall assembly [6].

right side, with a high vacuum pumping system and heating clothes for controlling the warm boundary temperature.

Test Methodology and Uncertainty Analysis: Calculation of k_e is highly sensitive to the thickness of the test specimen (the maximum is about 50 mm). The thickness is carefully measured or calculated, and any assumptions are considered. Materials can be blanket, clamshell, molded, or bulk-fill. Blankets can be applied in individual layers or in various layering combinations as desired. Multilayer insulation specimens can be installed in blanket, layer-by-layer, or continuously rolled fashion. The temperature sensors are Type E thermocouples, 30-gauge size, with vacuum sides at least 2 m long. For MLI specimens, five GN_2 purge cycles between 1 torr and 100 torr at a temperature up to 330 K are included. Without systematic controls to counteract this effect, at very low heat flux rates, the atmospheric pressure fluctuations can influence the results significantly and are eliminated by feeding the boil-off flow tubes into a custom plenum system set approximately 3 torr above the prevailing mean at atmospheric pressure and by controlling the back pressures of all three chambers within ±0.1 torr. The liquid level in the test chamber is at least 90% full. The total test duration may be hours to days. The total uncertainty in k_e is calculated to be 3.3% for the Cryostat-100, and the uncertainty in heat flux, q, is 3.2%. In most testing situations, for a given series of tests, the overall repeatability is demonstrated to be within 2%. Boil-off technology to measure ultralow heat flow ($k_e < 0.01$ mW/m-K and/or $q < 0.1$ W/m² for typical boundary conditions of 300 K/77 K in vacuum) is at the heart of efforts to develop and prove such advancements [6, 8, 13].

8.2.1.2 Selected Examples of Cylindrical Boil-Off Meter Calorimeters between 300 K and 77 K

Several 300 K–77 K cylindrical boil-off meter calorimeters (CBMCs) have been developed and utilized around the world. In terms of the design methodology, there are two types of CBMCs: (1) CBMCs with two end guard vessels, which may provide high accuracy of measurements, and (2) CBMCs without guard vessels, which have less complexity in construction and can also provide acceptable data with enough accuracy. We choose two of the CBMC types as examples to introduce their design features briefly.

Kamiya reports a CBMC with two guard vessels designed for evaluation of the electrically insulated MLI for the thermal shield of JT-60SA (a device being upgraded to a full-superconducting tokamak), shown in Figure 8.2 [8]. The thermal shields of JT-60A are cooled by pressurized helium gas between 80 K and 100 K. Most fusion facilities were not using multilayer insulation for the thermal shield due to plasma pulse operation. MLI is affected by the eddy current in the toroidal direction. MLI has a great advantage for saving refrigeration power by about 23 kW at 80 K in JT-60SA. To overcome the eddy current problem, its MLI needs to be electrically divided into 18 sectors in the toroidal direction. Each cryostat thermal shield (CTS) sector and MLI has to be insulated to withstand 50 volts at maximum, while the target heat load through the whole MLI must be less than 2.2 W/m². The CBMC is designed with three cryogen chambers and a water-cooled temperature boundary.

Fesmire introduces the Cryostat-200, which is a simplified and easily operating cylindrical test apparatus for measurement of the comparative k-value [14]. This apparatus includes only one 132-mm-diameter by 500-mm-long cold mass and accepts specimens up to 50 mm thick. An overall

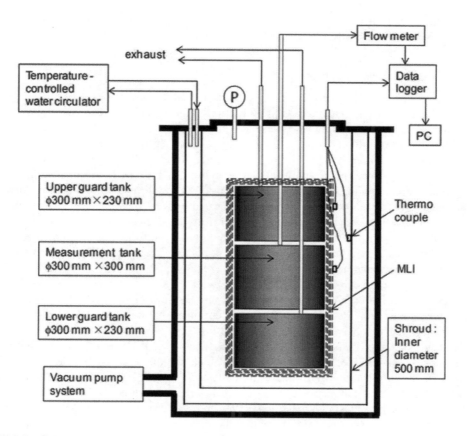

FIGURE 8.2 Schematic experimental set-up of a CBMC [8].

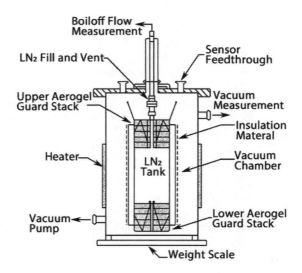

FIGURE 8.3   Simple design set-up of CBMC without guard vessels [14].

diagram of Cryostat-2 is given in Figure 8.3. The entire cold-mass assembly is easily removed and mounted on a wrapping machine to facilitate the testing of MLI and other layered types of insulation materials. The end effects are canceled by both aerogel stacks instead of guard LN_2 vessels. Cryostat-2 has also been successfully utilized to evaluate cryogenic thermal insulation performance at Kennedy Space Center (KSC), NASA, and has U.S. Patent No. 6,487,866.

8.2.2   CBMCs BETWEEN 77 K AND 4 K

There has not been enough experimental data for various MLI applications between temperatures lower than 77 K and 4.2 K. Therefore, several boil-off calorimeters have been designed, operated, and reported by Shu et al. separately for the qualification of MLI between 77 K and 4 K for different purposes around the world. Thermal radiation and contact heat transfer between adjacent insulation films are the dominant contribution of heat transfer through MLI at good vacuum conditions. The MLI is fabricated on a horizontally supported cylindrical cold body, such as SC magnets and SC cavities at accelerators [15]. To avoid excess tension in the film, contact pressure between the films is caused by the weight of the MLI itself. Thermal performance of the MLI without the use of spacer material between the aluminized polyester films has been evaluated experimentally in the CBMC. The calorimeter is used to examine how the thermal performance data obtained from the laboratory-scale calorimeter can be applicable to estimate the thermal performance of the MLI in the horizontal cryostat and how performance can be improved.

It is a double-guarded cylindrical calorimeter between 77 K and 4 K, which has a main boil-off tank, an upper guard tank, and a lower guard tank, as shown in Figure 8.3. They have two liquid nitrogen tanks completely surrounding them and eliminate the conduction heat through the fill and venting ports. The inner tanks with the sample film of MLI wound around their sidewalls are installed inside the outer liquid nitrogen shield tank. Thus, the MLI sample can be tested between the 4.2 K cold wall and 77 K hot wall. If the outer shield tank is removed and the inner tanks are filled with liquid nitrogen, an MLI sample can be tested between 77 K and the ambient temperature by keeping the same fabrication condition of MLI. The total number of layers of the reflective films is 40 in the tests. By using a mass flow meter, the heat flux through the MLI is measured by the evaporation rate of cryogen from the main tank. A correction factor of 1.157 for liquid helium was adopted to obtain the heat flux through the MLI from the reading of the mass flow meter.

The temperatures of the outer wall of the main tank and the reflective films were measured by 76-μm-diameter CR-Au/Fe type thermocouples. The vacuum chamber of the calorimeter is continuously evacuated by the 500-liters/sec turbomolecular pump and is maintained below 10^{-6} torr.

The MLI layers are wound around the vertical side wall of the inner tanks. As the gravity direction on the films is parallel to the axis of the inner tanks, the contact pressure is not affected by gravity, and homogeneous contact pressure among the layers can be realized. The layer density corresponds to the contact pressure. Then the circumferential length of each layer is calculated and used to wind the reflective films around the tank. The contact pressure between adjacent layers is found to be in the range of the pressure that is generated in the MLI fabricated around the horizontally supported cylinder.

Conclusion of the Tests. In order to improve the thermal performance of the MLI, the CBMC is effective to reduce the weight of the MLI by using thinner reflective film. The stack test of the MLI is a useful technique to get the relation between the layer density and the compressive pressure generated by the weight of the MLI.

8.2.3 CBMCs between 60 K and 20 K to 4 K

Between ambient temperature and 77 K, there is a rich amount of data for MLI thermal performance. However, there are serious demands to obtain thermal performance of MLI at 4.2 K with radiation from temperatures of 60 K and down to 20 K, especially in the fields of high-energy physics, fusion, aerospace, and so on [5, 16]. It is quite understandable that a 77 K–4 K CBMC is easy to use in the temperature region of 300 K to 77 K, but the reverse requires a redesign and reconstruction of the cryostat and instruments. Venturi introduced how a cylindrical boil-off calorimeter used to for 77 K–4 K has been successfully redesigned and modified as a calorimeter operated between 60 K and 20 K to 4 K [5].

The Original Test Cryostat. The original CBMC developed by Wroclaw University of Science and Technology is shown in Figure 8.4 and is used to measure performances of MLI between 293 K and 77 K and 77 K and 4.2 K. The cryostat is composed of an inner test vessel around which the MLI sample is wrapped. A double LHe guard vessel ensures insulation from other heat sources, and an external LN_2 tank provides shielding from room-temperature thermal radiation. The pipes of the guard and test vessel are double walled to aid in insulation. The boil-off gas in the test vessel is measured by a mass flow meter with an outlet to a helium recovery balloon.

Key Redesign and Modification. The modification to the test cryostat was made at CERN to allow intermediate temperature levels for the warm boundary to be fixed between 20 K and 60 K. A 1-mm-thick aluminum thermal screen (ATS) was designed and placed on a copper support in weak contact with the guard vessel. The aim is to establish, in steady-state conditions, an equilibrium temperature of the ATS between the inner tank at 4.2 K and the external tank at 77 K. Twenty K is reached with no additional heat applied. In order to set the temperature of the ATS, a 60-Ω electrical resistance heater that provides power in the 5–6 W range is placed to provide a heat source controlled by four CERNOX temperature sensors, which are distributed along its length.

Initial simulations of the system using ANSYS show that the optimal position for the heater is on the bottom part of the ATS in order to minimize temperature gradients. An important feature of the test cryostat is the measure of pressure underneath the MLI blanket through an integrated pipe connecting the sample vessel to the outside flange. This information is important to check the residual gas pressure from outgassing of the MLI sample that is directly related to molecular conduction within its layers.

Commissioning. Once the cryostat is closed, with the test blanket installed, it takes approximately three days to reach a vacuum level of 2E-7 mbar and a stable temperature of the thermal screen. The ATS reaches a temperature of 20 K with no additional power applied with a gradient of less than 1 K along its length. The power to be applied on the ATS heater to reach 60 K is approximately 6 W and the tubes are precooled by the exhaust flow. As a result, a systematic background error heat load of 20

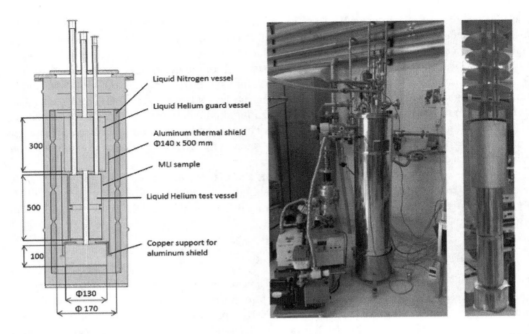

FIGURE 8.4 Left: Sketch of 77 K–4 K CBMC with double guard vessels and the special Al thermal shield on Cu support ring. Right: Cryostat [15].

mW was extrapolated. The cryostat's background heat load was measured by reducing the ATS temperature, via a special support, down to about 11 K, thus cancelling its radiation heat contribution. Stable conditions appear one hour after setting of the heaters. A sampling rate of every 20 seconds is sufficient for the time response of the system. A correction of ±2.5% is found on average.

8.3 FLAT PLATE BOIL-OFF CALORIMETERS

Comparison between Flat Plate Boil-Off Calorimeter (FPBC) and CBMC. Several cryostat instruments have been developed and standardized for laboratory testing of thermal insulation systems in a flat-plate configuration. Boil-off calorimetry is the measurement principle for determining the effective thermal conductivity (k_e) and heat flux (q) of a test specimen under a wide range of real-world conditions. Cylindrical configurations of CBMCs are better at minimizing (or even eliminating) unwanted lateral heat transfer or "end effects." Cylinders also align better with most applications, including tanks and piping. The vertically oriented cold mass assemblies of cryostats can cause some convection problems when tests are conducted at ambient pressure, but otherwise, these assemblies provide a stable platform for testing over a wide range of heat flows. Flat-plate configurations of FPBCs also offer a number of potential advantages: (1) the ability to handle small test specimens (when only a small piece can be obtained), (2) compression loading capability, (3) specialized ambient pressure testing with different purge gases, and (4) greater relevance to end-use application. Powder-type insulation testing is more difficult on flat-plate calorimeters but has been done successfully. Flat-plate cryostats are also easier to adjust for different cold boundary temperatures, as will be discussed.

The rate of heat transfer through the insulation test specimen, Q, is directly proportional to the cryogen boil-off flow data, V, as given by Equation 8.1, the same as with CBMC. The value of k_e is determined through a flat plate specimen, as given by Equation 8.5, and the definitions of symbols are given in Table 8.1. The heat flux $q = Q/A_e$.

$$k_e = Qx / A_e \Delta T = 4Qx / \pi \left(d_e\right)^2 \Delta T \tag{8.5}$$

8.3.1 FPBCs with Cryogen Guard Vessels

Fesmire et al. have separately reported the successful development and utilization of FPBCs. The design principle of these FPBCs has certain similarity. Cryostat CS-500 has been used in testing of insulation specimens for about 100 operations and is chosen as an example, as shown in Figure 8.5 [7].

With liquid nitrogen boil-off as the energy meter, the cold boundary temperature can be adjusted to any temperature between 77 K and approximately 300 K by the interposition of a thermal resistance layer between the cold mass and the specimen. A low thermal conductivity suspension system has compliance rods that adjust for specimen thickness and compression force. Material type, thickness, density, flatness, compliance, outgassing, and temperature sensor placement are important test considerations. Cryostat CS-500 is thermally guarded by a separate cryogen chamber to provide absolute thermal performance data and eliminate edge effects. Testing for thermal insulation materials (MLI, composites, foams, aerogels) is performed. Figure 8.6 is a 3D model of Cryostat CS-500.

This apparatus, which accepts round, 203-mm-diameter specimens, is guarded on the top and around its perimeter and can be easily adjusted to measure absolute thermal performance. An adjustable-edge guard ring enables calibration with a known material. With the use of liquid nitrogen (LN_2) vaporization as the energy meter, the FPBC can be adjusted to any temperature between 77 K and approximately 300 K.

The cold mass assembly, comprising the heat measurement vessel and thermal guard vessel, is suspended from the vacuum chamber lid, as shown in Figure 8.6. The cold mass assembly uses a

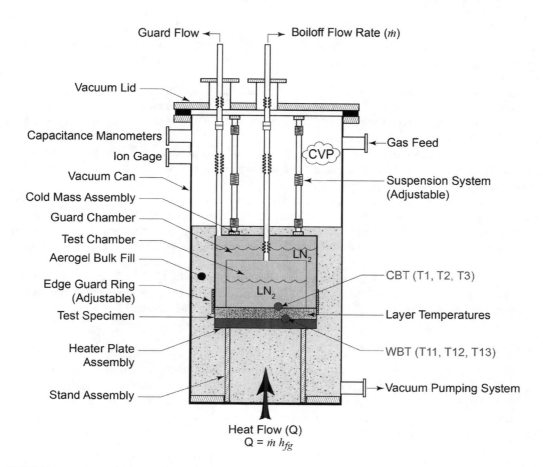

FIGURE 8.5 Cryostat CS-500: basic schematic [7].

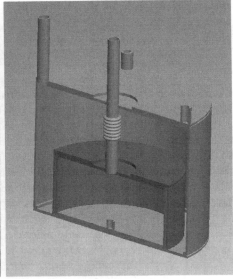

FIGURE 8.6   Cryostat-500 3D model: Overall system (Left) and cold mass assembly (Right) [7].

double wall as an interface between the side wall of the test chamber and the guard chamber. Such isolation is critical for thermal stability and the fine equilibrium necessary for an accurate boil-off measurement.

A low thermal conductivity suspension system includes compliance rod assemblies that can be adjusted for the thickness of a given test specimen between approximately 3 and 40 mm. Compression loading up to approximately 100 kPa (15 psi) can also be applied to the test specimen through the suspension system as required.

Cryostat CS-500 includes an external heating system for bakeout, a heating plate system for control of the warm boundary temperature, and vacuum levels as low at 5×10^{-5} torr. A LabVIEW data acquisition system is used.

The steady-state condition is reached when the boil-off flow rates from both chambers are stabilized, the temperature profile through the thickness is stabilized, and the liquid level in the guard chamber is at least 50% full (that is, covering the top surface of the test chamber). A stable state of the system is indicated by slight oscillation of the temperature sensors with no overall trend in their average value.

The total uncertainty in k_e is calculated to be 4.8% for the Cryostat CS-500. The mass flow meter automatically compensates for gas densities in the range of 273 K to 323 K and outputs the data in terms of a volumetric flow rate at STP. The overall repeatability is within 2%.

8.3.2   FPBCs without Cryogen Guard Vessels

In some practical cases, the needed thermal conductivities of insulation materials are not available, but the data accuracy required is not very restricted. In satisfaction of the requirements, an FPBC without a cryogen guard vessel (Cryostat CS-400) is developed, and the schematic of the simplified test apparatus without cryogen guard vessel is presented in Figure 8.7 [17]. In comparison with Compared to Cryostat CS-500, the Cryostat CS-400 is less complex and easier to operate and can

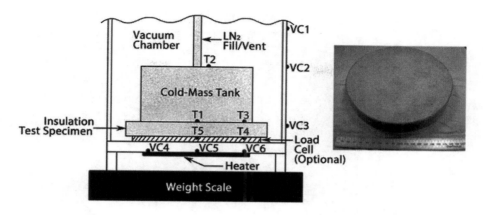

FIGURE 8.7 Left: Simplified schematic of Cryostat CS-400 showing typical locations of temperature sensors without LN$_2$ guard vessels. Right: insulation specimen [17].

be used for a wide range of materials and conditions in addition to the full cryogenic vacuum testing of high-performance materials. The cold-mass assembly can be configured for rigid or soft materials, with or without compressive loads applied. An optional load cell assembly is also provided. For monitoring the operational conditions and stabilizing state, the temperature sensors are allocated as follows: cold boundary temperatures (CBTs) T1, T2; warm boundary temperatures T4 T5, top of cold mass temperature T2, vacuum exterior temperatures VC1–3, heater temperatures VC4–6. The boil-off of LN$_2$ is measured associated with the sample's area, A_e; thickness, d; and the ambient T and P to determine the k_e of the materials.

8.3.3 MACROFLASH BOIL-OFF CALORIMETER (COMMERCIALLY AVAILABLE)

The Macroflash (Cup Cryostat) Boil-off Calorimeter is a specially designed and developed FPBC by Fesmire et al. which is easy to use, low cost, and commercially available now [18–20]. The Macroflash Calorimeter provides effective thermal conductivity (k_e) data for a wide range of materials from thermal insulation to structural composites to ceramics,

The apparatus provides a practical method and standardized way to measure heat transmission through materials under steady-state conditions at below-ambient temperatures and under different compressive loads, as presented in Figure 8.8. Another unique feature of this device is that it can provide test data at both large and small temperature differences. Using liquid nitrogen as a method to directly measure the heat flow rate, the device is applicable to testing under an ambient-pressure environment at a wide range of temperatures, from 77 K to 403 K. Test specimens (diameter 76 mm, thickness up to 10 mm) may be isotropic or non-isotropic, homogeneous or non-homogeneous. The Macroflash Calorimeter is currently calibrated in a range from approximately 10 to 1,000 mW/m-K using well-characterized materials. Reference data for hundreds of test specimens, including foams, powders, aerogels, plastics, composites, carbon composites, wood, glass, ceramic, metal, and multilayered composites, have been compiled from Macroflash testing. Macroflash is also under ASTM test standard C1774 Annex A4.

The cold boundary temperature is maintained by LN$_2$ at 77 K, while a heater assembly maintains a steady-state warm boundary temperature from ambient up to 403 K. Nitrogen or another gas is supplied to the instrument to establish a stable, moisture-free, ambient pressure environment. Compression loading is selected for 0, 14, or 34 kPa (0, 2, or 5 psi), with or without thermal grease, and the system can be readily adapted for other loads as required.

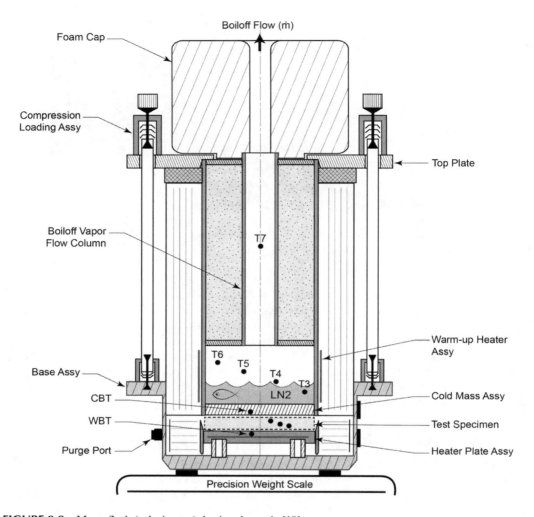

FIGURE 8.8 Macroflash (calorimeter): basic schematic [19].

Figure 8.9 (Right) shows the commercially available generation 2 Macroflash calorimeter. LN_2 fills to 220 g (or more) for initial cooldown and stabilization. Steady-state condition is between 50 and 100 g LN_2. The lower right displays some of testing samples. The Macroflash Calorimeter can automatically calculate the effective thermal conductivity of sample materials. The primary inputs are: thickness and diameter (mm), mass (g), WBT set point, interlayer temperatures (if used), and the primary outputs are: k_e calibrated, M_{dot} (mass flow rate), q (heat flow rate), Q (heat flux), WBT, and CBT. Total uncertainty in k_e is calculated to be 4.8% for the Macroflash. Overall repeatability for most test series is demonstrated to be within 0.5%, The method of direct calculation of the heat flow rate (Q) using boil-off calorimetry is shown in Figure 8.9 (Left). The Macroflash provides a cost-effective, field-representative methodology to test any material for below-ambient temperature applications to moderately elevated temperature conditions.

8.4 THERMAL CONDUCTIVE METER CALORIMETERS

The working principles of cryogenic calorimeters are based on two key mechanisms: cooling sources to the samples to control the cold boundary temperature and a meter to measure the heat load through the samples. A thermal conductive meter calorimeter utilizes a thermal conductive

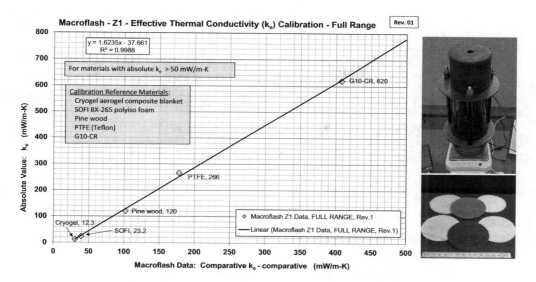

FIGURE 8.9 Left: Macroflash effective thermal conductivity k_e calibrated with full range. Right: Photos of commercial Macroflash Calorimeters and testing samples [20].

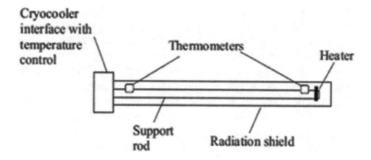

FIGURE 8.10 Calibration method of the thermal conductive meter.

heat rate meter to determine the heat load and mostly uses a cryocooler to control the boundary temperatures [21, 22]. Different from TCMCs, cryogenic boil-off meter calorimeters employ a liquid cryogen to control the cold boundary temperature and use the cryogen boil-off rate to obtain heat transfer data through the insulation specimen. The TCMC approach has three unique advantages: (1) the use of cryocoolers enables the user to set the boundary temperatures anywhere within the operating range of the refrigerators and therefore permits a wide range of temperature and temperature differences with the measurement. (2) The total heat transfer through the testing sample is by measuring the heat conducted through a cold support rod, which has a thermal conductivity as the standard reference. (3) The TCMC is safer than the CBMC in operation since volatile cryogens are not used. In order to obtain the absolute thermal conductivity, careful calibration is needed to eliminate the temperature-related effects on the support rod. There are also two types of the TCMC designs: TCMC for cylindrical samples and TCMC for plant samples.

Calibration of the Thermal Conductive Rod. As presented in Figure 8.10, the cold cylinder support rod is calibrated at various temperatures by applying a known heat to the rod to determine the calibration parameter in Fourier's Law:

$$Q = \theta \times \Delta T \qquad\qquad 8.6$$

where Q is the calculated heat; ΔT is the measured temperature difference; and θ is the calibration parameter, which includes the thermal conductivity, rod length, and rod cross-sectional area ($\theta = kA/L$, where k is the integral thermal conductivity of the rod between cold and warm temperatures). An error analysis of this calorimeter has shown accuracy of heat load measurements on the order of 1–2% depending on the temperature. Calibration of the support rod is performed separately [21]. A wire heater is placed at one end of the rod away from the thermometer. The heater is tightly wound in the threads and properly heat sunk using thermal grease. A four-wire scheme is used to accurately measure the voltage drop along the heater. The amount of current flowing along the heater wire is also measured by placing a known resister in series with the heater, which allows the accurate determination of the current. The product of the voltage and current yields the heater power. The rod is wrapped in several layers of MLI and placed inside a copper pipe, which acts as a radiation shield. One end of the copper pipe is connected to the same flange as the support rod. This configuration should minimize the temperature difference between the support rod and the radiation shield, which, in turn, minimizes the radiation load to the support rod. Based on the calibration, a Q vs ΔT table and chart are established for real insulation tests.

8.4.1 TCMCs with Cylindrical Insulation Specimens

At Glenn Research Center, a TCMC for measurement of MLI at low boundary temperatures is reported in detail by Johnson and Chato as follows and is shown in Figure 8.11 [22]. The calorimeter consists of cold and warm surfaces, which are a pair of nested cylinders with flat ends inside a cylindrical vacuum vessel. The cold (inner) cylinder has guarded top and bottom ends to minimize the effects of heat transfer at the ends of the test section. The guards are thermally separated from the test section by a 6.5-mm gap. They are structurally connected with thermally insulating G10 tabs. The inside of the warm (outer) cylinder and outside of the cold cylinder are painted with Aeroglaze Z306 (room temperature emissivity 0.90, solar absorptivity >0.95) to provide a black body surface for the insulation to view. The warm cylinder, test section of the cold cylinder, and a pair of guarded ends on the cold cylinder are each cooled by a cryocooler (the two guards are controlled by a separate cryocooler than the test section). The working portion of the cryocoolers is shown installed on the vacuum lid in Figure 8.11 (Upper). Each cryocooler-cooled subsystem is fitted with a trim heater to adjust the temperature of that subsystem. The MLI test specimen is wrapped around the sidewall of the cold cylinder. The portion of the heat flow from the warm cylinder through the MLI to the cold cylinder's test section must flow through the test section wall to a conductive plate (hub), then through a calibrated conduction rod (CCR), and last through a thermal strap to the cryocooler cold head like the configuration. In steady-state conditions, the heat flow through the MLI covering the test section is equal to the heat flow through the CCR. The thermal conductivity and cross-section area of each CCR were chosen to provide an optimal temperature drop in each respective CCR. The heat flow through the CCR is calibrated as a function of the temperature readings at two locations along the CCR.

Calibration is performed in the calibration rig, and accuracy is within ±3%. The calibrated heat conduction rod used for this testing has a 42.5-mm-wide square cross-section. The difference between predicted and actual heat flow plotted as a function of rod temperature differential is shown in Figure 8.12 (Left). A total of six rods are designed for different heat loads. The maximum thickness of insulation specimen is about 7 mm. Both the aluminized Mylar and the aluminum foil were attached to the top and bottom edge guards (G10 spacers designed to allow attachment of the MLI blanket without shorting the blanket to either the cold or warm surfaces) to eliminate contact with the cold boundary temperature. Room-temperature emissivity was measured for both coupons prior to installation on the calorimeter in accordance with ASTM E-408.

8.4.2 TCMCs with Flat Plate Insulation Specimens

Barrios and Van Sciver present a TCMC for flat-plate insulation specimen [23]. It is designed to accurately measure the thermal conductivity of spray-on foam insulation (SOFI) at temperatures

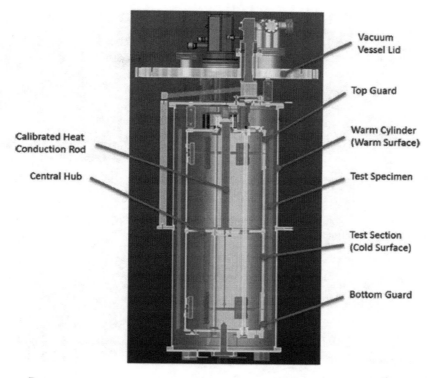

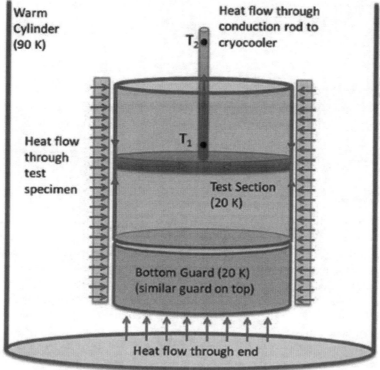

FIGURE 8.11 Upper: TCMC cross-section with cylindrical MIL specimen. Lower: Sketch of heat flows and conductive heat meter [21].

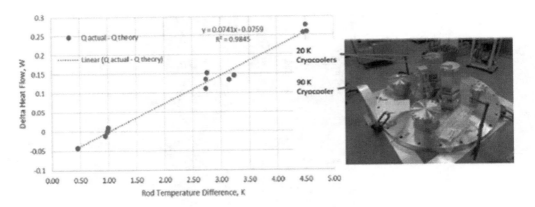

FIGURE 8.12 Right: Cryocooler mounted on vacuum vessel lid. Left: Difference between predicted and actual heat flow plotted as a function of rod temperature differential [22].

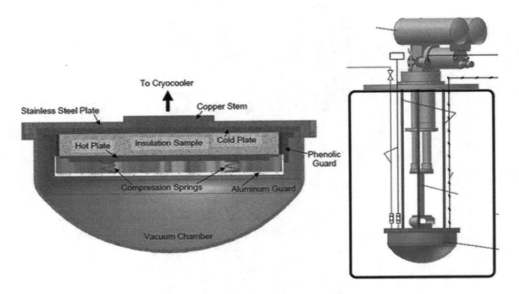

FIGURE 8.13 A schematic of the experimental chamber assembled with the cryocooler [23].

ranging from 20 K–300 K. The TCMC utilizes a single-sided guarded warm plate with only one cold plate and one specimen. The arrangement simplifies the cooling plate design, which is connected directly to the cryocooler as a heat sink. Figure 8.13 shows a sketch of the experimental chamber and the entire assembly. The estimated measurement error of the TCMC is less than 5% for insulation measurements.

Cold Plate and Hot Plate. The cold plate consists of a 6.35-mm-thick copper plate nested inside a larger-diameter 12.7-mm-thick stainless steel (SS) plate. The cryocooler (Cryomech model PT-810, 14 W at 20 K) is attached to the top of the cold plate with a copper stem and a braided copper thermal link. The overall temperature of the apparatus is controlled by a heater mounted to the head of the cryocooler. SS tubes are welded to the top of the cold plate to provide mechanical supports. The apparatus is placed inside a cryostat with a liquid nitrogen shield to isolate it from the ambient environment. The copper plate is used to provide a uniform temperature at the cold boundary of the specimen. Based on meeting the ASTM C177 requirement, the side of the copper plate facing the specimen is painted black. The hot plate consists of two copper plates bolted on either side of a foil heater. The side of the cold plate facing the sample is machined flat to ±127 μm. The compressibility of the insulation will compensate for the slightly higher flatness tolerance. The surface of the plate facing the sample is painted black to provide a hemispherical emittance greater than 0.8. Four SS compression springs (2.20 N/mm of spring rate) are mounted between the axial guard and the hot plate. The free length of the springs is 25.4 mm. The springs together provide an average pressure of 2.89 kPa on the surfaces of the sample.

Guards, Vacuum Chamber, and Thermal Insulation. There are a primary guard and a secondary guard. The primary guard consists of an aluminum guard attached to the cold plate by a phenolic ring. A foil heater is mounted to the aluminum guard. The heater makes the aluminum guard temperature match the hot plate and minimizes heat transfer to the aluminum guard. The vacuum chamber consists of a 3.40-mm-thick stainless steel end cap, which is welded to a SS flange. The chamber is maintained at high vacuum, serving as a secondary guard against heat leaks from the wall of the cryostat. Aerogel beads are filled into all the empty spaces within the experimental chamber to minimize heat leaks to the surroundings, including between the sample and primary guard and between the primary and secondary guards. An MLI shield (not shown in figure) is anchored to the first stage of the cryocooler to protect the experimental chamber from radiation from the cryostat wall.

Test Specimen. The specimen is machined to 222 mm in diameter and 25.4 mm in thickness. To keep a precise thickness and allow good contact with the hot and cold plates, the specimen should have flat, level surfaces. It is important to choose the proper specimens to be tested: Too heavy a specimen may overcome the force applied by the compression springs and cause poor contact with the cold plate. A porous specimen, such as SOFI, or surfaces that are not sufficiently flat, would also provide poor surface contacts. Aluminum backing, higher spring constant supports, or grease is applied to the SOFI specimen in order to enhance the surface contact. Specimens with high thermal conductivity may require more heater power to maintain the desired temperature difference than the heaters in the original design can provide. The powder/perlite type of insulation must be confined to a container to provide rigidity.

8.5 SPECIAL MULTIPURPOSE CALORIMETERS FOR MLI

8.5.1 FERMILAB SPECIAL MULTIPURPOSE CALORIMETER

Fermilab had the world's first superconducting accelerator of 6.5 km in operation in the 1980s, and then Fermilab was a key participant in the superconducting supercollider (SSC) program of the about-80 km superconducting accelerator. How to efficiently use multilayer insulation in keeping a huge cold mass cooled (around 4 K) is crucial in all of cryogenic thermal management. A special multipurpose calorimeter for MLI has been developed and utilized by Q.S. Shu and R. Fast to comprehensively study the performance of various MLI combinations, different shapes of various cold surfaces, and serious penetration through uncounted MLI in large applications [13, 24–26].

The calorimeter cryostat is about 3 m in height and 1 m in diameter. Its simplified schematic is shown in Figure 8.14 [26]. The center LHe boil-off vessel is surrounded by a primary ring-shaped cylindrical LHe guard vessel and then thermally protected by a secondary LN_2 guard vessel. There are two pairs of radiation shields of 4 K attached on the tops and bottoms of the LHe guard vessels and shields of 77 K are on the LN_2 guard vessel. All three cryogen vessels are connected and supported to the cryostat top flange by cryogen filling and vent tubes, shown in Figure 8.15. The 4 K center test copper plate (691 × 1,660 × 2 mm, black painted) is cooled and supported by a stainless-ness steel thermosiphon, which is welded on the plate and connected to the center test LHe vessel as shown in Figure 8.14.

The 77 K box (160 × 770 × 1,760 mm) surrounding the test plate is cooled and supported by two LN_2 thermosiphon stainlessness steel tubes, which are welded on the side pieces of the box and

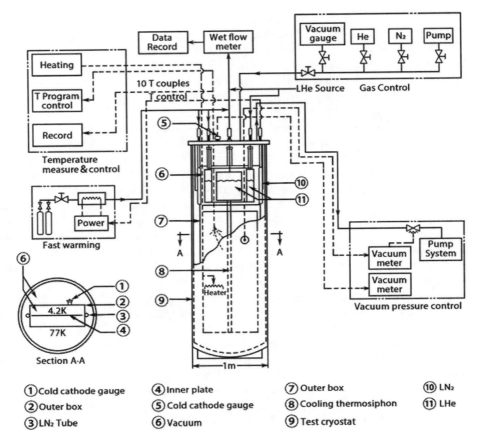

FIGURE 8.14 Simplified schematic of special multipurpose calorimeter for MLI [25].

connected to the LN$_2$ vessel. Eight thermocouples and 16 heaters are mounted on the large outer surfaces of the box to control the temperature of the warm sides. An additional eight thermocouples are used to measure the measure the temperature distribution through the MLI. Both large-side copper plates can be easily disassembled for workers to apply various MLI specimens and cut/repair MLI samples with penetrations, as shown Figure 8.15. Two cold cathode gauges are separately mounted on the cryostat top flange and on the 77 K copper box inside the vacuum space. The readings of the gauge on the 77 K box were used to monitor all the data presented.

77 K–4.2 K Test Arrangement. As described in papers [25, 26], the center copper plate (area 2.26 m^2) wrapped with the MLI specimen was refrigerated to 4.2 K by a thermosiphon tube connected to the liquid He boil-off vessel. The boil-off vessel was guarded by the primary liquid helium guard vessel and radiation shields. The outer copper box was maintained at 77 K by two thermosiphon tubes from a liquid nitrogen vessel. The accuracy of the wet test meter used for the helium gas boil-off is ±0.2%. The Q, K_e, and q are calculated by Equations 8.1, 8.2, and 8.3. Thermocouple wires (CR-Au/Fe) were provided with a heat sink by taping them to the LHe or LN$_2$ temperature surface, depending on which temperature was closer to the measured point, and installed in different locations and layers of the MLI specimen and copper box surfaces.

300 K–77 K Test Arrangement. The test arrangement made several adjustments from the test of 77 K–4.2 K. The inner copper plate was refrigerated to 77 K by a thermosiphon tube connected to the boil-off vessel, and it and the primary guard vessel both contained liquid N$_2$ instead of liquid

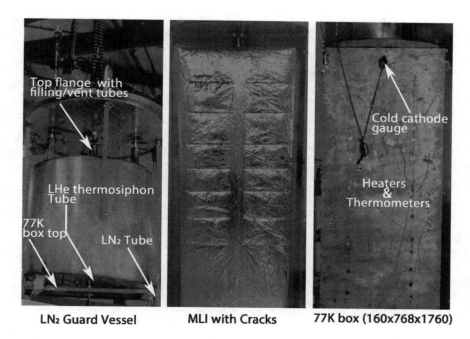

LN₂ Guard Vessel   **MLI with Cracks**   **77K box (160x768x1760)**

FIGURE 8.15   From left to right: LN$_2$ guard vessel (ring-shaped cylinder) with an LHe test vessel and LHe guard vessel inside. MLI wrapped on the center plate with cutting slots cooled by a thermosiphon tube from LHe test vessel. 77 K copper box cooled by two soldered thermosiphon tubes from LN$_2$ guard vessel. Source: *Q.S. Shu*

He. The second guard vessel was kept empty. To control the temperature of the outer (warm) box and to measure the temperature distribution in the MLI, copper-constantan thermocouples (diameter 0.07 mm) were taped to the box and plate and placed between some of the multilayers. To avoid measuring errors due to thermo-voltages, each thermocouple wire was led out of the test cryostat and to the scanner without joints or splices.

Test Arrangement for Penetration and Other Conditions. The calorimeter can test MLI performance in like-real application conditions such as (1) MIL blanks with cracks, slides, and penetrations; (2) simulation of cryostat leaks of He or N$_2$; and (3) center cold piece with shapes as in real applications (cooled by a modified thermosiphon from center boil-off vessel), which is surrounded by a box enclosure (cooled by a modified thermosiphon from secondary guard vessel).

8.5.2 CALORIMETERS FOR PENETRATION THROUGH MLI

In preparation for large-scale cryogenic upper stages, NASA had a need to more accurately predict the thermal loads that are transmitted to the large tanks through the insulation. As these issues are usually accounted for by increasing the thermal margin on a blanket, getting a full grasp on the heat loads due to integration will allow for predicting those heat loads with less uncertainty. Johnson et al. introduced a program to more fully understand the integration issues between various penetrations and multilayer insulation blankets both through experimentation and thermal modeling.

To perform testing on various penetration methods, a new type of calorimeter was needed. This calorimeter needed to be sensitive enough to measure very small changes in thermal performance of insulation blankets yet have enough capacity to take the much higher heat loads associated with placing a penetration through the blanket. Based on the work of Fesmire et al., a guarded liquid nitrogen boil-off calorimeter (known as Cryostat-600) was designed and fabricated with

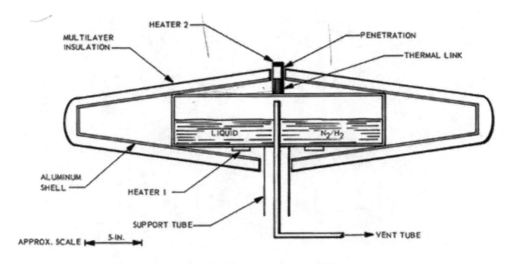

FIGURE 8.16 Cryostat sketch for study of insulation penetration [27].

built-in mounts for the penetrations, as shown in Figure 8.16. The built-in mounts were below the calorimeter to give a much more uniform cold boundary temperature for the insulation. A vacuum chamber was built for the calorimeter, and all tests were run at vacuum pressures in the 1×10^{-6} torr range. To determine the actual degradation around the penetration, the applied MLI and penetration loads must be known. Each of the six MLI blankets used was tested without a penetration (in accordance with Test 1 in Table 8.1) prior to being damaged for testing with a penetration. This allowed for subtracting out the baseline heat load. To calculate the strut thermal loads, temperature sensors were placed at known locations along the penetration (which was made of a known material and geometry) as shown in Figure 8.17 to allow for the calculation of the heat load down the penetration. These known heat loads were then subtracted from the measured load for each of the penetration tests.

Testing of various styles of integration of structural and fluid components into MLI blankets was completed at the Kennedy Space Center over the course of an eight-month test matrix spanning 22 different tests. Both temperatures and heat load data were gathered during the dedicated penetration calorimetry testing. The data from these tests were then used to verify a detailed thermal model, which was used to perform parametric analysis even beyond the testing. From that analysis, a simplified equation was generated to allow for the calculation of the integration heat loads from various penetrations into cryogenic tanks. The results from this experimental modeling study will allow for the quantification of integration losses for penetrations through MLI. This will decrease the uncertainty of the thermal performance of insulation systems applied to cryogenic tanks and vessels.

8.6 SPHERICAL CALORIMETRIC TANKS

8.6.1 1,000-LITER SPHERICAL-CALORIMETRIC TANKS

There are many spherical LH_2 and LO_2 tanks with capacities of more than hundreds of m^3 that have been and will be designed and constructed at Launch Complex at NASA, other space centers, and other large cryogen production and distribution centers around the world. Thermal performance as a function of cryogenic commodity (LH_2, LO_2, LN_2, etc.), insulation materials, vacuum pressure, insulation fill level, tank liquid level, thermal cycles, and lifetime are critical to these applications. Understanding the influences of these factors on cryogenic performance is vital both for improving

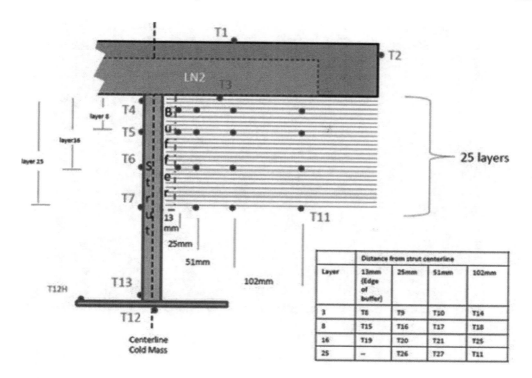

Layer	Distance from strut centerline			
	13mm (Edge of buffer)	25mm	51mm	102mm
3	T8	T9	T10	T14
8	T15	T16	T17	T18
16	T19	T20	T21	T25
25	--	T26	T27	T11

FIGURE 8.17 Penetration calorimeter with thermometers distributed along the MLI [27].

existing cryogen tanks and for the design of new cryogen tanks. In response to these demands, two 1,000-L spherical calorimetric tanks, identical scaled-down versions of the 323-m³ spherical liquid hydrogen tanks at Launch Complex 39 of KSC, were custom designed and built to serve as test calorimeters for the test projects, as shown in Figure 8.18 [28, 29].

8.6.2 CALORIMETER DESIGN AND INSTRUMENTATION

The inner tank has a 1,245-mm ID, 4.8-mm wall thickness, and annular space of 0.82 m³, and the outer tank has a 1,524-mm ID with 4.8-mm wall thickness. The cryogen capacity is 1,000 L. The unique low heat-leak inner tank support system and feedthrough is calculated to be 3 W. The as-built parasitic heat leak for the tanks is approximately equivalent to 4.2 W. The windows and fill ports are designed for insulation studies. Since the parasitic heat leak is a relatively small percentage of the overall system heat leak, the tanks are well suited as a calorimetric instrument to perform comparative testing of bulk-fill insulations. The measured k-value in the tank is consistent with the k-value obtained from Cryostat-100 (shown in Figure 8.1). This direct comparison holds for vacuum pressures from about 1 to 100 millitorr.

The calorimetry tank has a full complement of instrumentation, offering an extensive thermal and mechanical test capability, including the following elements partly shown in Figure 8.19 [29].

- Silicon diode temperature sensors for inner tank temperatures
- Thermocouples for temperature gradient through annular space and outer tank temperature
- Mass flowmeters for tank boil-off
- Cryo-Tracker for liquid level sensors
- Full vacuum range pressure transducers for tank annular space vacuum level
- Load cells for tank weight changes

FIGURE 8.18   View of the pair of 1,000-L spherical calorimetric tanks used for thermal and mechanical performance testing of LH$_2$ and LN$_2$ [28].

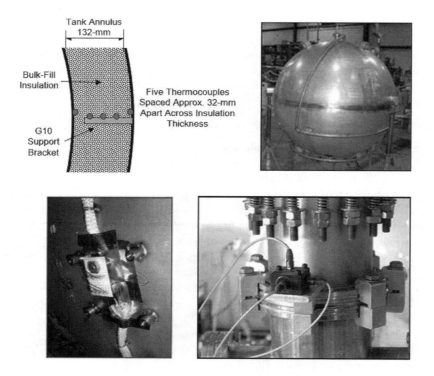

FIGURE 8.19   A: Five thermocouples along insulation in vacuum space. B: 12 thermocouples distributed along inner sphere. C: Silicon diode installation with radiant shield removed. D: Outer sphere tri-axis accelerometer installation [29].

- Tri-axial accelerometers for tank vibration levels
- Mechanical displacement indicators
- Pressure transducers for tank ullage and liquid supply pressures

8.6.3 TEST CAPABILITY AND KEY RESULTS

Fesmire et al. reported nearly 9,000 hours of steady boil-off data spanning 94 tests, which were collected over a period of 19 months using two calorimetric tanks, two insulations, and two types of liquids [29]. To minimize the effect of any inherent differences between the two tanks, testing was performed with nearly every combination of insulations and liquid possible, and it was verified that they have similar parasitic heat leaks. Keeping test conditions consistent over a two-year period of testing was challenging. The evaporative boil-off rate is affected by small variations in annulus vacuum pressure, liquid level, and ambient conditions. The tests and key results can be briefly summarized in the following points. Two representative results from a great amount of testing data are shown in Figure 8.20 as proving the efficiency of the spherical calorimetric tanks.

- Boil-off flow rate is the primary measurement, along with warm and cold boundary temperatures. System heat flux is calculated using the heat of vaporization
- Overall thermal conductivity of the actual field installation, k_{oafi}, is calculated using system geometry.
- LN_2 and LH_2 are separately utilized in tests. For the LH_2 test, special care is taken for an external vent stack system, hazard-proofing modifications, LH_2 supply, and GHe supply.
- Materials research: Thermal optimization of glass microsphere composites and perlite, evaluation of higher-strength bubbles for transfer line applications, granular physics modeling of glass bubbles.
- Demonstration testing: Life-cycle characteristics of field tanks and over-the-road tanks; aerogel bead evaluation in research tanks (for LO_2 storage); vibration and environmental effects evaluation; incorporating SMART tank concepts of passive devices, integrated cryocoolers, and shape memory alloy structures.

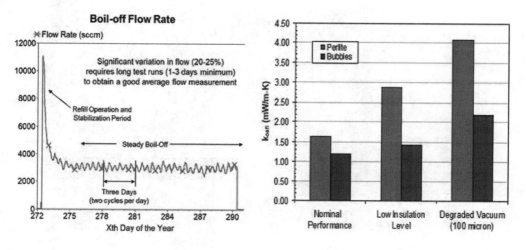

FIGURE 8.20 Right: Glass bubbles significantly outperform perlite with respect to normal operational, life-cycle considerations of low (settled) insulation and degraded vacuum pressure. Left: Boiloff rate as function of testing time [29].

- Field application and systems engineering: Implementation package for retrofit of existing tanks; design engineering approach for new large tanks, including instrumentation, monitoring, structural/mechanical/thermal.

8.7 CRYOGENIC HEAT MANAGEMENT WITH CALORIMETERS

SHIIVER is a large-scale cryogenic fluid management (CFM) test bed designed to scale CFM technologies for large, in-space stages. Technologies developed will play a critical role in enabling increasingly long-duration in-space missions beyond low Earth orbit (LEO). Initially, it is focusing on testing three cryogenic technologies:

1. Multilayer insulation will be applied to the tank domes to quantify the thermal performance of thick ($\geq$10 layers) MLI blankets at conditions and configurations representative of Space Launch System (SLS) upper-stage mission implementations.
2. Vapor cooling will be applied to the forward structural skirt to demonstrate the benefit of using the boil-off gas from the tank to reduce the structural heat load.
3. A radio frequency mass gauge (RFMG) will be installed inside the tank and tested.

In this section, discussions are focused on how the calorimeters are to be employed in the cryogenic heat management for the SHIIVER thermal test stack and preparation for the system test, as shown in Figure 8.21 [30–34]. The technologies from a 4-m-diameter test tank with skirt systems will be designed for an 8.4-m application, which might be found on a launch vehicle.

8.7.1 SMALL-SCALE TESTING OF MLI

To assess the design features of the SHIIVER insulation blanket, a calorimeter was developed to accurately perform small-scale testing of MLI down to 20 K [31]. This calorimeter (see Figure 8.22)

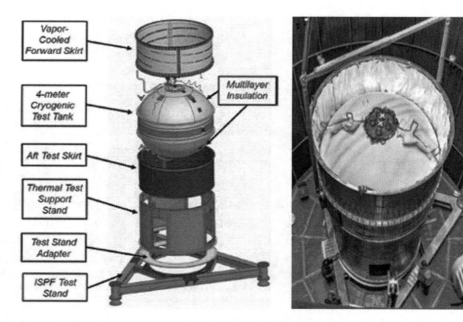

FIGURE 8.21 SHIIVER thermal vacuum test stack showing location of demonstration technologies (left). Preparation for system test of SHIIVER (right) [30, 34].

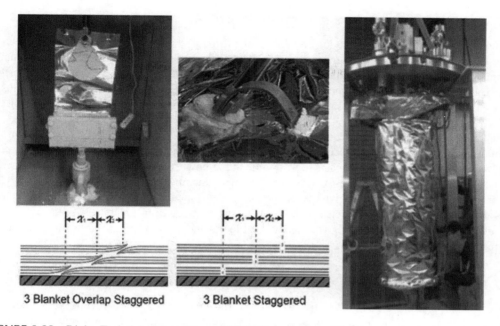

3 Blanket Overlap Staggered   3 Blanket Staggered

FIGURE 8.22 Right: Technicians installing a SHIIVER test coupon blanket onto the calorimeter. Testing includes studies of MLI blanket seams-joints, penetration to MLI blankets, and MLI blankets in five environments exceeding 498 N [31].

uses cryocoolers to maintain the cold boundary at temperatures close to 20 K and can control the warm boundary between 70 K and 300 K. This calorimeter is utilized to establish baseline data and principles for SHIIVER specific MLI design details, which include the number of layers of MLI and the preferred seam structure on the MLI blanket. Shu has reported MLI performance degradation due to cracks/slots and approaches to recover the thermal insulation property [24, 26]. However, there is still a lack of technical data showing how to seam large-scale MLI blankets. A series of seam structures of MLI blankets is proposed and tested in the project, such as blankets overlap staggered and blankets staggered.

8.7.2 Large-Scale Implementation and Testing of MLI

Ensuring the results from the multilayer insulation blankets scale to be relevant to NASA's current 8.4-m-diameter upper stage is important for SHIIVER. Since the current barrel of the hydrogen tank is the outer mold line of the vehicle, MLI cannot be placed there with existing technology, as the airflow during launch would rip it off. Thus, the MLI will only be placed on the domes for initial testing. However, SHIIVER will also have curtains that can drop over the outer diameter of the tank to help predict the improved performance if an advanced MLI system could be developed to survive that type of environment or if the stage could be placed in a shroud.

The final design is 30 layers of double aluminized reflectors, with each reflective layer separated by 2 layers of netting. The 30 layers are split into three sub-blankets made of 10 layers, each with outer, more durable cover sheets on each side. The mass for each dome is expected to be 19 kg, for a total mass of 38 kg. For the dome area of 17.4 m², the maximum loading on the blanket is approximately 11.6 kN. At a maximum acceleration of 5 g (49 m/s²), the total force due to the mass of the blanket is only 910 N, which is much less than the depressurization load. Each restraint will need to withstand approximately 284 N to support the total load.

8.7.3 Testing of Support Structure to the Propellant Tank

Although MLI will drastically reduce heating loads through the tank surface area, the large heat loads through the support structure to the propellant tank are also a great challenge. Most launch vehicles use skirt-type mechanical supports due to the location of the tanks in the structural design. The use of propellant boil-off vapor around skirts to reduce heating into the propellant tanks is desired. Small-scale testing has shown that a 50% reduction in heating from the skirt by using the boil-off vapor to intercept the heat being conducted down the skirts is a conservative estimate.

The heat intercepted by the cooling channel depends on the thermal conductance between the cooling tube and the skirt wall (see Figure 8.23, Left), the convection heat transfer coefficient, contact resistance, and contact area. In order to understand how these issues might affect the performance of a system, a sub-scale test was run on different attachment options (see Figure 8.23, Right).

The RFMG is a propellant quantity gauge being developed for low-gravity applications with possible use in long-duration space missions utilizing cryogenic propellants. A database of RF simulations of the tank containing various fluid fill levels and liquid configurations is generated for comparison to the measured data.

8.7.4 System Test

As shown in Figure 8.21 (Right), SHIIVER will go through three thermal vacuum test sequences plus reverberant acoustic testing. For all thermal vacuum testing, the environmental temperature will be approximately 290 K, and the vacuum pressure will be less than 1×10^{-5} torr. Initial thermal vacuum testing will develop a baseline system performance for a tank insulated with only spray-on-foam insulation. Both the thermal performance and the benefit of vapor cooling on the forward skirt will be evaluated without the MLI on the domes.

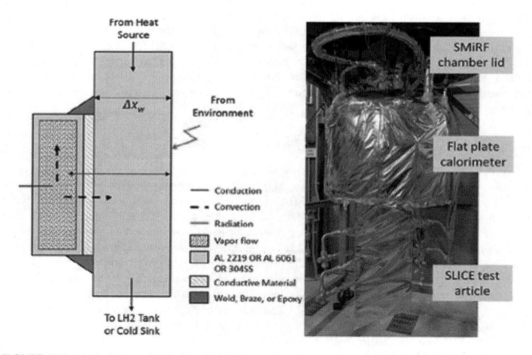

FIGURE 8.23 Left: Thermal path from boil-off vapor to skirt wall. Right: The test article hanging from a liquid hydrogen calorimeter [30].

A second thermal vacuum test will occur to investigate the thermal benefit of the MLI on the domes. Vapor cooling tests will also be run to assess the impacts of the dome MLI on the vapor cooling system benefits. An MLI curtain will be lowered around the SHIIVER tank to provide technical rationale for the performance benefit to insulating the entire tank in MLI if the tank can be contained within a shroud or if an MLI system could be developed to survive on the outside of a launch vehicle.

The SHIIVER test article will finally then encounter a third thermal vacuum test to determine the thermal effects of any damage the reverberant acoustic testing caused in the MLI. The lift-off acoustic test will be approximately 150 dB over a specified frequency spectrum for the duration of 40 seconds. The aero-acoustic environments are at a level of 159 dB for a duration of 20 seconds.

Conclusion. This demonstrates that calorimeters play important roles in cryogenic heat management for efficient storage and transport of cryogenic fluids in space. The demonstration of the scaling of the RFMG will help ready that technology for infusion into large tanks. The data developed using SHIIVER will be directly applicable to large upper stages.

REFERENCES

1. Kropschot R et al., 1960, Multiple-layer insulation *Adv. Cryog. Eng.* 5:579–586, Plenum, pp. 189–198.
2. Webb J, 1966, Apparatus for measuring thermal conductivity *U.S. Patent No.* 3,242,716.
3. Fesmire J, 2018, Aerogel based insulation materials for cryogenic applications, *ICEC/ICMC-2018*, Oxford, United Kingdom, September 3, 2018.
4. Shu QS, Fast R, and Hart H, 1987, Systematic study to reduce the effects of cracks multilayer insulation. Part 2: experimental results, *Cryogenics*, 27:298.
5. Venturi V et al., 2018, Qualification of a vertical cryostat for MLI performances tests between 20 K and 60 K and 4.2 K, *2019 IOP Conf. Ser.: Mater. Sci. Eng.* 502 012077.
6. Fesmire J and Johnson W, 2018, Cylindrical cryogenic calorimeter testing of six types of multilayer insulation systems, *Cryogenics*, 89, pp. 58–75.
7. Fesmire J, Johnson W, Swanger A, Kelly A, and Meneghelli B, 2015, Flat plate boiloff calorimeters for testing of thermal insulation systems, *Adv. Cryo. Eng. IOP*, 012057.
8. Kamiya K, Furukawa M, Hatakenaka R, 2014, Electrically insulated MLI and thermal anchor, *AIP Conference Proceedings*, 1573, p. 455.
9. Vanderlaan M, Stubbs D, et al., 2018, Repeatability measurements of apparent thermal conductivity of multilayer insulation (MLI), *Adv. Cryo. Eng.,* 278(1), p. 012195.
10. ASTM C1774 Standard Guide for Thermal Performance Testing of Cryogenic Insulation Systems 2013 (West Conshohocken, PA: ASTM International).
11. Fesmire J, et al., 2016, Cylindrical boiloff calorimeters for testing of insulation systems, *Adv. Cryo. Eng.,* 101(1), p. 012056.
12. Fesmire J and Dokos A, 2014, Insulation test cryostat with lift mechanism U.S. Patent No. 8,628,238 B2.
13. Shu QS, Fast R, and Hart H, 1986, Heat flux from 277 to 77 K through a few layers of multilayer insulation, *Cryogenics*, 26:671.
14. Fesmire J, Augustynowicz S, Heckle K, and Scholtens B, 2004, Equipment and methods for cryogenic thermal insulation testing, *Adv. Cryo. Eng.,* 49, 579.
15. Venturi V, et al., 2019, Experimental measurements on MLI performance from 20–60 K to 4.2 K, *IOP Conf. Series: Materials Science and Engineering*, 502, 012077 (ICEC/ICMC).
16. Funke Th, Haberstroh Ch, Performance measurements of multilayer insulation at variable cold temperature, *AIP Conference Proceedings*, 1434, p. 1279.
17. Fesmire F, 2019, Advanced thermal insulation materials and testing technology, Presented at A Joint Panel Meeting, July 17 2019, KSC, NASA.
18. Fesmire F, 2016, Below ambient and cryogenic thermal testing, *Presented at Exploration Research & Technology UB-R1*, KSC NASA, February 2, 2016.
19. Fesmire J, Bateman C, and Thomas J, 2020, Macroflash boiloff calorimetry instrument for the measurement of heat transmission through materials, *Adv. Cryo. Eng.,* 65, p. 012008.
20. Fesmire J, Coffman B, Meneghelli B and Heckle K, 2012, Spray-on foam insulations for launch vehicle cryogenic tanks, *Cryogenics*, DOI:10.1016, 2012.
21. Johnson W, Van Dresar N, Chato D, Demers J, 2017, Transmissivity testing of multilayer insulation at cryogenic temperatures, *Cryogenics*, 86, pp. 70–79.

22. Chato D, Johnson W, et al., 2016, Design and operation of a calorimeter for advanced multilayer insulation testing, DOI: 10.2514/6.2016-4775, 52nd AIAA/SAE/ASEE 2016.
23. Barrios M, and Van Sciver S, 2010, An apparatus to measure thermal conductivity of spray-on foam insulation, *AIP Conference Proceedings*, 1218, p. 938.
24. Shu QS, Fast R, and Hart H, 1988, Theory and technique of reducing the effect of cracks in multilayer insulation from room temperature to 77 K, *Adv. Cryo. Eng.*, 33.
25. Shu QS, Fast R, and Hart HL, 1987, Systematic study to reduce the effects of cracks multilayer insulation. Part 2: Experimental results, *Cryogenics*, 27, p. 298.
26. Shu QS, Fast R, and Hart H, 1988, Crack covering patch technique to reduce the heat flux from 77 k to 4.2 k through multilayer insulation, *Adv. Cryo. Eng.*, 33, pp. 299–304.
27. Johnson W, Kelly A, and Fesmire J, 2014, Thermal degradation of multilayer insulation due to the presence of penetrations, *Adv. Cryo. Eng.*, 59, p. 471.
28. Sass J, Fesmire J, Morris D, et al., 2008, Thermal performance comparison of glass microsphere and perlite insulation systems for liquid hydrogen storage tanks, *Adv. Cryo. Eng.*, 53, p. 1375.
29. Fesmire J, Morris D, et al., 2006, Vibration and thermal cycling effects on bulk-fill Insulation materials for cryogenic tanks, *Adv. Cryo. Eng.*, 51, pp. 1359–1366.
30. Johnson W, Ameen L, Koci F et al., 2017, Structural heat intercept, insulation, and vibration evaluation rig (SHIIVER), *Engineer NASA GRC*, July 6, 2017, Space Cryogenics Workshop.
31. Johnson W, Van Dresar N, et al., 2017, Transmissivity testing of multilayer insulation and cryogenic temperatures, *Cryogenics*, 86, pp. 70–79.
32. Johnson W, Oberg D, Frank D, Mistry V, and Koci F, 2020, Testing of SHIIVER MLI coupons for heat load predictions, *Adv. Cryo. Eng.*, 755(1), p. 012151.
33. Johnson W, Balasubramaniam R and Westra K, 2020, Testing of heat flux sensors at cryogenic temperatures, *Adv. Cryo. Eng.*, 65, p. 012074.
34. Johnson W and Chato D, 2020, Performance of MLI seams between 293 K and 20 K, *Adv. Cryo. Eng.*, 65, p. 2020.

9 Cryogenic Heat Switches for Thermal Management

Quan-Sheng Shu, Jonathan A. Demko, and James E. Fesmire

9.1 INTRODUCTION

Cryogenic heat switches (CHSs) are novel instruments with controlled variable heat conduction, working in a certain temperature region about from 50 mK up to 400 K. Heat switches can alternatively provide high thermal connection or ideal thermal isolation to the cold mass [1].

Various cryogenic detectors in space are normally cooled by a running cryocooler with a second redundant cryocooler as backup. Such configurations rely on one heat switch to provide a highly thermally conductive connection to the running cryocooler and another heat switch to thermally isolate the redundant cryocooler. For many scientific efforts in condensed matter physics, optical/laser, and radio frequency experiments, novel heat switches were developed to thermally connect and/or isolate the objects to the LHe bath or cryogen-free cooling stage.

Heat Switch Ratio. The heat switch ratio is a non-dimensional parameter used to compare the performance of heat switches and strongly depends on the properties of materials of construction. A simplified work principle of CHS is shown in Figure 9.1. With the switch closed, heat flows from the cooled object to the cryocooler at temperature T_c. For simplicity, the transition process is ignored so that the $\Delta T (D)$ is constant, and D is assumed to be much smaller than T_c. When the switch is opened, a degree of isolation between the cooled object and the redundant cooler is provided. The heat flow rate between the object and the two coolers is represented in Equation 9.1.

$$\dot{Q}1 = \int_{Tc}^{Tc+D} K(1)dT \quad \text{and} \quad \dot{Q}_2 = \int_{Tc+D}^{Th} K(0)dT \tag{9.1}$$

Common definitions of the heat switch ratio are given in Equation 9.2. One option (left) is the thermal conductance ratio based on the physical properties of the switch and allows evaluation of a potential switch prior to the development of the detailed design. The second option (right) is the ratio of heat flows representing the switch configuration, including various parasitic losses.

$$R_k = K(1)/K(0) \quad \text{or} \quad R_h = \dot{}_1/\dot{}_2 \tag{9.2}$$

Solid Heat Switches (SHSs). SHSs include three groups: (1) the working principle is based on the dramatic change of the thermal conductivity during its phase change; (2) the phase change of material will cause a significant change of its mechanical property, introducing the thermal contact or disconnect; and (3) the mechanical property of material changes with temperature to create a thermal contact on/off. Currently, superconductors, magneto-resistive materials, shape memory alloys, SC maglevs, differential thermal expansion, and bimetal and magneto-strokes are separately employed or used in combination for SHS development. The advantages of SHS are that it covers wide cryogenic T; is compact, robust, and cost efficient; its working parameters are adjustable; and it does not require a vacuum-tight enclosure for operation.

Gas Heat Switches (GHSs). Gas heat switches use gas to carry heat flows between surfaces that have different temperatures and also have on-off functions to control the heat flows. GHSs can

DOI: 10.1201/9781003098188-9

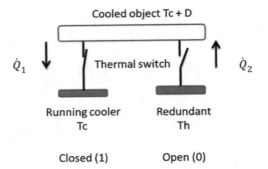

FIGURE 9.1 Simplified cryogenic heat switch arrangement: instrument being cooled and redundant cryo-cooler off.

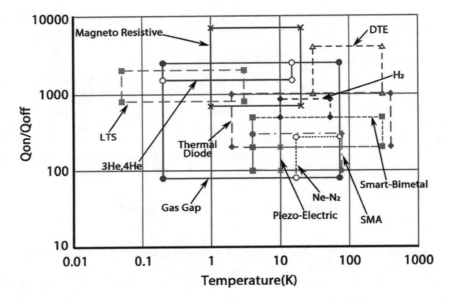

FIGURE 9.2 Idealized and simplified performance chart of cryogenic heat switches [1].

be categorized into two groups: (1) cryogenic thermal diode heat switches (CDHSs), which are based on heat pipes, and (2) cryogenic gas gap heat switches (GGHSs), which rely on conventional gas heat transfer through narrow gaps between two enlarged surfaces in different temperatures. Generally, the on and off functions of the switches are performed by a liquid trap with absorber materials to control the gas volume and pressure inside switch working spaces. With different working gases, GHSs can transport large heat flows at T from 200 mk up to 400 K. GHSs can be very strong and robust and last a lifetime.

Numerous CHSs have been developed and implemented, but some CHS technologies are still in their infancy. One CHS can work well in a particular application but may underperform or even fail in another situation. Shu et al. briefly summarize the performance and working ranges of various CHSs with a chart as presented in Figure 9.2.

9.2 SUPERCONDUCTING CRYOGENIC HEAT SWITCHES

9.2.1 THERMAL CONDUCTIVITY OF SUPERCONDUCTORS

Superconducting cryogenic heat switches (SCHSs) are based on the fact that the thermal conductivity, λ_n, of a superconductor in the normal state (induced by a magnetic field exceeding the critical field below the critical temperature) is much greater than its thermal conductivity, λ_s, in the superconducting state. The phase diagram of a superconductor in the H (magnetic field) and T (temperature) planes is shown in Figure 9.3 (T_c is critical temperature, and H_c is critical magnetic field). Bardeen–Cooper–Schrieffer theory indicates that the paired electrons (the so-called Cooper pair) are superfluid and do not carry entropy while moving in the superconductor. Therefore, Cooper pair electrons do not contribute to thermal conductivity. At a finite temperature, the ratio of densities $n_e(T)$ of the normal electrons to the densities $n_c(T)$ of superfluid Cooper pair electrons is presented in Equation 9.3.

$$n_e / n_c \approx e^{-\left[\frac{\Delta(T)}{KT}\right]} \tag{9.3}$$

$$e^{-\left[\frac{\Delta(T)}{KT}\right]} \qquad \text{Proportional to Boltzmann factor}$$

In pure metals, the thermal conduction in the normal state is always entirely electronic, and the lattice conduction term (phonon conductivity) can be neglected. As the metal becomes superconductive, its thermal conductivity, λ_s, falls below the value of the normal phase, λ_n. This decrease in thermal conductivity is caused by the gradual disappearance of the thermal distribution of the free electrons. Therefore, the ratio of the electronic thermal conductivity in the normal state to the phonon thermal conductivity in the superconducting state can be larger than 10^5. When the grain dimensions are about 0.1 cm, $\lambda_n/\lambda_s \approx 6T^{-2}$; for $T = 0.1$ K, then, λ_n/λ_s is about 600. For high-purity wires of Pb, In, and Sn, the ratio may be from $10T^{-2}$ to $500T^{-2}$, as indicated by Lounasma [2]. Table 9.1 presents the T_c and H_c of several superconductors as potential SCHS candidates.

9.2.2 DESIGN AND APPLICATION OF SCHSs

There are three general techniques used to achieve 50 mK cooling in research labs and in space: multi-stage adiabatic demagnetization refrigerators (ADRs), closed-cycle dilution refrigerators, and hybrid sorption-ADR coolers. In the last cooling stage for all refrigerators (around 200–50 mK), SCHSs are needed to manage the heat transport on or off. Based on the thermal properties of

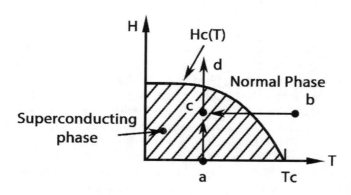

FIGURE 9.3 Phase diagram of temperature vs magnetic field for a superconductor.

TABLE 9.1

T_c and H_c of Several Superconductors in SCHSs

Material	T_c (K)	H_c (mT)
Zn	0.85	5.3
Al	1.2	10.5
In	3.4	29.3
Sn	3.7	30.9
Pb	7.2	80.3

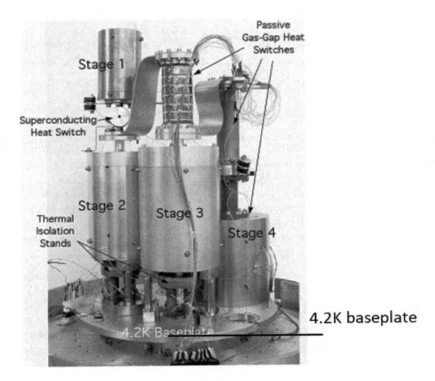

FIGURE 9.4 Four-stage continuous adiabatic demagnetization refrigerators (CADR) structurally and thermally connects to a mechanical cryocooler through a single interface plate [3].

superconductors, several SCHSs have been designed, tested, and implemented at these temperature ranges. The arrangement of one SCHS (50–200 mK) and three gas gap heat switches (200 mK–4.2 K) in a multistage ADR by Shirron et al. [2] are shown in Figure 9.4.

Krusius developed an SCHS with zinc foil for large heat flow below 50 mK [4]. Zinc was chosen because it is easy to solder to a copper base and has good thermal cycling. To obtain a large cross-sectional area of the switch and also to avoid magnetic flux trapping, the switch is thus composed of many parallel foils or wires. He used nine 0.17-mm-thick zinc foils, which were indium-soldered to copper end posts. The switch is operated by a small superconducting magnet with 65 mA to close the switch between a He_3–He_4 dilution refrigerator and an ADR in the precooling process. The heat switch has served in routine use for precooling a nuclear demagnetization refrigerator to below 18 mK. The typical thermal resistance was $740T^{-1}$ (mK)^2/μW. Schuberth designed and tested several

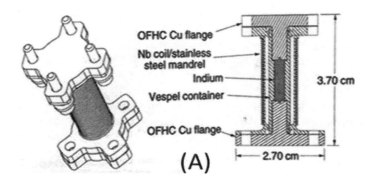

FIGURE 9.5 The SCHSused in a CADR for temperatures below 0.35 K [6].

SCHSs, which consisted of tin and indium wires pressed between machined silver rods and oriented parallel to the operating magnetic field [5]. As the design is so compact, they have a low heat capacity, low stray field of the switching magnet, and low eddy current heat due to the operating field. The area to length ratio in the design is higher. The indium wire is 1.0 mm in diameter with purity 99.99% and tin 0.5 mm, with 99.999%. The Ag rod is 6 mm with residual resistance ratio (RRR) on the order of 1,500. The magnet has 2,600 turns in 20 layers of 0.1-mm-diameter NbTi wire with 200–360 mA to provide fields higher than 30 mT.

More challenges in recent high-resolution detectors in space and laboratory applications are for targets cooled to below 50 mK with much smaller heat loads (10 µW) and using mechanical coolers (instead of dilution refrigerators) at a higher precooling base T (4 K–10 K). The success of the continuous ADR also depends on having suitable heat switches.

Shirron et al. [6] reported a prototype SCHS made from high-purity indium (99.99+%) with oxygen-free high thermal conductivity (OFHC) copper end pieces. The design is different from most previous SCHSs, as shown in Figure 9.5. The SCHS is expected to have an on/off ratio of about 2,000 at 50 mK and on-stage conductance of 8 mW/K and may transport 3.5 µW back to the continuous stage. In the continuous stage, the switch must efficiently transfer heat at a temperature difference of only 5–10 mK. The SCHS has a much higher on/off conductance ratio than an MR switch and was chosen for Stage 1. Metal In (99.99+%) with OFHC copper end pieces was used, while Sn and Al were design alternatives.

Dipirro and Kittel presented comprehensive reviews of SCHSs [7, 8]. One of the highest temperatures for the conductance ratio and off conductance to be applicable is about 0.5 K. Lead is the active element in the heat switch. This switch construction is very rugged and is able to completely support the cold ADR stage. The CADR requires the lowest temperature heat switch. The superconducting heat switch needs to be on at the lowest temperature and off at the highest temperature. Nb has the highest T_c of 9.1 K, but its type-II behavior trapping magnetic fields is not good for SCHSs.

9.3 MAGNETO-RESISTIVE HEAT SWITCHES

9.3.1 CHANGE OF THERMAL CONDUCTIVITY

At low T, electron thermal conductivity is a linear function, while phonon thermal conductivity drops off. Electronic heat conduction in compensated elemental metals (Ga, Cd, Be, Zn, Mo, and W) at low T can be suppressed so thoroughly by a several-Tesla magnetic field that the heat is effectively carried only by phonons. Therefore, magneto-resistive materials were chosen to build heat switches due to their high switching ratio (potentially 10,000 at 4 K), simplicity, reliability, short switching time, and tunable advantages in different magnetic fields [9, 10]. Magneto-resistive heat switches (MRHSs) have no moving parts and no enclosed fluid, as well as a wide T range. MRHS can be used for all four stages of ADR.

In 1-mm-diameter single crystal samples of Tungsten (W), the ratio of zero field to high-field thermal conductivity can exceed 10,000. Bartell indicates that the thermal conductivity of a single tungsten crystal (RRR about 20,000) may be calculated based on Equation (9.4), where the constants $b_0 = 0.0328$, $a1 = 1.19 \times 10^{-4}$, $a2 = 3.57 \times 10^{-6}$, $a3 = 1.36$, $a4 = 0.000968$, and $n = 1.7$. The calculated thermal conductivity of the MRHS has been validated by comparing the experimental results of the miniature ADR with modeled predictions. When turning the heat switch from on to off, the thermal conductivity is reduced by about 1,000 with 0.5 T.

$$K(T) = b_o T^2 + \cfrac{1}{\cfrac{a1 + a2T^2}{T} + \cfrac{B^n}{a3T + a4T^4}} \tag{9.4}$$

9.3.2 MRHS Development

Duval et al. [11] built and tested several MRHSs with tungsten crystal by vacuum brazing for thermally connecting to the ADR post. It has a very short switching time, depending only on the ramping speed, less than 30 s, of the magnet used. They were tested in the 2 K–10 K range, with magnetic fields up to 3 T. The main drawback of the standard MRHS is the mass and complexity of the magnet system for the controlling field to switch off the MRHS. Canavan et al. [9] present a technique of minimizing this mass and complexity by using the MRHS in continuous adiabatic demagnetization refrigerators (CADR). For any pair of stages in the CADR, the magnetic field of the higher-temperature stage is low when the switch needs to be closed, and it is high when the switch needs to be open. The switch would then act passively, opening and closing at the proper phase in the cycle without a separate control circuit.

The MRHS developed by Bartlet et al. for the milli-kelvin cryocooler (mKCC) is shown in Figure 9.6 [10]. A single W crystal (99.999%) is wire electrical discharge machining cut from a sample (12.5 mm in diameter and 32.3 mm in height [12 mm excluding mounting flanges]) shown in Figure 9.6 (Left). The cylinder (z) axis is aligned with the 001 direction. The other crystal directions are unknown, but conductivity in the xy plane should be nearly isotropic. Due to the slotted design, the MRHS has an effective 1.5-mm^2 cross-section and a free path length of 31 cm. The residual resistivity ratio has been measured and is about 32,000, and the resistivity measured 2.60×10^{-12} 'Ωm at 4.2 K and 8.35×10^{-8} 'Ωm at room T. The thermal performance of the MRHS was tested, and the results are very promising. However, as in the mKCC design, two of these miniature ADRs connected to a common cold stage via additional MR heat switches can be used to make a small tandem continuous ADR, as shown in Figure 9.6 (Right): one ADR recycles and is isolated from the cold stage, while the other is connected to the cold stage and provides cooling. A change in thermal conductivity of a factor of 1,000 can be achieved within 5 s.

9.4 SHAPE MEMORY ALLOY HEAT SWITCHES

9.4.1 Shape Memory Alloy

Shape memory alloy (SMA) can recover large strains (up to 8%) by undergoing a temperature-induced phase transformation [12]. This strain recovery can occur against large forces, resulting in their use as actuators. Although research into potential cryogenic heat switch applications of low-temperature SMA materials has been explored (such as cryogenic tanks, habitat walls, space radiators, etc.), the science and understanding of phenomena in the cryogenic realm is still in its initial stages.

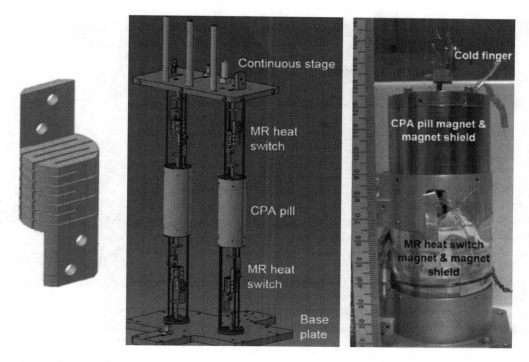

FIGURE 9.6   Left: Single-crystal tungsten MR heat switch for the mKCC. Center: The single milli-kelvin cryocooler (SMKCC) ADR insert. Right: Complete miniature ADR [10].

The two main types of shape-memory alloys are copper-aluminum-nickel and nickel-titanium (NiTi) alloys. NiTi-based SMAs are preferable for most applications due to their stability, practicability, and superior thermo-mechanic performance. There are two shape memory effects:

One-way shape memory. The alloy can be easily stretched using an external force. After removal of the force, the alloy shows permanent deformation. It can recover its original shape upon heating. Subsequent cooling does not change the shape unless it is stressed again.

Two way shape memory. The material remembers two different shapes: one at low temperatures and one at the high-temperature shape. Two-way shape memory lies in training. Training implies that a shape memory alloy can "learn" to behave in a certain way.

9.4.2 SMA Training for Cryogenic Applications

Training for Cryogenic Applications. Swanger et al. presented apparatus and methods for low-temperature training of SMAs [13, 14]. Most commercially available alloys exhibit actuation properties at temperatures above 0°C. The key target of SMA materials being developed by the NASA Glenn Research Center (GRC) is for providing two-way actuation at about 4 K to moderate 400 K. The SMA were fabricated from NiTi-based ternary alloys with elemental addition including Fe, Cr, and Co (1–5%). The mechanical training regimen was focused on the controlled movement of rectangular strips with S-bend configurations at around 30 K.

The training specimens are 40 mm long, 10 mm wide, and 1 mm thick, with an out-of-plane S-bend of 10 mm. With the transformation temperatures of some specimen alloys below 77 K,

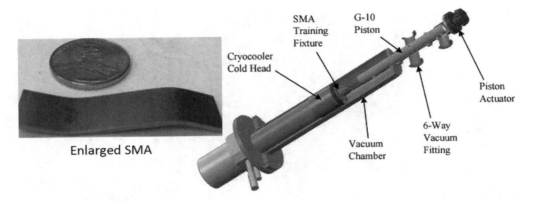

FIGURE 9.7 Left: S-shaped SMA specimen. Cut-away of low-temperature training of materials. Right: Hardware configuration [13].

the Cryomech AL230 cryocooler (25 W of at 20 K) was used to provide the cryogenic temperature. Procedurally, the specimens undergo high-temperature annealing in the S-shape, followed by numerous rounds of cool-down, inelastic deformation via compression until flat while cold, and then controlled warm-up unloaded in order to achieve two-way actuation in the desired direction.

Training Apparatus. An apparatus for low-T training of materials (ALTM) was designed, built, and tested at the Cryogenic Test Laboratory (CTL) of KSC NASA, as shown in Figure 9.7. With the test specimen and instrumentation in place, the cold head, holding fixture, and aluminum piston were wrapped with 20 layers of MLI. The vacuum chamber was then carefully slid down over the G10 cylinder. The test specimen and piston face were allowed to thermalize for approximately two hours. Compression was then applied using the actuator until the SMA specimen was pressed completely flat. The specimen was held in this position for one hour, after which the piston was raised and the system was allowed to warm to room temperature under vacuum. This process was repeated four additional times, producing a total of five training cycles per SMA specimen. Various S-shaped experimental SMAs were trained using the ALTM at temperatures as low as 30 K and resulted in successful two-way actuation in several cases.

9.4.3 DESIGN AND DEVELOPMENT OF SMAHS

The temperature can approximately vary between –233°C and 127°C during lunar day/night cycles. Benefan et al. introduced their development of a shape memory alloy-activated heat pipe-based thermal switch for cryogenic use in future Moon and Mars missions [15, 16]. The shape memory alloy heat switch (SMAHS) rejects heat from a cryogen tank into space during the night cycle while providing thermal isolation during the day cycle. The design of the actuation mechanism of the SMAHS was based on a biased configuration where the SMA actuators provide motion in one direction and biasing force springs in this case provide motion in the opposing direction. As shown in Figure 9.8, a custom $Ni_{47.1}Ti_{49.6}Fe_{3.3}$-based SMA (2.16-mm-diameter wires by New Hartford, New York) with a reversible transformation was used as the sensing and actuating elements, while thermomechanical actuation was accomplished through an antagonistic spring system, resulting in strokes up to 7 mm against bias forces of up to 45 N. The system thermal performance with a variable-length, closed two-phase heat pipe gave heat transfer rates of 13 W using pentane as working fluids (and 10 W using R-134ᵃ). The SMA exhibited a reversible phase transformation between a trigonal R-phase, and a cubic austenite phase was used as the sensing and actuating elements. The actuation system was tested for more than 30 cycles, equivalent to one year of operation on the Moon.

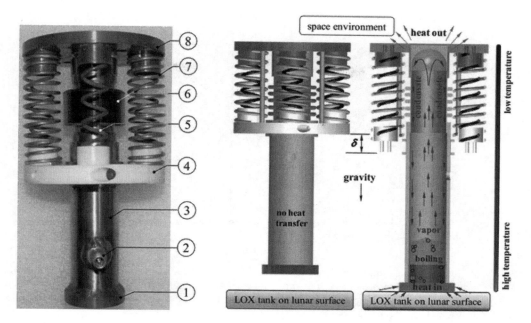

FIGURE 9.8 Left: Final thermal switch assembly. 1. Copper evaporator, 2. evacuation/charging valve, 3. heat pipe containment, 4. Teflon (moving) plate, 5. $Ni_{47.1}Ti_{49.6}Fe_{3.3}$ SMA actuators (×4), 6. pipe extension bellows, 7. stainless steel bias spring (×4) and 8. copper base and condenser. The bias spring in the front was removed to show the position of the SMA spring. Right: Principle of thermal switch operation: The open (or off) position during lunar day (hot) and the closed (or on) position during lunar night (cold) [16, 27].

During a lunar day cycle (~14 days) when temperatures are near 127°C, the SMA actuators are in their open or off position and exist in their high-temperature phase, where they overcome the force from the bias springs. The switch provides thermal isolation. During a lunar night cycle when temperatures are near −233°C, the temperature of the surrounding space environment drops, resulting in a phase transformation in the SMA springs. The phase transformation results in the R-phase, which is now less stiff and overcome by the bias springs. The resulting actuation (facilitated by bellows in the heat pipe) takes the switch to its closed or on position, wherein it is in thermal contact with the LO_2 tank. The final SMAHS assembly is shown in Figure 9.8 (Left), and the bias spring in the front was removed to show the position of the SMA spring.

The SMA springs, with a free length of L_o, are pre-loaded in the off configuration using the bias springs, yielding an initial deflection of $\delta_{initial}$. Once cooled, the SMA springs expand due to the stiffness change and reach a working deflection, δ_{work}, where the switch makes contact with the heat sink in the on configuration. The reverse motion is accomplished as the actuators are heated back to the initial conditions. Figure 9.9 presents some of test results of the stroke–temperature response and the heat transfer rate as a function of contact pressure.

9.5 MAGLEV-SMART BIMETAL HEAT SWITCHES

Shu et al. reported the development of an energy-efficient cryogenic transfer line of 6 m employing magnetic suspension (maglev) operated by a smart bimetal heat switch (SBMHS) [17, 18]. The prototype transfer line exhibited cryogen-saving potential of 30–35% in the suspension state as compared to its normal support state. Key technologies advanced include novel magnetic levitation using multiple-pole HTS with rare earth permanent magnets (REPMs) and the SBMHS as the

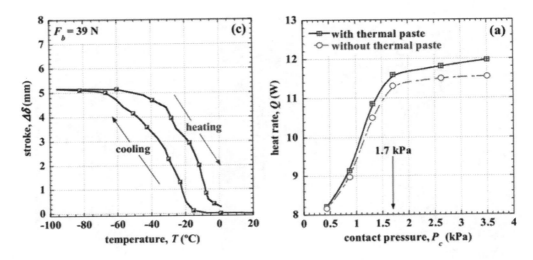

FIGURE 9.9 Left: Stroke–temperature response of the SMAHS under varying bias loads. F_b denotes the total bias force from the superposition of bias springs, bellows, and weight of the heat pipe. The maximum strokes obtained were 5.1 mm. Right: Heat transfer rate as a function of contact pressure at the interface between the evaporator and the heat source (right) [15].

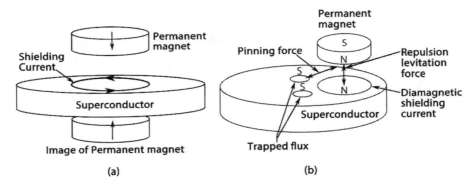

FIGURE 9.10 Levitation and stabilization mechanisms used in the maglev transfer line: A: diamagnetic response for levitation; B: flux pinning for stabilization.

passive warm support actuator. Although still in the prototype stage, it provides potential applications for zero boil-off (ZBO) cryogenic tanks, transfer lines, and SC magnetic bearings. The SBMHS can also be employed in space and other cryogenic areas.

9.5.1 MAGLEV WITH HIGH-TEMPERATURE SUPERCONDUCTOR-PM

The levitation as well as the horizontal–vertical stability provided by a high-temperature superconductor (HTS) and PM are illustrated in Figure 9.10. The levitation force develops because a shielding current forms immediately beneath the PM. The non-superconductive regions that exist in type II superconductors, referred to as pinning centers, trap the flux from the PM. This flux-pinning phenomenon creates horizontal stability of the levitation element. When the PM moves, its flux line will move with it, but the trapped flux lines do not move. The magnetization of these residual flux acts to attract the PM back to its starting position.

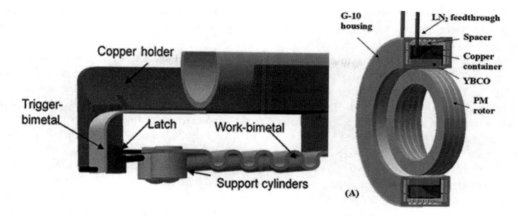

FIGURE 9.11 Right: Cylindrical multi-PM ring design. Left: Scheme of the bi-metal heat switch [20].

The relative positions and shapes of the HTS and PM can be designed to enhance the force and meet different configurations. Three performance indices are emphasized in all maglev configurations: sagging distance, final levitation gap, and levitation forces. The HTS is initially cooled at a certain cooling height and due to the gravity of inner line and cryogen; the HTS will approach the PM to accumulate enough levitation force to balance the loads. In a compact maglev transfer line design, 2–3 mm sagging travel is allowable, but larger sagging of the inner line may compress the superinsulation and/or cause solid contacts between the inner and outer vessels.

Although levitation forces between a PM and a HTS can be evaluated analytically [19], various configurations of HTS–PM combinations have been experimentally investigated and optimized.

The Multiple-Pole Levitation System. The multiple-pole levitation system, as illustrated in Chapter 3 was initially considered to support the inner tube by use of a few discrete HTS-PM blocks as four poles to form the support system. The multiple-pole design also allows for the freedom to alter the shapes and structures of the HTS and PM. For example, the HTS can be tiles with curvature or simply rectangular blocks. With one pole at a displacement of 2–3 mm, a levitation force of 20–40 N was easily achieved.

Cylindrical Multi-PM Rings. The multi-ring (desk) maglev system, as presented in Figure 9.11 (Right), has strong axial and radial holding forces [20]. Multi-PM rings with Fe shims to enhance the magnetic fields are surrounded by a cylindrical HTS ring inside a LN_2 cryostat. The PM ring-block is 200 mm OD, 15 kg weight, NdFeB-H33?, Br = 1.2 T, and its maglev force can reach a few thousand N.

9.5.2 SMART BIMETAL HEAT SWITCHES

The smart bimetal heat switch as a passive cryogenic actuator is introduced by Shu et al. and originated from developing a cryogenic transfer line with maglev. The main advantages of SBMHS are: (1) it is passively controlled by temperature variations, and (2) actuator parameters, such as switching T and net displacement, can be readjusted with ease [17, 21].

The unique SBMHS is a crucial technique issue in the maglev transfer line. The maglev system functions only when the line is cooled to below the T_c of the HTS. Therefore, a warm support system is needed to keep the tubes/pipes concentrically in a proper position, while the inner line is warm (without cryogen). Normally a bimetal actuator continuously changes following the change of adjacent temperature. The second critical tech issue is how to keep the bimetal actuator inactive when the transfer line and the bimetal actuator are being cooled before reaching the T_c of the HTS (from 300 K to above 77 K). The SBMHS should start withdrawn and disconnect mechanically from the outer vacuum jacket only when the HTS-PM functions. The SBMHS should not gradually change its shape as temperature changes. The retreat/resumption of support should be automatic, and a net displacement of nearly 7 mm is necessary.

Several prototypes for demonstrating the working principles of the SBMHS were built and tested. One assembled in the 6-m cryogenic pipe is shown in Figure 9.11 (Left). The 3D model shows SBMHS in the warm state: the main work-bimetal arm is stretch down and locked in the position by a second trigger-bimetal key in the latch. The upper photo shows the SBMHS in the cold state: the work-bimetal arm is withdrawn up and released from the previous position by the trigger-bimetal key out from the latch. The bimetal is made of Invar alloys, FeNi36, with a linear thermal expansion coefficient $<5 \times 10^{0}$/K as a passive component and copper ($> 15 \times 10^{-6}$/K) as active components. The copper in between the active and passive bimetal components provides excellent thermal properties necessary for low-temperature functions. The physical parameters of the bimetal are: specific thermal bending 24.6×10^{-6}/K and bend strength 150 N/mm². The cryogenic tests prove that the SBMSH is switched at 80 K while cooling down and regaining the warm state position successfully supporting 6 kg weight, as shown in Figure 9.12.

9.5.3 Design and Test of 6-m Cryogenic Transfer Line with Maglev and SBMHS

PM rings with ideal radial magnetization produce quite high levitation force, but the cost is quite high. To obtain strong magnetic excitation, the field intensities of various PM and iron combinations are analyzed using finite-element codes. It is found that to balance the cost, weight, and total length of the PM subassembly, the most powerful and economical magnetic design is given by the PM-Fe-PM design of "sandwiched" multi-poles, as shown in Figure 9.12. HTS can be tiles with curvature or simply rectangular blocks.

The HTS block is glued to the copper conductive holder, and the holder is brazed to the inner SS pipe. YBCO sizes are 65 mm length, 35 mm width, and 13 mm thickness. Testing proved that two-pole maglev units can meet the tech requirements of 40 N support force with <2 mm sagging of the inner pipe. In total, three maglev supports and three SBMHSs are evenly mounted at 1.0, 3.0, and 5.0 m along the 6-m inner pipe. The initial clearance of 7 mm is maintained between the multilayer insulation (MLI) layer and the inner wall of the outer pipe. The 6-m prototype maglev cryogenic transfer line was successfully tested several times at CTL, NASA KSC [22].

9.6 DIFFERENTIAL THERMAL EXPANSION HEAT SWITCHES

9.6.1 DTE-HS Working Principles

Differential thermal expansion-based heat switches (DTE-HSs) rely on the different coefficient of thermal expansion (CTE) of components. In the most common design, a small gap separates two solid components, one of which has a high CTE compared to the other. When the temperature decreases, the high CTE material "shrinks" and closes the gap between the two solids. Below a certain "switching temperature," the gap is fully closed, thus providing a heat conduction path [13–15, 23–25]. One DTE-HS disadvantage is that it requires a relatively small gap on the order of several μm. In order to enlarge the tiny gaps for ease of construction and operation, thermoplastics with a high CTE, such as polytetrafluoroethylene (PTFE) and ultra-high molecular weight polyethylene (UHMWPE), are used. While most metals have a CTE of 10 K–206/K, PTFE, for example, has an order of magnitude higher CTE compared to copper at room T.

A trade study is briefly summarized [25]: The desired properties of the thermally isolating support components are low thermal conductivity, large CTE at cryogenic temperatures, high elasticity and yield strength, low/zero creep susceptibility, and low density. The desired properties of the end pieces include high thermal conductivity, small CTE at cryogenic temperatures, low susceptibility to cold welding, ability to polish gap surfaces, and low density.

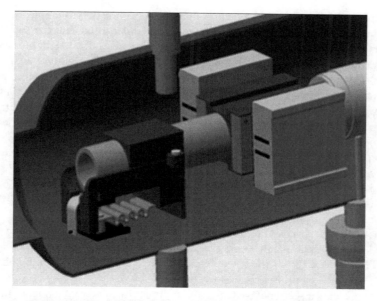

FIGURE 9.12 Upper: Scheme of a section with the maglev-PM block, a BMHS, and additional mechanical solid support (for comparison). Lower: Photo of testing BMSH [21, 22].

9.6.2 DESIGN AND TEST OF DTE-HS

Early Development. Marland et al. [26] and Bugby et al. [27] reported their work on DTE-HS, which was developed to examine components and potential applications. The first generation of DTE-HS was two beryllium end pieces separated by a thin-walled stainless steel tube. A narrow, flat gap is created between the end pieces. When the smaller end piece is cooled, the stainless steel tube shrinks, closing the flat gap until contact is made between the beryllium pieces. As the smaller end piece continues to cool, the switch is actuated on. In order to actuate the switch off (without warming the cryogenic component), a heater located halfway along the length of the stainless steel tube is used to expand the tube and separate the gap surfaces. Then came the so-called second generation: the polymer-aluminum DTE-HS has a larger gap, conductance ratio, and repeatable and time-invariant performance. The Ultem 1000 solid polymer rod, with a large CTE at cryogenic temperatures, was selected for the thermally isolating support tube/rod to replace the thin-walled tube. 99.999% aluminum, which has high thermal conductivity and low hardness, was elected for use in the end pieces. The temperature range of the warm side during off operation extended from >150 K (first generation) to >80 K (for cold-side temperatures 35 K). It has a larger gap dimension, 0.13 mm for the tested unit compared to 0.05 mm for the first generation. The weight reduced from 250 to 80 g and the length from 75 to 58 mm. Tested off: 700–2,300 K/W with 230 K for smaller end piece; 1,100–1,200 K/W with 300 K for smaller end piece. Tested on: 0.94 K/W at 34 K and 0.28 K/W at 60 K–90 K. Ratio: 2,000–5,000.

Later Development. Dietrich et al. developed two DTE-HSs and further improved a lightweight version [23, 28]. While most metals exhibit a coefficient of thermal expansion of $10^{-20} \times 10^{-6}$/K, PTFE has a CTE of 500×10^{-6} at room temperature. UHMW-PE has an even higher thermal contraction than PTFE between room temperature and 77 K.

Figure 9.13(a) presents a sectional drawing, and Figure 9.13(b) is a 3D model of the single cylindrical switch design. The part connected to the heat load (detector side) consists of an inner shaft made of a solid copper cylinder (10 mm diameter) with a flange on one end. The part connected to the cold head consists of a copper flange with four integrated copper jaws that are separated from the inner cylinder by the gap. A non-reinforced hollow UHMW-PE cylinder is put around the jaws to act as the high-CTE element. The two copper parts are held together by four thin stainless steel tubes (2 mm diameter by 150 μm wall thickness), which mainly determine the off-state thermal resistance. The total height of the switch is 30 mm. The contact pressure of the jaws to the shaft at 100 K was estimated to be 1.4 MPa, while the maximum tensile stress in the UHMW-PE was estimated to be 5 MPa. The switch showed a state change around 220 K and on/off-state conductivity of more than 1 W/K and 3 mW/K, respectively.

To reduce the mass and to increase the stability simultaneously for the intended cooling of infrared detectors in space, capillary tubes were replaced by a single thin-walled tube made of titanium alloy. To further reduce the thermal conductance in the off state, the shaft holder material was changed from copper to a low outgassing epoxy. To lower the total mass, the gap width had to be reduced from 80 to 70 μm. Measurements of the contact conductance were performed using

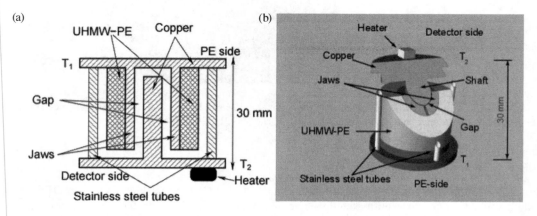

FIGURE 9.13   Design of DTE-HS. (a): Cross-section, (b): 3D model [23].

a pulse-tube cryocooler (model PT08) as shown. Finally, the mass of the new switch is only 25 g compared to 250 g for the previously realized switch.

9.7 PIEZO HEAT SWITCHES

9.7.1 PRINCIPLES OF PIEZO ACTUATORS

A simple oxide ceramic made from PZ material (Pb, Zr, Ti) is commonly used and shows much better piezo-electrical and piezo-mechanical efficiency than quartz and single crystals. Piezo stack actuators mainly act like an expanding element to generate a compressive force. The complete motion cycle is nearly proportional to the input voltage signal [29]: Achievable maximum strains of a piezo stack range between 0.1% and 0.15% of length. For example, a piezo stack (5×5 cm/20 cm) strokes 20 μm under unipolar operation 0–150 V. The maximum force generated depends on the stack's cross-section area. The force limits are in the range of 7–8 kN/cm².

Piezo actuator advantages include working with extended environment (vacuum, cryogenics, etc.), compact design, high mechanical power density even in miniature structures, power consumption only when motion is generated, and no stand-by/no sustainer energy consumption. Piezo actuators are widely used for precision positioning tasks. The motion is reversed due to discharge.

9.7.2 PZHS DESIGN AND TEST

The existing GGHSs and SCHSs have demonstrated the performance required at $T \le 1$ K. However, the hermetic joints required to contain the ³He exchange gas within the switch housing have shown vulnerability to the mechanical stresses associated with thermal cycling and vibration. The piezo heat switch (PZHS) is thus an attractive alternative technology, since this device has an essentially unlimited range of cryogenic operating T, is mechanically robust, and is free from hermetic sealing requirements [30].

When the piezoelectric positioner (an attocube, ANPz101) is positively energized, the lower plate moves upward until mechanical contact is established with the upper plate. After the desired heat transfer is complete, energizing the positioner with negative voltage moves it downward until the switch opens. To minimize the off conductance, the columns were constructed from G-10 hollow rods (L = 4.3 cm, OD = 6.3 mm, ID =3.2 mm), and the plates were mounted on Vespel SP1 insulators (7 mm thick) [30]. To maximize conductance, the upper and lower plates were made from ultra-high-purity (99.999%) copper with a contact area of 1.45×1.45 cm. The plate surfaces must be as flat and as parallel as possible. Once the surfaces were immaculate, the switch plates were cleaned and gold plated with ~1 μm of Au. The plating prevents tarnishing of the copper and also acts as a "cushion" to the switch surfaces, further enhancing the effective contact area.

During testing, the PZHS was connected to a cryocooler with a base temperature of ~3 K. All closed switch data were obtained with the positioner applying its maximum specified force of 8 N. Thermal conductance of the PZHS was measured between 4 K and 10 K. The PZHS on-conductance is 2.8 mW/K at 4 K, and on/off conductance ratios were ~100 at the lowest temperature of 4 K and ~200 at the highest measured temperature of 10 K with the positioner applying its maximum force of 8 N. The electromechanical behavior of multilayer piezo-actuators for fuel injector applications at cryogenic T and a cryogenic heat switch with stepper motor actuator were reported in [31, 32].

9.8 CRYOGENIC HEAT PIPES

9.8.1 CRYOGENIC LOOP HEAT PIPES

Heat pipes (HPs) only allow heat to flow from the evaporator to the condenser. HPs are able to provide heat transport efficiency over 100 times that of copper and consequently allow minimal T

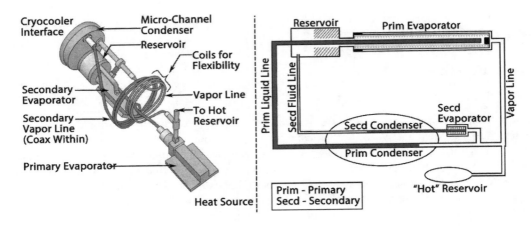

FIGURE 9.14   Right: Schematic of an advanced CLHP. Left: Miniaturized CLHP, Bugby [37].

drops for heat transport across significant distances with less weight. The flexibility of the heat pipe allows for improving integration and assembly in space orbit and pointing transport [33–35, 36]. The most common Cu-water HP is for electronics cooling. Below 100 K, the pairs include stainless steel envelopes with a working fluid of nitrogen, oxygen, neon, hydrogen, or helium.

The cryogenic loop heat pipe (CLHP) has an envelope with two pipes, wicks, and a working fluid. The conductive evaporate end is thermally connected to instrument to be cooled, and the conductive condenser end is thermally connected to the cryocooler or cryogen container. A schematic diagram of the advanced CLHP is shown in Figure 9.14 (Right). Standard heat pipes are constant conductance devices; the temperature drops linearly as the power or condenser temperature is reduced. For some applications, such as satellites, the electronics will be overcooled at low powers or at the low sink temperatures. Variable conductance heat pipes (VCHPs) are used to passively maintain the temperature of the electronics being cooled as power and sink conditions change.

Bugby et al. developed several miniaturized CLHPs with short and long transport lengths [37]. Both CLHPs utilized neon as the working fluid and operated in the temperature range of 30 K–40 K. The detailed structure of the short transport length neon-charged CLHP is presented in Figure 9.14 (Left). Preliminary test results show that the short transport length CLHP is able to transport a heat load of 0.1–2.5W applied to the primary evaporator. The on conductance was about 1 W/K, and off resistance was 1,400 W/K. The long transport 3-m-length CLHP is able to transport a heat load of 0.1–0.8 W applied to the primary evaporator. The on conductance is about 0.7 W/K, and the off resistance is estimated to be over 5,000 K/W.

9.8.2 Pulsating Heat Pipes

Pfotenhauer et al. [36, 38] and Natsume et al. [39] introduced several pulsating heat pipes (PHPs), which transfer heat by two phase flow mechanisms through closed loop tubing and have the advantage that no electrical pumps are needed to drive the fluid flow. For studying superconductivity material MgB2 around 30 K, a hydrogen PHP is made up of a long capillary pipe without a wick that is bent into many turns. The experimental setup includes a cryocooler, pulsating heat pipe element, heater, and thermometer, as shown in Figure 9.15A–B. An appropriate size for the capillary pipe is required in order to form the alternating distribution of vapor slugs and liquid plugs, which are the basic requirement for the oscillation movement. The selected value for the inner diameter of the capillary pipe is 2.3 mm. When the filling ratio is 35% and the heating power is 5 W, this PHP provides its highest effective thermal conductivity of 18.7 kW/m·K, and the temperature difference between condenser and

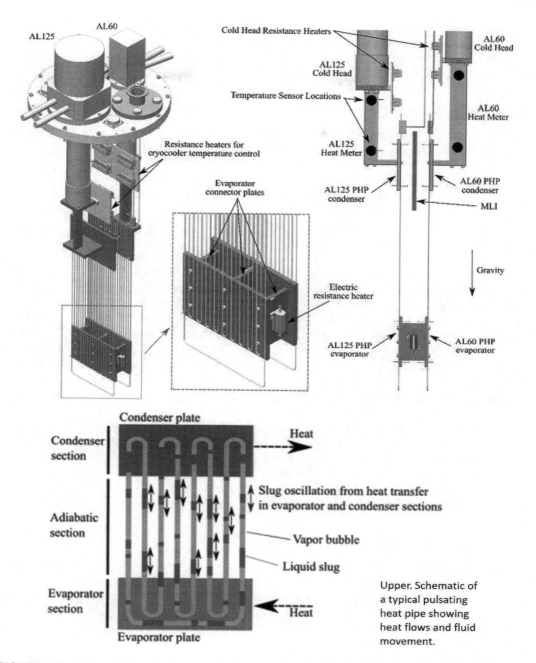

FIGURE 9.15 Lower: Schematic of a typical pulsating heat pipe. Upper: Experiment facility CAD assembly in an isometric view (left) and profile view (right). Temperature sensor location with black circles [36].

evaporator is 1 K. Another pulsating heat pipe operating at temperatures ranging from 77 K to 80 K and using nitrogen as the working fluid is also reported [38]. The experimental data show that the heat transfer between the evaporator and condenser sections can produce an effective thermal conductivity up to 35 kW/m-K at a 3.5 W heat load.

A proof-of-concept, two-condenser nitrogen pulsating heat pipe for use as a passive thermal switch in applications with redundant cryocooler installations was developed [36]. Two independent

PHP condensers were each thermally linked to an independent cryocooler cold head, with the two associated PHP evaporators attached to a common cold plate heat load. This is accomplished by the thermal isolation provided by dryout in the PHP associated with a non-operating cryocooler. Mechanical actuation is not required, as the PHP is fully passive.

9.8.3 SPACECRAFT APPLICATIONS OF CHPs

The spacecraft thermal control system has the function to keep all components on the spacecraft within their acceptable temperature range. This is complicated by the following factors:

1. Micro-g environment
2. Thermal radiation is the only approach to remove heat from the spacecraft
3. Widely varying external conditions
4. Long lifetimes, with no possibility of maintenance
5. Electrical power availability is limited

Some spacecraft are designed to last for 20 years, so heat transport without electrical power or moving parts is desirable. Large radiator panes (multiple m²) are required to reject the heat. CHP and CLHP pipes are used extensively in spacecraft, since they do not require any power to operate near isothermally and can transport heat over long distances. In the satellite Orbiting Carbon Observatory-2, Nakornpanom et al. introduced a cryocooler with beginning-of-life (BOL) heat load of approximately 3 W at 110 K. With the thermal mechanical unit (TMU) cold block controlled to 110 K, the cryogenic subsystem (CSS) maintains the three focal plane arrays (FPAs) near the desired temperature of 120 K. The waste heat from the cryocooler is removed by a heat rejection system (HRS) that utilizes two variable conductance heat pipes and a space viewing radiator. Grooved wicks are used in spacecraft instead of the screen or sintered wicks used for terrestrial heat pipes, since the heat pipes don't have to operate against gravity in space. This allows spacecraft heat pipes to be several meters long, in contrast to the roughly 25-cm maximum length for a water heat pipe operating on Earth. The condenser sections of the VCHPs are bonded and fastened to the radiator, and the evaporator sections are attached to an aluminum thermal plane, which interfaces to the cryocooler. A closed-loop heater controller, with the heater located at the reservoir and a temperature sensor at the evaporator, control the cryocooler heat rejection temperature to within ±0.3 K.

9.9 CRYOGENIC DIODE HEAT SWITCHES

When a heat pipe/thermal diode is modified by using small-diameter stainless steel tubing to connect it to a liquid trap (LT) cooled by a small secondary cooler with on-off switch function, the device is called a cryogenic diode heat switch. Normally the secondary cooler is thermally isolated from the primary cooler. During normal operation, a small heater keeps the LT filled only with vapor. To effectively turn off the heat pipe, the primary cooler is heated, and the small heater is ramped down; then the LT captures the working fluid.

Brandon [40] and Cepeda-Rizo/Bugby [36, 41] have developed several CDHSs separately. The two devices employ a methane heat pipe with a liquid trap for on-off actuation. It allows heat to flow in only one direction (forward mode) to thermally manage two CCD cameras on the NASA/JPL SIM Lite telescope, as presented in Figure 9.16 (A CCD camera contains a charged-coupled device (CCD), which is a transistorized light sensor on an integrated circuit). The LT is positioned on the condenser end and has its own cooling source. During normal operation, the small LT heater keeps the LT warm enough so that it is filled only with vapor while the heat pipe is on. To turn off the heat pipe, the LT heater is turned off, and all working fluid migrates to the LT. With the heat pipe in the off condition, only a small amount of heater power is required on the evaporator end to achieve a significant temperature rise for decontamination. The heat pipe can be turned back on by simply

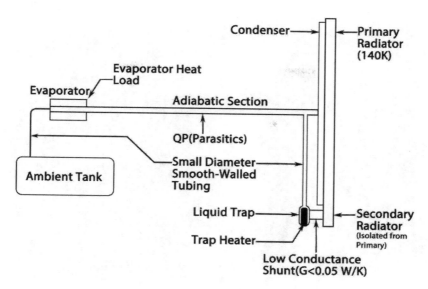

FIGURE 9.16 Schematic of the CDHS with methane and a liquid trap. An ambient tank is used to maintain the pressure of the heat pipe at safe levels [41].

repowering the LT heater. At a hot-side temperature of 150 K, the heat load is 6–12 W. The cold-side cryo-radiator is at 140 K with a transport length of 1.4 m. Periodically, the hot side is heated to 293 K with minimal heater power for decontamination. The CDHS can also work for cryocooler redundancy applications. A suitable working gas should be chosen for each particular application depending on the operational temperature extremes. The CDHS can be used with cooling systems with redundant cryocoolers or multiple cryogenic radiators. This heat pipe technology is highly suited for the decontamination of cryogenic focal planes with minimal heater power required.

Glaister developed a flexible CDHS under micro-gravity conditions for space flight. The heat pipe working fluids were oxygen with an operating temperature range of about 60 K–145 K and methane about 95 K–175 K. The composite wick (with fine screen mesh surrounding a coarse wick) provides over five times more transport capacity than a homogeneous wick once the pipe is primed. To provide the switching function, a LT is assembled with a stainless steel wire cloth (200 mesh)-wrapped cylindrical core inside. The on/off thermal resistance is about 0.05 K/W and >1,000 K/W at 85 K–120 K and 0 gravity. The cooling power is about 5 W at 100 K. Na-Nakornpanom et al. [42] introduced the heat reject system used in flight for the primary science objective of the Orbiting Carbon Observatory-2 (OCO-2) to collect space atmospheric carbon dioxide, as shown in Figure 9.17. The waste heat from the cryocooler is removed by means of a heat rejection system that utilizes two variable conductance heat pipes and a space viewing radiator, as illustrated in Figure 9.17. The condenser sections of the VCHPs are bonded and fastened to the radiator, and the evaporator sections are attached to an aluminum thermal plane, which interfaces to the cryocooler. A closed-loop heater controller, with the heater located at the reservoir and a temperature sensor at the evaporator, controls the cryocooler heat rejection temperature to within ±0.3 K.

Dipirro et al. developed a compact, high-conductance CDHS switch using liquid helium and wick material. As a heat pipe, its operation is similar to that of capillary pumped loop heat pipes. The device used an attaching getter that can be heated. The helium desorbed from the getter while heating to about 35 K by a small power (<3 mW) with the switch on. The on state conductance is over 400 mW/K at 1.8 K. Unheated, the getter quickly cools (to about 24 K) and absorbs enough helium so that the switch turns off. The off conductance is 3.5 μW/K at 1.5 K. This device is designed to be used in adiabatic demagnetization refrigerators.

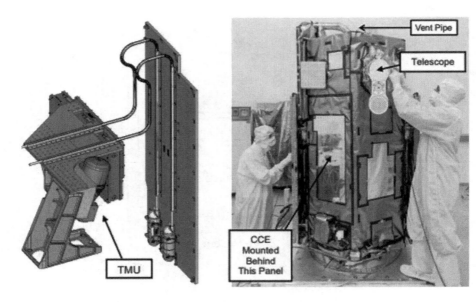

FIGURE 9.17   Left: Cryocooler heat rejection system. Right: Orbiting carbon observatory-2 (OCO-2) [42].

9.10   CONCEPT OF GAS GAP HEAT SWITCHES

The gas gap heat switch has been widely and effectively implemented for thermal management over large temperature ranges. These devices rely upon adding or removing gas from the interior of the hermetical switch body to thermally link or unlink portions of the switch. The gas characteristics and adsorption properties must be taken into account to determine a functioning temperature range. For example, below about 0.2 K, GGHSs are not usable since the saturated vapor pressure of even He^3 is too low to provide much conduction.

Inside the hermetically sealed shell are two conductive fins (or other shapes) that are attached respectively to one cold end or the warm end, and separated by narrow gaps. If gas is removed from the switch interior by cold getter, the switch is off. The heat leak from one end of the switch to the other is dictated by the conductance of the shell. When gas is refilled in by heating the getter, the switch is on and heat flows through the gas between the fins. In the on state, the switch must have a large surface area and a small gap between the warm and cold surfaces. Usually Cu is arranged either in concentric cylinders or interleaved fingers, separated by a gap limited by manufacturability or robustness to vibration. Shell materials must have a low permeability to helium or other working gas.

In general, the larger the surface area (gas contacted) and the thinner the gap between warm and cold surfaces, the higher the heat transport obtained. However, there are several important factors that limit choices of design and performance of heat switches.

Mechanical design improvements can lead to optimized on or off characteristics of a GGHS. Their on conductance is mainly determined by the gas properties and the gap geometry. However, their operational T range is limited by the gas-sorbent pair adsorption characteristics. Traditionally the gas chosen is helium, since it has the best thermal conductivity below 100 K, and the sorbent used is activated charcoal. Such a switch is limited to be used at cold-end temperatures below ~15 K. In order to obtain a customizable device working in the whole range below 100 K, Martins et al. extensively characterized hydrogen, neon, and nitrogen under different sorption conditions [43]. T_{cold} is the cold finger temperature as used, K_{on} W/m-K, on/off is the ratio of conductance between both states, and T_{off} stands for the sorb temperature required to produce an off state.

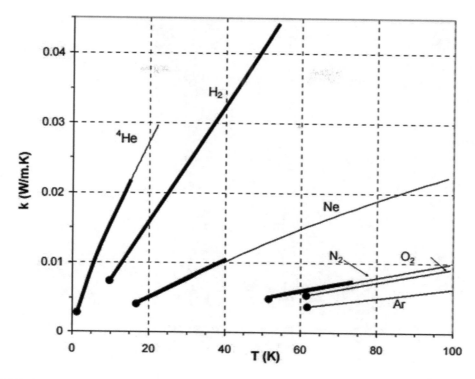

FIGURE 9.18 Conductivities of gases at low temperatures superimposed with a bold line that represents the range of temperatures required for a heat sink to the sorb. Source: *http://webbook.nist.gov/chemistry/fluid/*

H_2 extends the temperature range, making it possible to use the switch from 8 K to 53 K with very good conductance. Ne is a gas suitable for a shorter range of T (17 K to 40 K) for its lower actuating on-T. N_2 may be an interesting gas for extending the operating T range up to 73 K, but the very high sorb temperature for reaching the on ($T_{ON} > 100$ K) is a possible drawback. For lower temperatures, He_3 is used instead of He_4. In Figure 9.18. thermal conductivities of gases at low T are superimposed as a bold line that represents the range of T required for a heat sink to the sorb. Lower points correspond to a condensation point at the minimum pressure that still produces a viscous regime.

9.11 H_2, N_e, AND N_2 GGHSS

GGHSs using H_2 and Ne as heat exchange gases were developed over time [45, 46]. As presented in Figure 9.19 by Catarino, the GGHS consists of two cylindrical copper blocks (the hot and cold blocks) with a 100 µm gap between them. Those blocks are kept in place by a thin stainless steel supporting shell (SSSS) of 100 µm thickness. Such a shell also ensures a closed volume for the gas to be kept. A cryopump (or sorb) weakly thermalized to the cold block is the actuator for the switching. Such a sorb pump uses 30 mg of activated charcoal. Cooling down the sorb to a low enough temperature leads to gas adsorption, and then no thermal conduction exists between the copper blocks: the off state is reached. A gas gap heat switch in the off state is still, to some extent, a conducting device. The cylindrical supporting shell dominates this heat conduction. On the contrary, heating up the sorb to a certain temperature switches to the on state. Those temperatures are dependent on the gap width, the gas properties, the amount of gas in the switch, and the amount of sorb (and also its properties).

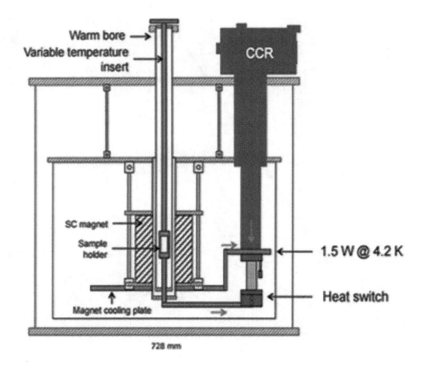

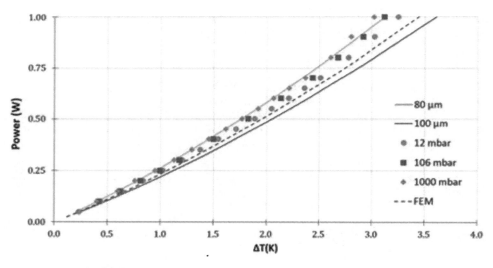

FIGURE 9.19 Upper: GGHS with GMC: GGHS thermally coupled to the GMC and to the VTI through a thermal link [50]. Lower: Heat flow power transferred through the switch vs ΔT in the on state with helium for different charge pressures [50].

The role of the gas in resistance is dependent on its pressure. Actually, as the pressure reaches a high enough value to produce a viscous regime in the gap, the gas conductance becomes essentially pressure independent. When this viscous regime is reached, a GGHS is considered to be in its on state.

For neon, the minimum temperature to actuate the switch ranges from 17 K to 40 K; for hydrogen, it's from 9.5 K up to 55 K. In order to easily compare the conductance with the various gases, a linear fit was adjusted for small temperature differences. The measured values for the thermal on conductance were 110 mW/K at 11 K for H_2, 74 mW/K at 20 K for Ne, and 85 mW/K at 70 K for N_2. Measured thermal off resistances were 4,000 K/W at 11 K for H_2, 3,000 K/W at 20 K for Ne, and 1,150 K/W at 70 K for N_2. Radiation contributes less than 2% of the off conductance at 70 K for temperature differences up to 15 K. An on/off conductance ratio of about 440 is obtained for the H_2 GGHS at 11 K; subsequently, a 220 ratio was measured for Ne at 20 K and a 100 ratio for N_2 at 70 K.

9.12 ⁴HE AND ³HE HEAT SWITCHES

GGHSs using ⁴He and ³He have also been developed. One distinctive application for He GGHSs is in magneto-instruments (optical or electrical) with a variable temperature sample cooled by cryogen-free cryocooler or by a liquid helium bath. Reported by Berryhill, the GGHS allows the sample temperature to be varied from 4 K to 300 K while maintaining the magnet at 4.2 K [47]. Kimball and Shirron presented a GGHS with low activation power and quick-switching time for other low-temperature applications for soft X-ray spectrometer instruments and others [48, 49].

9.12.1 GGHSs for Cryogen-Free Magnet Systems

A helium GGHS is well able to work with a cryogen-free cryocooler system as well as with an LHe cooling system. Discussion here will focus on cryogen-free systems since the design and methodology can logically adapt to LHe system with ease.

Helium GGHS with Pulse Tube Cryocooler (PTC). Berryhill et al. introduced a C-Mag Optical magneto-optic instrument, which is a versatile material characterization system [47]. It allows the researcher to simultaneously control the applied magnetic field and *T* of a sample while studying its electrical and optic properties. The GGHS allows the sample temperature to be varied from 10 K to 300 K while maintaining the magnet at 4.2 K or below. The system consisted of a GGHS, a 6-T (Tesla) split pair magnet, a 0.7-W Cryomech cryocooler, an SI model 9700 T controller, a TM-600 *T* monitor, and a cryostat with integrated heat switch and KRS-5 optical windows.

The heat switch design was determined by the allowable space in the bore of the magnet as well as the distance from the 4.2 K cold plate of the cryocooler to the magnet centerline. The sample space was approximately 13 mm, and the overall length of the switch from sample end to 4.2 K cold plate was slightly more than 127 mm. Internal heat exchange surfaces were gold-plated OFHC copper, and the outer shell was stainless steel. A 3-mm-diameter SS tube extended through the rear of the switch and connected to an activated charcoal getter. Small copper wires were epoxied onto the getter back to the 4.2 K cold plate for cooling. The sample tip was non-inductively wound with a 10-ohm resistor for control of the sample *T*. Two temperature sensors were used to monitor the sample tip temperature.

With 100 mW into the getter, the sample end *T* was approximately 6 K, and increasing power to 250 mW, the sample end *T* was reduced to 4.5 K. Once the getter was turned off and the temperatures allowed to stabilize, the sample end temperature was approximately 35 K, and the upper temperature limit of the sample end was 300 K with the sample heater powered. The overlapping of the temperature ranges provided sufficient room for control of the getter power for a continuous ramp from lowest to highest obtainable temperature.

Helium GGHS with G-M Cryocooler (GMC). Barreto et al. developed several GGHSs for cryogen-free magnet systems, which can also be easily modified for various applications with liquid helium cryostats [50]. The idea relies on a parallel cooling path to a variable temperature insert (VTI) of a magnetic property measurement system. A Gifford-McMahon cryocooler (1.5 W @ 4.2 K) would serve primarily as the cold source of the superconducting magnet, dedicating 1 W to this cooling, under quite conservative safety factors. The remaining cooling power (0.5 W) is to be diverted toward a VTI through a controlled GGHS that was designed and built with an 80-μm gap width. Cryogen-free superconducting magnet systems (CFMSs) including a VTI along with a CCR

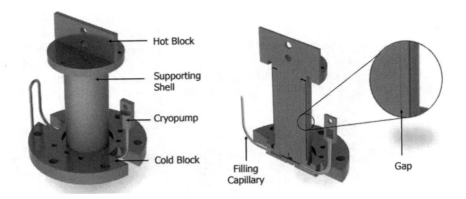

FIGURE 9.20   Section drawing and photo of entire assembly of the Ne GGHS prototype [50].

are now commercially available. The inclusion of a gas gap heat switch would eliminate the need for a helium open circuit. The superconducting coil is cooled with a CCR, and the same cryocooler can be used in parallel to cool the sample holder, as shown in Figure 9.19. The GGHS height is 8.5 cm, with a gas volume of 3.2 cm³.

The VTI must operate from 4 K up to room temperature, drastically reducing the conductance from the VTI to the CCR; that is, turning such a switch off will allow the VTI to be opened to room temperature while the superconducting magnet is still working. The necessary condition to avoid quenching is to provide 1 W at 4 K under conservative safety factors. The remaining cooling power (0.5 W) is aimed to be diverted toward the heat switch, which is thermally coupled to the sample holder through a thermal link.

The entire assembly of the GGHS prototype is illustrated at Figure 9.20. At gap width of 80 μm of the GGHS, an on thermal conductance of 290 mW/K was measured with helium and an off state of 90 μW/K, resulting in an on/off ratio of 3.2×10^3 at 4 K. The experimental results are in close agreement with both a simple thermal model and with the finite-element method. The on/off heat transport is a function of the T difference. The upper covers present heat flow power through the switch vs ΔT in the on state with helium for different charge pressures. Lines represent the analytic curves obtained with our thermal model, dots are for experimental results, and dashed lines the numeric approximation using the finite-element method (FEM) for a gap width of 100 μm.

9.12.2   GGHSs BELOW 4 K

Thermal noise (Johnson noise) is the random white noise generated by thermal agitation of electrons in a conductor or electronic device. The minimum signal to be detected by the device is limited by the noise level. The noise power is proportional to the absolute temperature of the conductor.

$$P / B = K_b T \tag{9.5}$$

where P is the power (in watts), B is the total bandwidth (Hz) over which the noise power is measured, K_b is the Boltzmann constant (1.381×10^{-23} J/K, joules per Kelvin), and T is the noise temperature (K).

In order to read out the targeted signals from the noise, more astronomic detectors work at lower cryogenic T. For instance, for X-ray astrophysics detectors and SpicA far-infrared instruments, the signal energy level is very low. Therefore, operating the instrument below 100 mK and using a highly sensitive sensor becomes essential. Multistage adiabatic demagnetization refrigerators and

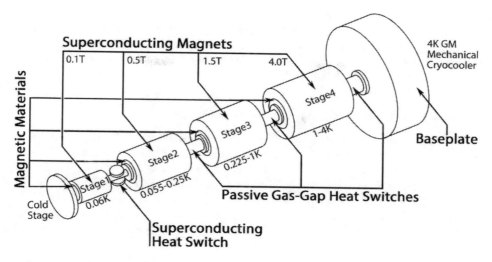

FIGURE 9.21 Simplified concept of a four-stage CADR with four SC magnets and four heat switches.

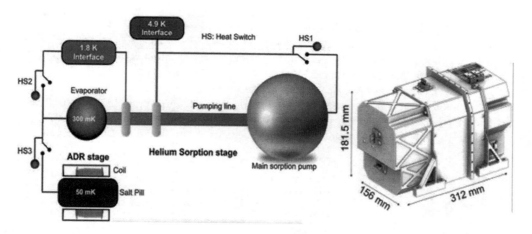

FIGURE 9.22 Safari hybrid cooler thermal design showing two GGHSs and one SCHS. The entire dimensions of the hybrid cooler are very compact [51].

hybrid ADR coolers (hybrid sorption-ADRs or dilution refrigerator-ADRs) have been successfully developed and implemented for various space flight applications [51–53].

Because the ADR uses a cyclic process to obtain cooling, alternately magnetizing and demagnetizing a salt pill at high and low temperatures, efficient heat switches are needed. In practice, heat switches are located at several different locations between different cooling stages of the ADR and hybrid ADR. A four-stage ADR normally uses three GGHS and one SCHS, as shown in Figure 9.21. There are two GGHSs and one SCHS utilized for the 50-mK hybrid sorption-ADR cooled for SPace Infrared telescope for Cosmology and Astrophysics—SpicA FAR-infrared Instrument, shown in Figure 9.22 [51]. The entire dimensions of the hybrid cooler are very compact.

Dipirro and Shirron systematically summarized the design and application of heat switches for ADRs [53]. It is mentioned that switching ratio (the relative on and off conductance) of a heat switch is a commonly used metric, but this is usually inadequate to capture the different temperature boundaries in the on and off states. Rather, it is important to focus on the average heat flow or, even better, the average entropy flow in the on and off states of the heat switch. Considerations in the design of heat switches for ADRs include size, mass, structural integrity, actuation method, reliability, speed of actuation, and on/off conduction ratio (including all associated parasites). As a practical matter, it is also very useful if the switch is very conductive at higher temperatures in order to speed cooldown. For ADRs operating below 10 K, the gases of choice are usually ^{3}He or ^{4}He.

9.12.3   Low-Power, Fast-Response Active GGHSs below 4 K

Low Power, Fast-response Active GGHSs. Kimball et al. developed a low-activation power and quick-switching time active GGHS for low-temperature applications [48–49]. Heat switches tend to be designed as close as possible to the limits of material strength and machinability, using materials that have the lowest thermal conductivity to strength ratio. In addition, switching speed is important for many applications, and many designs and switch types require a compromise between the power used for actuation and on/off transition times. The soft X-ray spectrometer instrument developed for the Japanese Astro-H mission requires less than 0.5 mW of power to operate, has on/off transition times of <1 minute, and achieves a conductance of > 50 mW/K at 1 K with a heat leak of <0.5 μW from 1 K to very low temperature. Details of the design and performance are presented in Figure 9.23A–B, which shows a cut-away cross-section of the GGHS with enlarged views of the orbital-welded joints that allow the reentrant titanium shell. The tapered copper fins are connected to either end of the heat switch shell but are separate from one another by a gap of 0.36 mm. The getter material is bituminous charcoal and is located at the top of the switch.

Helium GGHSs are robust enough to survive the brutality of a rocket launch and maintain the necessary on and off conductance to meet the requirements of a cooling system designed to hold a detector array at 0.050 K for more than 24 hours yet recycle within a one-hour time frame.

GGHSs with ^{3}He in Ultra-Low T Systems. Several portable, cryogen-free ultra-low-temperature cooling systems using a continuous ADR were introduced by Dipirro et al. [53, 54] and Hepburn et al. [55]. The system can continuously cool to 50 mK with a cooling power of about 20 μW at 100 mK. The GGHS with ^{3}He is a crucial device to manage the heat flow in the system, as presented

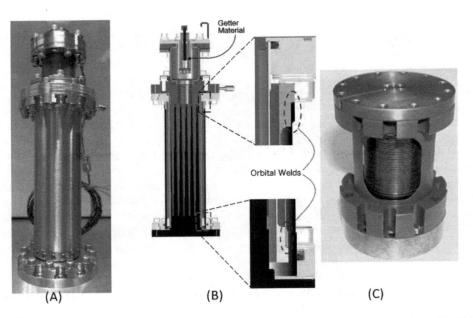

FIGURE 9.23   A: Complete GGHS ready to be integrated into the ADR used to cool the Astro-H soft X-ray spectrometer detector array. B: Cut-away view [48]. C: The ^{3}He GGHS with a bellow shell [53].

in Figure 9.23C. Its shell is made of bellows sealed with external Vespel support and 38-mm outer flanges. The bellows extend the thermal path and reduce the thermal conductance through the shell. GGHS shells have been made from polymers (for instance, Vespel) or composites which are lined or overlapped with a low-conductance, defect-free metal foil and bonded in place with epoxy. The design life time is five years, and total loss is no more than 25% of the charged gas. It requires a leak/permeation rate of $<5 \times 10^{-9}$ standard cm^3/s. A comparable switch of Vespel without foil was over three orders of magnitude worse than required. However, ^{4}He can form a superfluid film trapped between the foil and the polymer shell. Alternatively, ^{3}He was successful in solving this problem on the X-Ray Spectrometer project for imagine in universe. As an alternative to polymers and composites, all-metal shells of titanium alloy Ti (15-3-3-3) are used for gas containment. To speed the pumping and reduce the heat from the getter, a heat sink is added midway down the pumping line. The amount of gas and getter material in a GGHS is balanced to allow a turn-on T that is not too high as well as a turn-off T that is high enough to be quickly reached when the getter heater is turned off.

Titanium Reentrant Shell. To fit into the pre-determined space allocated for the heat switches, the thermal length of the shell must fit into the spatial length of the composite-tube heat switches. Thus, for two of the four switches needed in the ADR, a reentrant design was developed that nested three tubes into the equivalent length of a single tube. The metal chosen for the tubes needed to have the lowest thermal conductivity possible. Therefore, Ti (15-3-3-3) was used, as it has one of the lowest thermal conductivities for a metal at temperatures below 10 K.

Two other heat switches are used in the ADR assembly to connect a third ADR stage between the helium tank and a Joule-Thomson cryocooler. It is not as critical to limit parasitic conduction through these switches because of the relatively warm temperatures involved.

9.13 PASSIVELY OPERATED GGHSS

Passively operated GGHS can be passively turned off without the need for a separate, thermally activated getter. They have the following advantages: passive GGHSs do not need a heater to activate the getter and a thermometer to measure the getter temperature then subsequently simplify the sample control sytem. No additional heat is added to the system for activating the getter, which makes the thermodynamic efficiency higher. It has a quicker response time, and since there is no external getter housing, the structure is more compact and light.

Vanapalli et al. reported a passive, adaptive, autonomous GGHS around 250 K–310 K [44]. and Shirron et al. seperately introduced several passive GGHSs near 0.2 K–13 K used in multi-stage ADRs for space applications [49, 56–59]. The switch on/off state is controlled by the getter's temperature. The getter is attached to the ADR; therefore, the switch state is often controlled by the ADR stage temperature. The choice of getter material is dictated by temperature range of the stage, the "high" temperature stage using charcoal, near 1 K using a sintered stainless steel puck, and near 0.16 K using gold-plated copper fins provided by innards. Off-state conductance is dominated by the hermetic outer shell. Typical "low-temperature" switches use titanium shells (superconducting below ~3.4 K), and higher-temperature switches may use stainless steel.

Performance of GGHSs at <13 K relies on the strong temperature dependence of the vapor pressure of ^{4}He adsorbed onto neon or copper substances when the coverage is less than one monolayer. Difference in the binding energies of ^{4}He to the neon or copper give rise to different temperatures where the switches transition between on and off. For passive operation, the switch must operate in the molecular limit so that a change in vapor pressure has an appreciable effect on conductance. Equally important, the vapor pressure must be a very strong function of temperature near the desired on/off point. Where the switch links the refrigeration stage to a fixed heat sink, rapid turn-off is critical for minimizing the parasitic heat flow that will occur when the stage cools below the sink temperature.

The properties of ^{3}He are almost ideal for passive switch operation at very low temperatures. Its saturated vapor pressure (SVP) varies as an exponential over temperature. Over the range from

0.15 K to 0.20 K, the SVP changes by over 1,000, providing the means for a very high switch ratio. Dipirro et al. also employed ^{3}He condensed as a thin film on alternating plates of copper. The switch is thermally conductive above about 0.2 K and is insulating on either end of the switch cooled below 0.15 K. The on conductance is 7 mW/K at 0.22 K.

REFERENCES

1. Shu QS, J A Demko JA, Fesmire JE, 2017 Heat switch technology for cryogenic thermal management, *CEC IOP Conf, Materials Science and Engineering* 278, 012133.
2. Lounasma OV, 1974 *Experimental principles and methods below 1 K*, Academic Press, London.
3. Shirron PJ, 2003 A portable, cryogen-free ultralow temperature cooling system using a continuous ADR, *2003 Cryogenic Engineering Conference*, September 22–26.
4. Krusius M, Paulson D, 1978 Superconducting switch for large heat flow below 50 mK *Rev. Sci. Instruments*, 49(3), pp. 396–398.
5. Schuberth E, 1984 Superconducting heat switch of simple design, *Rev. Sci. Instrum.*, 55(9), p. 1486.
6. Shirron P, Canavan E, Dipirro M, 2000 Multi-stage continuous-duty adiabatic demagnetization refrigerator, *Advances in Cryogenic Engineering*, 45, p. 1629.
7. DiPirro M, Shirron P, 2014 Heat switches for ADRs, *Cryogenics*, 62, pp. 172–176.
8. Kittel P, 2002 Heat switch limitations on multi-stage magnetic refrigeration, *Adv in Cryo Engr,* 47B, p. 1167.
9. Canavan E, Dipirro M, Jackson M, Panek J, Shirron P, 2002, Magnetoresistive heat switch for the continuous ADR, *AIP Conference Proceedings*, 613, p. 1183.
10. Bartlett J, Hardy G, Hepburn ID, 2015 Performance of a fast response miniature adiabatic demagnetisation refrigerator using a single crystal tungsten magnetoresistive heat switch, *Cryogenics*, 72, pp. 111–121.
11. Duval J, Cain B, 2003 Design of miniature adiabatic demagnetization refrigerator *Advances in Cryogenic Engineering,* 49B, p. 1729.
12. Benafan O, 2012 Deformation and phase transformation processes in polycrystalline NiTi and NiTiHf high temperature shape memory alloys, PhD Dissertation University of Central Florida, Orlando.
13. Swanger A, Fesmire J, Trigwell S, Williams M, Gibson T, Benafan O, 2015 Apparatus and method for low-temperature training of shape memory alloys, *IOP Conference Series*, 102.
14. NASA Technology Transfer Program, 2016 Adaptive thermal management system, KSC-TOPS-44.
15. Benafan O, Notardonato W, 2013 Design and development of a shape memory alloy activated heat pipe-based thermal switch, *Smart Materials Structure, Smart Mater. Struct.*, 22, 105017 (17pp).
16. Notardonato WU, 2008 Recovery of lunar surface access module residual and reserve propellants, *Cryogenics*, 48, pp. 210–216.
17. Shu QS, Cheng G, Yu K, Hull JR, Demko JA, Britcher CP, et al., 2003 Low thermal loss cryogenic transfer line with magnetic suspension, *Adv Cryog Eng*, 49B, pp. 1869–1876.
18. Shu Q, Susta J, Hull J, Fesmire J, Demko J, 2006 Magnetic levitation technology and its applications in exploration projects, *Cryogenics,* 46(2–3), p. 105.
19. Hull JR, Cansiz A, 1999 Vertical and lateral forces between a permanent magnet and a high-temperature superconductor, *J. Appl Phy.*, 86, pp. 6396–6404.
20. Shu QS, Werfel FN, et al., 2006 SBIR phase II report: Smart cryogenic heat switch and energy efficient magnetic levitation cryogenic transfer line, *AMAC report, 2006.*
21. Shu QS, Hull JR, Demko et al., 2005 Development of an energy efficient cryogenic transfer line with magnetic suspension, *Presented at the Cryogenic Engineering Conference 2005.*
22. Fesmire J, et al., 2006 Testing of prototype magnetic suspension cryogenic transfer line, *Advances in Cryogenic Engineering*, 51, p. 539.
23. Dietricha M, Eulera A, 2013 Compact thermal heat switch for cryogenic space applications, *Cryogenics*, 59, 2014, pp. 70–75.
24. NIST cryogenics technologies group. Material properties data base. Internet address; 2013. [accessed 01.06].
25. Kirby RK, 1956 Thermal expansion of polytetrafluoroethylene (Teflon) from 190 to 300 K, *J Res Natl Bureau Stand*, 57(2), pp. 91–94.
26. Marland B, Bugby D, Stouffer C, 2004 Development and testing of an advanced cryogenic thermal switch and cryogenic thermal switch test bed, *Cryogenics,* 44, pp. 413–420.
27. Bugby D, Stouffer C, 2005 Cryogenic thermal management advances during the cryotool program, *Advances in Cryogenic Engineering,* 51B, p. 1799.

28. Dietricha M, et al., 2017 A light weight thermal heat switch for redundant cryocooling on satellites, *Cryogenics*, 83, pp. 31–34.

29. First Steps towards Piezoaction, Piezomechanik Dr. Lutz Pickelmann GmbH.

30. Jahromia A, Sullivan D, 2016 Piezoelectric cryogenic heat switch, *Rev of Sci Instruments* 85 065118.

31. Shindo Y, Narita F, 2011 Cryogenic electromechanical behavior of multilayer piezo-actuators for fuel injector applications, *Journal of Applied Physics*, 110, 084510.

32. Melcher BS, Timbie' PT, 2015 Cryogenic heat switch with stepper motor actuator, *Review of Scientific Instruments*, 86, p. 126112.

33. Timinger H, 2008 Design parameters for cryogenic thermosyphons, *CEC*, 53, pp. 1307–1314.

34. Pfotenhauer JM, 2015 Experimental investigation on a pulsating heat pipe with hydrogen, *CEC 2015 IOP Conf. Series: Materials Science and Engineering*, 101, p. 012065.

35. Na-Nakornpanom A, 2015 In-flight performance of the OCO-2 cryocooler CEC 2015 IOP Publishing, *IOP Conf. Series: Materials Science and Engineering*, 101, p. 012024.

36. Mueller1 BW, Pfotenhauer1 JM, Miller1 FK, 2020 A two-condenser pulsating heat pipe for use as a passive thermal disconnect in redundant cryocooler implementations, *CEC 2019, IOP Conf. Series: Materials Science and Engineering*, 755, p. 012031.

37. Bugby D, et al., 2004 Development of advanced tools for cryogenic integration, *Advances of Cryogenic Engineering*, 49, pp. 91914–1922.

38. Pfotenhauer J, 2015 Design and Operation of a Cryogenic Nitrogen Pulsating Heat Pipe, CEC 2015, *IOP Conf. Series: Materials Science and Engineering*, 101, p. 012064.

39. Natsume K, 2013 Development of a flat-plate cryogenic oscillating heat pipe for improving HTS magnet cooling, *Physics Procedia*, 45, pp. 233–236.

40. Brandon R, Paulsen J, 2016 Cryogenic thermal diodes, *AIP Conference Proceedings*, 504, p. 785.

41. Cepeda-Rizo J, 2012 Methane cryogenic heat pipe for space use with liquid trap for on-off switching, *AIP Conference Proceedings*, 1434, p. 293.

42. Na-Nakornpanom A et al., 2015 In-flight performance of the OCO-2 cryocooler, *IOP Publishing IOP Conf. Series: Materials Science and Engineering*, 101, p.012024 (CEC).

43. Martins D, Catarino I, 2010 Customizable Gas-Gap Heat Switch, *AIP Conference Proceedings*, 1218, P-1652.

44. Vanapalli S, Colijna B, 2015 Passive, adaptive, autonomous gas gap heat switch, *Phys Proc*, 67, p. 1206.

45. Vanapalli S, Keijzer R, 2016 Cryogenic flat-panel gas-gap heat switch, *Cryogenics*, 78, pp. 83–88.

46. Catarino I, Bonfait G, Duband L, 2008 Neon gas-gap heat switch, *Cryogenics*, 48, pp. 17–25.

47. Berryhill A, et al., 2008 Novel integration of 6T cryogen-free magneto-optical system with variable temperature sample using a single cryocooler, *Advances in Cryogenic Engineering*, 53, p. 1523.

48. Kimball1 M, Shirron P, 2015 Low-power, fast-response active gas-gap heat switches for low temperature applications, *IOP Conf. Series: Materials Science and Engineering*, 101.

49. Kimball1 M and Shirron P, 2012 Heat switches providing low-activation power and quick-switching time for use in cryogenic multi-stage refrigerators, *Advances in Cryogenic Engineering*, pp. 853–858.

50. Barreto J, Borges P, Kar S, et al., 2015 Gas gap heat switch for a cryogen-free magnet system, *IOP Conf. Series: Materials Science and Engineering*, 101, p. 012144.

51. Duval J-M, 2015 Qualification campaign of the 50 mK hybrid sorption-ADR cooler for SPICA/SAFARI, CEC 2015 IOP Publishing, *IOP Conf. Series: Materials Science and Engineering*, 101, p. 012010.

52. Fukuda H et al. Properties of a two stage adiabatic demagnetization refrigerator, CEC 2015 IOP Publishing IOP Conf. Series: Materials Science and Engineering 101 (2015) 012047 afwa3286@chib-u.jp

53. DiPirro M, Shirron P, 2014 Heat switches for ADRs, *Cryogenics*, 62, pp. 172–176.

54. Shirron P, Dipirro M, 2003 Portable cryogen-free ultra-low temperature cooling system using continuous ADR, *Advances in Cryogenic Engineering*, 49B, p. 1746.

55. Hepburn I, et al., 2003 Space engineering model cryogen-free ADR for future ESA space missions, *Advances in Cryogenic Engineering*, 49B, p. 1737.

56. Shirron P, et al., 2002 Passive gas-gap heat switches for use in adiabatic demagnetization refrigerators, *Advances in Cryogenic Engineering*, 613, p. 1175.

57. Dipirro M, Shirron P, 2003 Design and test of passively operated heat switches for 0.2 to 15 K, *Advances in Cryogenic Engineering*, 49, B, p. 436.

58. Kimball M, et al., 2017 Passive Gas-Gap Heat Switches for use in Low Temperature Cryogenic Systems, *IOP Conf. Ser. Mater. Sci. Eng.*, 278, p. 012010.

59. Tuttle J, Canavan E, et al., 2017 Development of a space-flight ADR providing continuous cooling at 50 mK with heat rejection at 10 K, *CEC 2017 IOP Conf. Series: Materials Science and Engineering*, 278.

10 Current Leads for Superconducting Equipment

Jonathan A. Demko, Quan-Sheng Shu, and James E. Fesmire

10.1 INTRODUCTION

A current lead is used to provide an efficient means of transmitting electrical power between ambient-temperature electrical connections to low-temperature connections. This chapter contains a discussion of current lead development and an overview of current lead analysis. This is followed by sections that discuss many applications of current leads, such as research magnet systems, MRI magnets, high-energy physics (HEP) applications, fusion magnet systems, superconducting power applications, and specialty current lead designs.

In all current lead designs, an electrical connection to a power source is made at room temperature, with the warm end maintained around 290 K. The cold end temperature of the lead must interface to a component at temperatures in the 2 K to 45 K range in many superconducting magnet applications, such as in fusion magnets, MRI magnets, and accelerator magnets. This allows for the possibility to design the current lead with a conventional portion that starts at room temperature and interfaces with a high-temperature superconductor (HTS) section at around 80 K. This section will span the remaining temperature range to the lowest temperature and finally connect to a low-temperature superconductor (LTS). For high-temperature superconducting applications, the low end temperatures are much warmer, reaching 80 K. This is typical of current leads for power applications such as in HTS cables, transformers, fault current limiters, and motors. These would be fabricated entirely from conventional materials such as copper, brass, or aluminum.

Current leads range in capacity from fractions of an ampere for instrumentation to 60,000 amps for large fusion magnet systems. The transition from the warm end to the cold end must be accomplished in a manner that minimizes the refrigeration load. The load on the refrigerator can in general come from two sources, conduction and joule heating. Depending on the application, there are other constraints that may be placed on the current lead such as high voltage and the ability to operate for short-duration overcurrent faults or loss of coolant flow that is often referred to as a loss of flow accident or LOFA.

In general, current leads may be classified into a few basic designs such as conduction cooled, which typically pass through a vacuum, and all heat is removed through conduction at the cold end. These types of lead designs are frequently used in superconducting power applications where the replenishment of the cryogen is to be avoided.

There are two types of convectively cooled leads. Vapor-cooled leads rely on the heat conduction down the lead to boil a liquid cryogen that generates vapor that is used to cool the lead. This type of design is used in many research applications. Forced-flow cooling designs are used in many large-scale magnet systems, such as particle accelerators and fusion machines such as the International Thermonuclear Experimental Reactor (ITER). In these situations, the controlled flow of a cryogen is used to keep the lead from burnout. For each of these design concepts, the use of a high-temperature superconductor at the low-temperature end provides an additional alternative that the lead designer may select depending on whether all constraints required by the application can be met.

The thermal optimization of current leads in a vacuum is the simplest approach. Briefly, this can be accomplished using the equations derived by McFee [1] for the optimum shape parameter and minimum heat load to the low temperature end. The set of equations solved is based on the one-dimensional

DOI: 10.1201/9781003098188-10

heat conduction equation with internal heat generation and temperature-varying properties. The equations can be solved numerically to obtain the optimal shape factor, IL/A, and heat load, $[Q_L]_{min}$, between any two temperatures. Figures 10.1 and 10.2 present calculated values for the optimal shape factor for a lead in a vacuum.

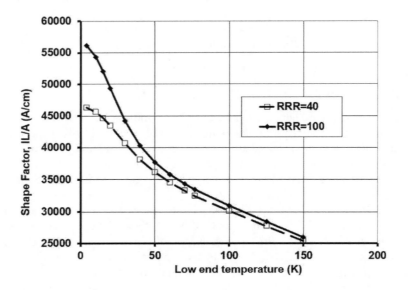

FIGURE 10.1   Optimal shape factor for a conduction-cooled lead.

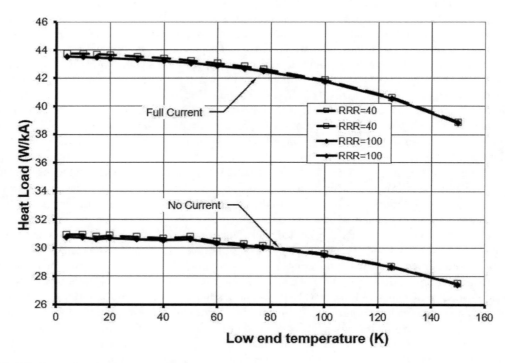

FIGURE 10.2   Conduction-cooled lead optimized heat loads as a function of cold end temperature from 300 K warm end temperature.

One of the earliest vapor-cooled lead designs was given by Efferson [2]. It was constructed using bundles or tubes of fine copper wires. The number of tubes could be adjusted to provide the optimal operating conditions for any design current. The copper wire tubes provided a large surface area for the helium gas to cool the wire. The cold end of the lead was submersed in the bath of liquid helium, and in its initial application, the heat load would boil liquid helium to produce helium vapor that cooled the copper wires. The optimal operating current of vapor-cooled leads was experimentally determined by Efferson was shown to be given at the current where:

$$\left(\frac{dI}{d\dot{m}}\right)_{I_{op}} = 0 \tag{10.1}$$

An early analytical optimization of vapor-cooled current leads was performed by Lock [3]. In his approach, he uses an average value of the electrical resistivity, ρ, of the lead based on the type of copper for different residual resistivities, ρ_0, to calculate the optimal heat leak, Q_0 / I, and geometry parameter, $\left(IL/A\right)_{opt}$. The paper compared results assuming an average ρ taken at 300 K and 80 K for varying ρ. There is a significant difference based on the choice of the average properties.

Another more extensive compilation of lead designs and formulae for design calculations was given by Buyanov [4]. These formulae were useful when computer methods were not readily available. Presently, most analysis can be conducted using numerical methods [5].

The analysis of vapor-cooled current leads can be accomplished with a one-dimensional energy balance for the conductor and the cooling flow, as given in the following equations. In these equations, γ is the density, λ is the thermal conductivity, C is the specific heat, A is the cross-sectional area, h is the convective heat transfer coefficient, P is the heat transfer surface area per unit length (or perimeter), e_v is the vapor enthalpy, and ρ is the electrical resistivity of the conductor. The subscripts c and v stand for the conductor and vapor, respectively.

$$\frac{\partial\left(\gamma_c A_c C_c T_c\right)}{\partial t} = \frac{\partial}{\partial x}\left(\lambda_c A_c \frac{\partial T_c}{\partial x}\right) + \frac{\rho I^2}{A_c} + hP\left(T_c - T_v\right) \tag{10.2}$$

$$\frac{\partial\left(\gamma_v A_v C_v T_v\right)}{\partial t} = \dot{m}_v \frac{\partial e_v}{\partial x} + hP\left(T_v - T_c\right) \tag{10.3}$$

For analysis of conduction-cooled leads in a vacuum, the last term of the first energy balance equation for the lead itself is eliminated as well as the second equation, which is the energy balance for the cooling fluid. A discussion of a numerical method for solving these equations is provided in [5]. The optimization of a helium-cooled lead is discussed in [6,7]. Peterson [7] notes that the numerical solution agrees closely with the analytical presentation of Lock [3] and actual operating experience.

The design of current leads is in practice reduced to the selection and optimization of length and cross-sectional area. The conventional metal portions of current leads have been made in several configurations to allow for high heat transfer efficiency as well as compact current carrying capability. Typical constructions include bundles of small-diameter wire [2] and pipe, thin conducting foils, spiral fin heat exchangers [8], and zig-zag flow fin configurations [9].

The basic requirements for leads are that they minimize the load to the refrigerator and that they operate stably under adverse conditions such as fault currents, which can be 10 to 20 times the maximum design operating current. In the case of convectively cooled leads, the loss of cooling flow can cause instability if the current stays on. Refrigeration loads depend on the design of the lead. These loads are composed of a combination of heat loads by conduction to the cold space and, when convective cooling is used, the provision of cryogen to cool the lead. For low-temperature systems at liquid helium temperatures, there are two basic approaches. The first is when boil-off from the helium is used to balance the heat conducted through the cold end. This type of design has been discussed by several researchers [2,3,4].

Forced-flow cooling is an alternative in magnet systems that use high pressure and supercritical helium. For these situations, the optimization is a bit more complicated because both the liquefaction load and heat conduction can be present. If the Carnot efficiency for both the refrigeration and liquefaction may be assumed to be equal, then the ideal Carnot power may be considered the parameter to be minimized. It may be calculated by:

$$P_{Carnot} = \frac{T_H - T_L}{T_L} \dot{Q}_{cold} + \Delta h_{liq} \dot{m} \tag{10.4}$$

where T_H is the temperature of the warm end of the lead, which is usually around 300 K for an all-conventional copper lead; T_L is the cold end temperature, which is around 4.5 K for most low-temperature magnets; $\dot{Q}_{cold}$ is the heat conducted through the cold end; and Δh_{liq} is the ideal amount of energy required to liquefy the helium. For helium at 0.1 MPa, the ideal liquefaction is $\Delta h_{liq} = 6819$ J/g , and $\dot{m}$ is the cooling flow in g/s.

10.1.1 Short-Duration Overcurrent Heating

In many power grid applications, such as power cables, transformers, or fault current limiters, fault currents that can be more than ten times the maximum operating current frequently occur. Some of the HTS devices have fault current-limiting capabilities because of the non-linear increase of resistance of the superconductor with currents that are greater than the critical current [10]. For these devices to be designed to withstand these types of events, the current lead temperatures must remain stable (no thermal run away) and below the melting temperature of the lead or any solder joints in the conventional portion. There must also be enough thermal mass in any HTS portion to prevent damage to the HTS wire. The temperature rise during a fault of the conductor in the lead can be estimated by an energy balance, assuming that there is no cooling. This results in the electrical resistive heating being balanced by the increase in energy stored in a section of the lead. The one-dimensional equation is:

$$\int_{\tau=0}^{\tau=t_{fault}} I^2 \rho \frac{L}{A} d\tau = \int_{T=T_i}^{T=T_f} ALc\,dT \tag{10.5}$$

where L is the length of the lead, A is the cross-sectional area, ρ is the electrical resistivity, and c is the volumetric heat capacity. The equation can be rearranged to collect the parameters that are functions of temperature on one side and parameters that are a function of time on the other side of the equals sign to give:

$$\int_{\tau=0}^{\tau=t_{fault}} \frac{I^2}{A^2} d\tau = \int_{T=T_i}^{T=T_f} \frac{c}{\rho} dT \tag{10.6}$$

This equation allows one to calculate the temperature rise at a location along the lead that arises from the resistive heating produced during a current pulse applied during a fault. On one side of the equation is the integral of the current density over time, and the other side are the temperature-dependent properties of the material. As an example, the cryogenic properties from 5 K to 300 K properties for a copper alloy with a residual resistivity ratio (RRR) of 40 are provided in Figures 10.3a–c.

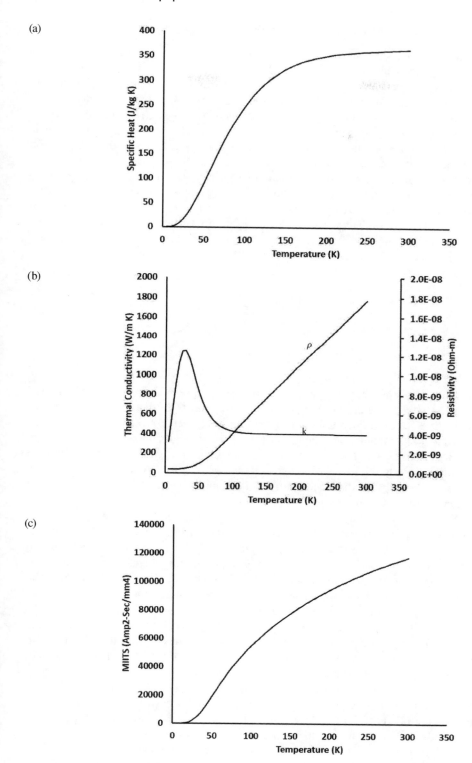

FIGURE 10.3 Cryogenic properties of a copper alloy with a RRR = 40. (a): Specific heat. (b): Thermal conductivity and electrical resistivity. (c): 10^{-6} times the integral of the current squared with time, known as the MIITS integral.

10.2 CURRENT LEADS FOR HIGH-ENERGY PHYSICS MAGNETS

In the field of high-energy physics, several types of magnet systems are employed for accelerator magnets and detectors. Typical accelerators have hundreds of magnets, which may require currents exceeding 10 kA. These magnets and current leads are located around the magnet ring, which may extend for several kilometers. For example, the Fermilab Tevatron has a 6.86-km ring, and the CERN Large Hadron Collider (LHC) is 27 km in circumference. As discussed in an earlier section, the heat load to 4 K per kA per lead is around 45 W for a conduction-cooled lead. With hundreds of current leads supplying power to the magnets of a superconducting accelerator, the refrigeration requirement would be prohibitive. For this reason, the current leads of superconducting accelerators are cooled by force-cooled mode. Forced-cooled operation significantly simplifies the design of the magnet system and does not depend on the orientation of the current lead (horizontal or vertical).

Forced-flow current leads used in many accelerator installations were made from conventional materials such as copper or brass and attached to a low-temperature superconductor at the cold end. This type of lead has similar operating characteristics as those shown in [11], which can be summarized by dividing the heat leak and cooling flow rate by the design operating current. This gives the following performance ratios for this type of lead:

$$1.0 < \frac{\dot{Q}}{I}\left(\frac{W}{A}\right) < 1.2 \tag{10.7}$$

$$0.049 < \frac{\dot{m}}{I}\left(\frac{g}{sec - A}\right) < 0.06 \tag{10.8}$$

Commercial leads are available for this type of operation from various sources [12]. Most large accelerator facilities have taken the task to design custom current leads for their own special needs.

The Superconducting Supercollider Laboratory (SSCL) developed 6.6 kA leads [6] for the accelerator dipoles and 10 kA magnet test laboratory [8]. These were all copper construction with a helical fin heat exchanger design. The helical fin pitch was selected such that there was enough surface area for helium vapor to remove the heat generation along the lead. The single helical flow path ensures that the cooling will be distributed uniformly along the lead. The spiral fin or helical flow path heat exchanger is shown in Figure 10.4.

Following the Superconducting Super Collider (SSC), the next-generation collider, the CERN Large Hadron Collider, required more than 3,000 current leads transferring around 3 MA of current to and from the superconducting magnets [13,14]. There is a wide range of leads, as listed in Table 10.1 The largest-capacity leads used a copper warm portion with a high-temperature superconducting lower section. The low current capacity leads, 120 A or less, were all copper. From the number of leads and the ratings in the table, it can be inferred that the current leads are a significant portion of the refrigeration/liquefaction load of the LHC.

The design of the leads was performed at CERN, and several prototypes were tested. For the HTS portion of the lead, Bi-2223 tapes with a gold-doped silver matrix were stacked together in sufficient numbers to provide the desired critical current. A schematic of the 13-kA lead is shown in Figure 10.5. The cold end of the HTS is connected to a Nb-Ti bus bar.

The resistive part was cooled by helium gas supplied at around 20 K and 1.3 bar. The HTS part was self-cooled by vaporization of helium at 4.2 K. The 20 K helium flow was controlled using a warm valve that was controlled off the interface temperature of the warm end of the HTS section. The flow was regulated to maintain the warm end of the HTS element at around 50 K.

Each lead had a minimal amount of instrumentation to provide for safe operation and protection. This consisted of two temperature sensors at the warm end of the HTS section and one at the room temperature terminal. There were also eight voltage taps along the lead for monitoring the resistive, HTS, and Nb-Ti sections of the lead. For the resistive section, a 100-mV threshold is set, and a 3-mV threshold is set to trigger a power abort of a circuit if these are reached.

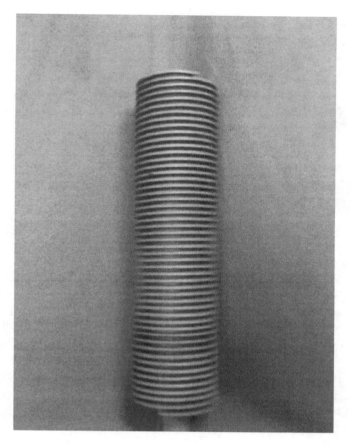

FIGURE 10.4   Spiral fin heat exchanger design used at the SSC.

TABLE 10.1
Lead Count for the LHC Machine

Quantity	Current Rating (A)	Lead type
64	13,000	Copper and HTS
298	6,000	
820	600	Copper and HTS
2104	60 to 120	Copper

An upgrade of the LHC accelerator is allowing for modification of the current leads to include some new innovations. One concept uses an auxiliary flow of helium, which will keep the superconductor cold but allow the upper conventional portion to warm up and prevent frost buildup when there is no current applied. An illustration of this concept taken from Brandt et al. [15] is provided in Figure 10.6.

The flow paths are illustrated in Figure 10.7 for the main cooling flow and the auxiliary flow. This auxiliary flow is used to cool the HTS portion of the lead but is diverted from the current lead and then cools the thermal shields surrounding the helium vessel in the LHC distribution feed box. The main helium lead flow passes through the fins of the central heat exchange section, as shown in

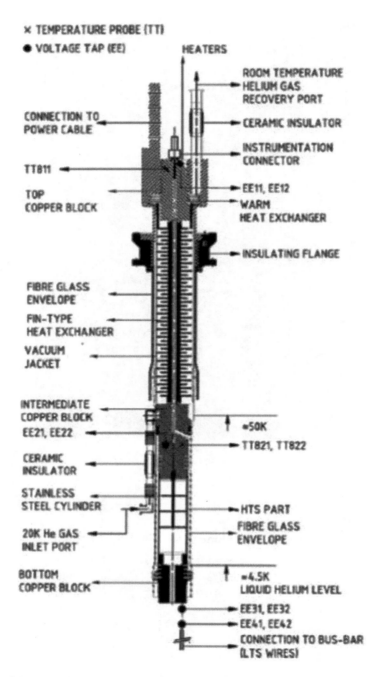

FIGURE 10.5   Schematic of the LHC 13,000-A current lead.

Figure 10.8. It is warmed up to near room temperature and then connects to a bayonet that interfaces with the 300-K 1-bar helium return line.

Separate control valves on each helium stream are used to regulate the flows. The goal is to control the temperature of the exposed current lead upper section so that it remains warm enough to prevent frost buildup.

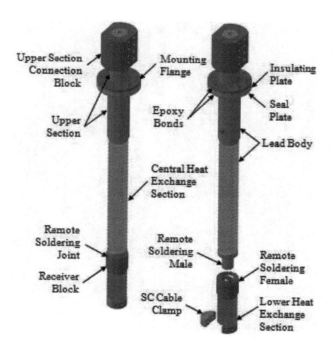

FIGURE 10.6   13-kA current leads for the LHC upgrade [15].

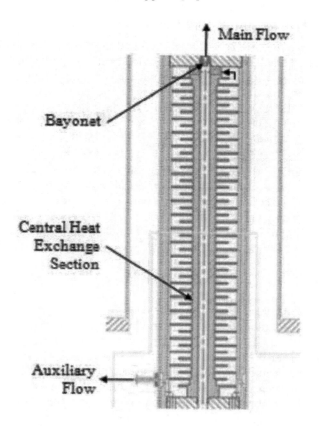

FIGURE 10.7   Helium flow paths for the LHC upgrade 13-kA current leads.

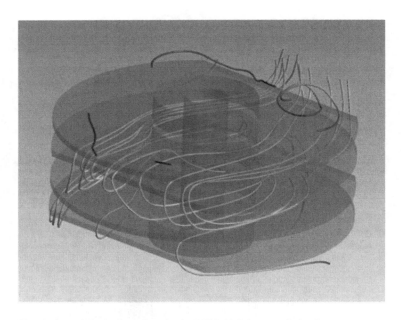

FIGURE 10.8   Simulation of the helium flow in the LHC 13-kA upgrade leads.

10.3   CURRENT LEADS FOR MRI MAGNETS

Magnetic resonance imaging (MRI) is a diagnostic tool used largely in medical applications to produce figures of the anatomy and physiological processes of a subject to diagnose illnesses. Other applications involve chemical analysis. These systems are in wide use around the world. Most MRI magnets use low-temperature superconductors and must operate at liquid helium temperatures (2 K to 5 K). Iwasa [16] discusses benefits that may be obtained if MRI systems were manufactured with HTS instead. The thermal loads in these systems are mainly from two sources, the cryostat and current leads. Energy-efficient designs seek to minimize the heat load from the current leads.

The French-German project Iseult-Neurospin is a 11.75-T MRI magnet system [17]. The current leads have a design requirement of 1,500 A, and they should be able to withstand a three-hour-long slow current dump without any active cooling. Under this scenario, the leads must remain under 100°C. The final design is a braid-in-tube type using a brass composition of 90% Cu and 10% Zn. There is no superconductor in this particular lead design. The selection of brass was made because of the higher thermal stability in the brass. Some of the main design parameters for construction and predicted operating conditions of this lead are given in Table 10.2. The leads are shown in Figure 10.9a and b.

A pair of leads were tested at Saclay, and the predicted operating parameters are compared to the measured steady state values in the following table. The tests demonstrate that the performance of current leads met the steady-state design goals.

The designers also tested the leads under the design fault scenario where the leads are operated at I_{op} = 1,483 A steady state was reached with helium gas supplied for cooling the lead. The flow was stopped, and the full current remained applied for 15 seconds, at which time a slow dump (drop in current) took place. Temperature sensors were placed at distances of 0.615 and 0.715 m from the cold end on both leads. The measured temperature rise was below the 100 K limit.

The design and performance results of a pair of vapor-cooled MgB2 current leads for a 1.5-T MRI magnet developed by Kiswire Advanced Technology Ltd. in Korea are discussed in [18]. The magnet would operate at a current of 500 A. To reduce the liquid helium (LHe) consumption of

(a)

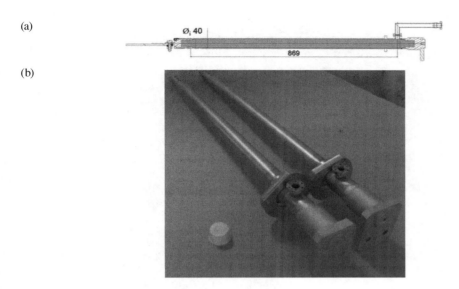

(b)

FIGURE 10.9   (a) and (b): Current leads for the Iseult-Neurospin, a 11.75-T MRI magnet system [17].

TABLE 10.2
Design Features for the Iseult-Neurospin 1,500-A Current Leads

Design Features

Design current	1,800 A
Nominal operating current	1,483 A
Tube design	
Tube material	Stainless Steel 304
Tube diameter	43 mm
Tube wall thickness	1.5 mm
Current carrying braid design	
Braid wire diameter	0.5 mm
Number of wires	3,068
Length of braid	1.05 m
Cryogenic conditions	
Cold end temperature	4.45 K
Warm end temperature	293 K
Helium gas pressure	1.25 bar
Helium mass flow	0.095 g/s
Cold end heat load at 1,483 A	1.82 W
Warm end heat load at 1,483 A	33.9 W
Cold end heat load at 0 A	0.65 W

---

the MRI system, the current leads were designed to be retractable and would be detached from the magnet operated under a persistent-current mode.

The lead construction consisted of a conventional copper portion, which was optimized assuming that the copper had a RRR = 157. The optimum length was 0.450 m long with a cross-section of 1.96×10^{-6} m². The construction utilized magnesium diboride (MgB_2) superconducting wire at the cold end. The MgB_2 was in a normal metal matrix, forming a wire with a diameter of 1.03 mm. The

TABLE 10.3

Comparison of Design and Measured Operating Parameters at I = 1,483 A

Parameter	Design	Lead A	Lead B
Operating current	1,483 A		
Voltage drop at 1,483 A	75.1 mV	69.9	68.6
Helium gas pressure	1.25 bar	1.05 bar	1.05 bar
Helium bath temperature	4.45 K	4.25 K	4.25 K
Helium mass flow	0.095 g/s	0.097 g/s	0.096
Cold end heat load at 1,483 A	1.82 W	1.99 W	1.96 W

normal conductor to superconductor ratio is 1.78. The MgB2 wire specification calls for the critical current of 400 A at 30 K and 1 T, so at least two wires are needed for each lead. The leads were designed to allow for cold He vapor to flow through the leads. The design used three MgB2 wires twisted and soldered to five copper wires of similar diameter. The multi-contact connection is made at the cold end of the lead.

The leads were tested using a Nb-Ti low temperature superconductor connecting the lead pair. These leads did not perform well due to a higher-than-expected joint resistance of 80 $\mu\Omega$ in the multi-contact socket. This resulted in excessive Joule heating at 300 A, which caused the leads to quench. This also produced a higher-than-expected heat load at the cold end. This design demonstrates the high degree of difficulty of having a demountable joint in a current lead with an acceptable joint resistance.

From cryogenic evaluation, the cold-end heat input of the vapor-cooled current leads was calculated using the LHe consumption measured by a LHe-level sensor and mass flow meter. Furthermore, thermal analysis of the current leads was carried out using the finite-element method. During operation tests at various operating currents, the voltage and temperature of the current leads were measured to evaluate the effects of employing the MgB2 wire on the thermal performance of the leads. In addition, conventional vapor-cooled copper current leads were also examined, and the results were compared with the test results of the proposed vapor-cooled MgB_2 current leads [18].

MRI systems are working at removing helium as a coolant by the use of high-temperature superconductors [16,19]. Tosaka et al. discuss the development of a 9.4-T MRI system for whole-body measurements. The elimination of helium as a coolant reduces complexity on one hand, but then requires much more careful consideration in the design of the magnet coils and the current leads. The system must be mainly conduction cooled. The systems described in [19] have a small helium circulation loop to provide cooling to the coils, but the cooling is still mostly by conduction. It is assumed that in this type of MRI arrangement, the leads would be conduction cooled as well by the use of a heat exchanger at the cold end of the lead.

10.4   CURRENT LEADS FOR FUSION MAGNETS

There are many machines used in fusion research that rely on very high current magnets to control the plasma formed in different experiments. Wendelstein 7-X (W7-X) is one of the largest stellerator fusion experiments and is located at the Max Planck Institute for Plasma Physics [20]. The machine was successfully operated with He plasmas in 2015 and H_2 plasmas in 2016. To confine the plasmas, the W7-X machine relies on a superconducting magnet system with 50 non-planar and 20 planar coils. These are grouped into five equal modules that are connected electrically in seven circuits containing ten coils of each type. One special feature of the current leads [21] is that they are installed upside down in the bottom of the cryostat.

The ITER tokamak will be the world's largest fusion machine and plans to harness the energy of fusion (www.iter.org/mach). There are several LTS magnets systems in this device, including the toroidal field (TF) coils, poloidal field (PF) coils, the central solenoid (CS), and 18 superconducting correction coils inserted between the TF and the PF coils. The current leads for these magnets will be designed to carry current as indicated in Table 10.4 [9].

The current leads for ITER have standard operational requirements. This includes high efficiency, which means minimal heat load to the cold end and minimal helium flow. This is achieved as previously described through the optimization of the length and current carrying cross-section of the lead. The leads must have enough heat capacity to withstand a thermal runaway from a loss of flow accident. The leads are forced-flow cooled, so the heat exchanger design must have a minimal pressure drop. Finally, a low resistance connection to the warm bus and low temperature superconductor must be provided. At the very high currents at which these leads operate, joint resistance can cause heating that would quench the magnets. For example, when the TFC leads are at 68 kA, a joint resistance of 0.1 nΩ would result in a heat load of 0.46 W per lead, which must be removed at the low-end operating temperature of between 2 K and 5 K.

The design selected for the ITER leads uses a conventional portion, made from copper, between the room-temperature connections at around 300 K to a high-temperature superconducting section at 65 K. The conventional or upper portion of the lead is forced-flow cooled using helium vapor supplied at 50 K and 0.4 MPa. The design is scaled up from the LHC HTS leads. The heat exchanger design for the copper portion is a fin type with a zig-zag flow configuration. The design is shown in Figure 10.10. A comparison of the geometries for the various magnets that are in ITER is presented in Table 10.5.

At this point, a joint is made between the copper section and a high-temperature superconductor portion that would bridge the temperature range to 4.5 K. The HTS material is a Bi-2223 superconductor tape with a gold-doped silver matrix. The HTS section is conduction-cooled by a helium heat

TABLE 10.4
Summary of Current Leads for the ITER Tokamak

Magnet Type	Number of Leads	Operating Current (kA)
Toroidal field coil	18	68
Poloidal field coil	12	45
Central solenoid	12	46
Correction coils	18	10

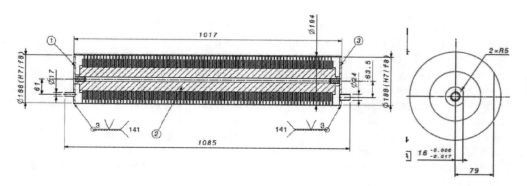

FIGURE 10.10 Design of the heat exchanger for the 68 kA TFC lead for ITER [22].

TABLE 10.5

Dimensions of the Heat Exchangers (HXs) of the ITER Leads. The cut is the segment cut from the edge of the fin to allow helium to pass. Cuts are on alternating sides of the HX

Lead Type	Design Current	Length (mm)	Fin Diameter (mm)	Core Diameter (mm)	Cut (mm)	Central Hole Diameter (mm)
CC	10 kA	903	110	38	10	10
PF/CS	55 kA	951	174	82.6	13	16
TF	68 kA	951	188	92.4	15	16

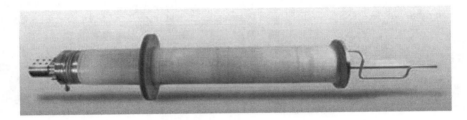

FIGURE 10.11 Completed HTS current lead for JT-60SA TF coil system. Source: *www.jt60sa.org/wp/ picture-gallery/, last accessed 17 June 2021*

FIGURE 10.12 Development of 52 kA HTS current lead for the ITER CS magnet test application [20].

exchanger at the cold end. The completed lead for the ITER TF coil is shown in Figure 10.11. The completed lead for the ITER CS magnet test is shown in Figure 10.12.

General Atomics (GA) successfully tested a pair of 52-kA currents in the GA CS feeder system at the Institute of Plasma Physics, Chinese Academy of Sciences. The tests provided comparisons on the performance with the design conditions. Table 10.6 lists some of the results.

The 52-kA HTS current leads have a room-temperature interface connection that is water cooled and heated. The conventional copper or resistive portion of the lead is the zig-zag cooling flow configuration. The HTS portion is made using 90 Bi-2223 gold-doped silver matrix tapes. For the test configuration, the LTS joint has a twin-box topology. Results of the test are given in Table 10.6.

The significant operating characteristics met the ITER design goals as provided in the table, with one exception: the LOFA time was less than the 600-second design time. The design requirement of 15-kV Paschen voltage tests was also met.

TABLE 10.6

Comparison of Design Operating and Test Conditions for a pair of ITER CS Current Leads

Items	Test Results	Items
Mass flow rate	3.2~3.3 g/s	3.8 g/s for 50 kA
HEX pressure loss	0.6 bara	1 bara
Joint resistance of four twin-box joints	Max. 0.6 nOhm	2 nOhm
Joint resistance of HTS tops	5~6 nOhm	10 nOhm
LOFA time	600 s	300 s
Vacuum	<1E-4 Pa	1E-3 Pa
Thermal shield	<90 K @ 7 days cool	100 K
Pulsed current	Yes. No voltage in high temperature superconducting current lead is larger than 1 mV, which guides the threshold setting for GA operation in future	HTS top control stability induction voltage is small
HTS conduction heat load	10.3 W per lead	12 W per lead

10.5   CURRENT LEADS FOR SUPERCONDUCTING POWER APPLICATIONS

The application of superconducting devices requires that electric current, ranging to several thousand amperes, be brought into the cold region of the cryostat. Large superconducting magnets and particle accelerator magnets are typically made using low-temperature superconductors. The application of high-temperature superconducting leads to superconducting magnetic energy storage (SMES) devices.

Recent developments in high-temperature superconductors have resulted in several demonstration projects for HTS cables requiring that the utility make electrical connections to ambient [10, 23, 24]. Power system components such as cables and transformers operate under alternating current, which is typically 50 to 60 Hz. The effective operating currents, as typically specified, are the root mean square (rms) currents, which are related to the maximum or peak current in a sinusoidal wave form as:

$$I_{eff} = \frac{I_{peak}}{\sqrt{2}} \cong 0.707 \, I_{peak}$$

Current leads can be sized using the rms current for the maximum normal operation. Since these devices operate on AC, the leads are not always solid copper or brass but are frequently designed using braided conductor to reduce losses that arise from the alternating currents.

The structure housing the current leads for HTS cables is often referred to as the termination. The American Electric Power Company (AEP) HTS cable was designed for 13.8 kV three-phase operation. This termination makes the transition for all three phases in a single termination. Other projects are at higher operating voltages, such as the Long Island Power Authority (LIPA) demonstration, which is a 600-meter-long, 138-kV demonstration. This cable design used separate terminations for each phase. The larger insulator bushings seen in the LIPA figure are required due to the higher operating voltage of the current leads.

The terminations for the Southwire Co.–American Electric Power Bixby substation of a high-temperature superconducting power transmission cable are shown in Figure 10.13. The structure must provide for the ambient connection to the substation. The design withstands seasonal

FIGURE 10.13 Termination containing the current leads for a high-temperature superconducting cable operating around 77 K.

fluctuations of the ambient temperature and varying weather conditions. Heat losses through the current leads are one of the most significant thermal loads and for this application are concentrated at the ends of the cable. This installation was 200 meters long, and other existing projects had longer lengths, extending to a kilometer or longer. An additional function of the terminations is to provide for the thermal contraction of the cable, which can be tens of centimeters for very long cables. One way is to have the terminations mounted on a rolling platform which moves as the cable shrinks during cooldown. The LIPA project terminations are shown in Figure 10.14.

Current leads for power system applications are most often designed to be conduction cooled with no coolant stream along the lead. So, another consideration in the design is that the leads not develop ice balls under varying ambient conditions. The HTS cable demonstrations had their terminations outside in conditions that can be warm during the summer and below freezing in the winter. The leads use commercial high-voltage bushings at the warm end, which have been proven to operate under these conditions.

HTS transformers have many advantages, including a smaller size and, very significantly, the removal of transformer oil as the dielectric. This makes the application of HTS transformers much safer since the fire hazard of the burning oil is not present. Current leads for these devices will be at a high voltage (HV) and low current (LC) at one side and then at a higher current but lower voltage. Typical utility-sized units operate with three-phase electric power, so there will be three HV, LC and three low-voltage, high-current leads as well as a neutral or ground connection.

High-temperature superconductors have a current dependent resistance characteristic that has been explored for use in current limiting devices known as fault current limiters (FCLs). These FCL devices limit the available short-circuit current to values significantly lower than they would be if they were not present. This provides an extra level of protection to the power grid, improving the grid resilience and reliability [10]. These fault current limiting devices can be either in a cable or as a separate device in the circuit.

Many HTS power devices operate at liquid nitrogen temperatures. For these cases, there is no superconducting portion, only a conventional lead made from copper or brass. One unique

FIGURE 10.14 Long Island Power Authority—American Superconductor HTS cable terminations.

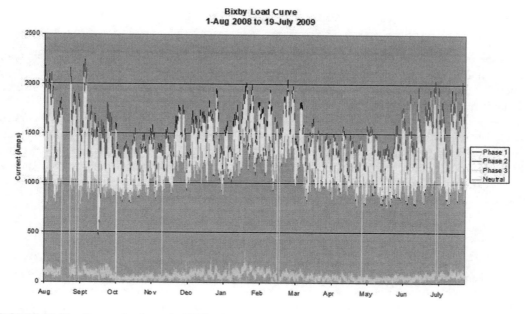

FIGURE 10.15 Current loads on the HTS cable [30].

difficulty for these applications is that the operating current is determined by the demand of the users. Figure 10.15 shows the current on the Southwire Co. HTS cable demonstration for a one-year period. The current fluctuates daily and seasonally. This installation was designed for a 2 kA maximum, which had been reached for a short time in multiple instances over the year.

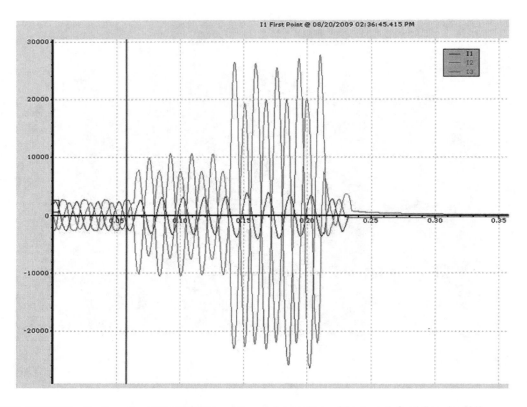

I1 First Point @ 08/20/2009 02:36:45.415 PM

FIGURE 10.16 Fault on the AEP HTS cable with a peak transient—27 kA peak for 4.5 cycles [31].

In addition, these installations are subject to fault currents, which may be ten times or more than the maximum operating current. These devices are required to be able to return to service in a short time after the fault is cleared. An example of a fault experienced by the Southwire Co. AEP cable is shown in Figure 10.16.

The design must allow for a maximum operating current to be sustained for several hours in steady-state operation. The design must also accommodate the possibility of a short-duration fault current that can be ten times or more than the operating current. This is typically done by selecting the lead features to have a length sufficient to provide a thermal mass that will not overheat during expected fault conditions. This can be done using the methods described in earlier sections of this chapter.

10.6 LEADS WITH SPECIAL FEATURES

The development of current leads has led to research in special features. One of these is the Peltier current lead [26], which utilized n-doped and p-doped Peltier tablets at the warm end of the lead in the current carrying circuit. These thermoelectric elements carry heat away when a current is applied. This results in the potential of cooling the ends by 50 K to 60 K below room temperature. These Peltier cascades are solid-state devices that have been made with Bi_2Te_3. The devices can operate under the application of AC or DC currents. The heat flow that results from a combination of conduction and the thermoelectric effect can be expressed as [27]:

$$\dot{Q} = -kA\frac{dT}{dx} \pm \alpha TI$$

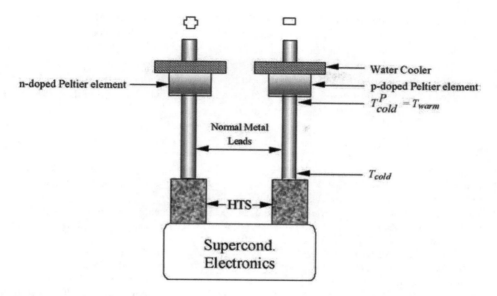

FIGURE 10.17   Principle of using the Peltier effect to cool current leads.

where k is the thermal conductivity, and α is the Seebeck coefficient. For a p-type thermoelectric element, $\alpha > 0$. When there is no current applied, thermoelectric cooling is not present. Also, the direction of current flow determines the cooling. As shown in Figure 10.17, n-doped thermoelectric material would be used on the positive current lead and p-doped thermoelectric element on the negative lead.

After a magnet system is charged, the heat load from a current lead remains even though it is not bringing current to the magnet [28] discusses one option for disconnecting the lead using a multi-contact connector at the thermal intercept where the HTS portion of the lead connects to the HTS portion. This seems like a logical choice because it will greatly reduce the driving temperature difference for heat transfer to the magnet, thus reducing the heat transfer. In addition, this allows the conventional portion of the lead to warm up since it is detached from any cold connection, preventing frost formation on the warm connection of the lead.

An analysis was performed in [27] using ANSYS of a tapered, cone-like lead configuration.

10.7   SUMMARY AND CONCLUSIONS

The design of current leads for cryogenic applications depends on many factors, which depend on the application. Small superconducting magnet systems may be able to use conduction-cooled leads. Large superconducting magnets, such as in fusion machines like ITER or particle accelerators, rely on vapor-cooled leads because of the extremely high currents that must be supplied to the magnets.

The warm end of the current lead is always made using conventional conductors such as copper, aluminum, or brass. If the low-temperature end of the lead is at liquid helium temperatures, it is frequently advantageous to use high-temperature superconducting wires below 75 K to reduce the heat load.

Superconducting power devices such as power distribution and transmission cables, transformers, and fault current limiters are being developed that operate on HTS materials. Therefore, the leads will be designed from conventional materials. Since these systems are frequently alternating current, the leads are often designed using braided copper. This also provides a means for managing thermal contraction of components.

The analysis of high current leads is most often conducted with numerical models, and results obtained by those experienced in the field frequently match the final design. The designs being used are very efficient at minimizing the heat load to the cold component and the coolant flow needed. Optimal operating design depends on the refrigeration and liquefaction system operational requirements as well as stability to off-design or fault conditions. As the applications for cryogenic equipment increase, so will the variety of designs of current leads that will be needed. But these will be based on the design principles discussed in this work.

REFERENCES

1. R. McFee, "Optimum Input Leads for Cryogenic Apparatus," *Rev. Sci. Instr.*, Vol. 30, No. 2, 1959.
2. K. R. Efferson, "Helium Vapor Cooled Current Leads," *Rev. Sci. Instr.*, Vol. 38, No. 12, pp. 1776–1779, 1967.
3. J. M. Lock, "Optimization of Current Leads into a Cryostat," Cryogenics, December 1969, pp. 438–443.
4. Yu. L. Buyanov, "Current Leads for Use in Cryogenic Devices. Principle of Design and Formulae for Design Calculations," *Cryogenics*, Vol. 25, Feb 1985.
5. J. Demko, W. Schiesser, R. Carcagno, M. McAshan, "A Method of Lines Solution of the Transient Behavior of the Helium Cooled Power Leads for the SSC", SSCL-Preprint-107, Also Proceedings of Seventh IMACS International Conference on Computer Methods for Partial Differential Equations, Rutgers University, June 1992. www.iaea.org/inis/collection/NCLCollectionStore/_Public/26/073/26073694.pdf last accessed 16-September-2017.
6. J. A. Demko, W. E. Schiesser, R. Carcagno, M. McAshan, and R. McConeghy, "Thermal Optimization of the Helium Cooled Power Leads for the SSC," Supercollider 4, Plenum Press, 1992, p. 635.
7. T. Peterson, "Magnet Current Leads," Presentation at the U.S. Particle Accelerator School (USPAS), January 2017, http://uspas.fnal.gov/index.shtml.
8. Q. S. Shu, J. A. Demko, R. Domam, D. Finan, T. Peterson, I. Syromyatnikov, and A. Zolotov, "The thermal optimum analyses and mechanical design of 10 kA vapor cooled power leads for the SSC superconducting magnet tests at MTL," *IEEE Trans. Appl. Supercond,* Vol. 3, No. 1, 1993, p. 408.
9. A. Ballarino, P. Bauer, B. Bordini, A. Devred, K. Ding, E. Niu, M. Sitko, T. Taylor, Y. Yang, and T. Zhou, "Qualification of fin-type heat exchangers for the ITER current leads," 2015 IOP Conf. Ser.: Mater. Sci. Eng. 101.
10. C.M. Rey, R.C. Duckworth, J.A. Demko, D.R. James, M.J. Gouge, "Test Results of a 25-m Prototype Fault Current Limiting HTS Cable for Project Hydra," Advances in Cryogenic Engineering: Transactions of the Cryogenic Engineering Conference—CEC, American Institute of Physics, Vol. 55A, pp. 453–460, 2010.
11. P. Seidel (Editor), "Applied Superconductivity: Handbook on Devices and Applications," Wiley-VCH, 2015.
12. www.americanmagnetics.com/datasheets/current_leads.pdf, Last accessed December 20, 2017.
13. A., Ballarino, "Large-capacity current leads," Physica C 468 (2008).
14. A. Ballarino, "Operation of 1074 HTS Current Leads at the LHC: Overview of Three Years of Powering the Accelerator," IEEE Trans. Appl. Supercond. 23(3) (2013).
15. J.S. Brandt, S. Cheban, S. Feher, M. Kaducak, F. Nobrega, and T. Peterson, "Current lead design for the Accelerator project for upgrade of LHC," IEEE Trans. Appl. Supercond. 21(3) (2011), p. 1066.
16. Y. Iwasa, "HTS and NMR/MRI magnets: Unique features, opportunities, and challenges," Physica C, 445–448, 2006.
17. F.P. Juster, C. Berriaud, P. Bouziat, P. Brédy, H. Lannou, L. Quettier, T. Schild, V. Stepanov, "Iseult-Neurospin 1500 A Currents Leads: Conceptual and Experimental Results," MT-25 Poster A0, August-September 2017.
18. J. Kim, Y.H. Choi, Y-G. Kim, I. Shin, S. Yoon and H. Lee, "Design and Performance Results of Optimal Vapor-Cooled MgB2 Current Leads for a 1.5 T MRI Magnet," Presented at the Applied Superconductivity Conference, 2017 Aug. 27—Sep. 1, Netherlands, Amsterdam; Session: Current Leads, Links, and Bus Bars; Program I.D. number: Mon-Af-Po1–09 [553]MT-25.
19. T. Tosaka, H. Miyazaki, S. Iwai, Y. Otani, M. Takahashi, K. Tasaki, S. Nomura, T. Kurusu, H. Ueda, S. Noguchi, A. Ishiyama, S. Urayama, H. Fukuyama, "Project Overview of HTS Magnet for Ultra-High-Field MRI System," Physics Procedia 65 (2015) 217–220.

20. K. Ding, T. Zhou, K. Lu, Q. Du, B. Li, S. Yu, X. Huang, C. Liu, K. Zhang, K. Jing, Q. Ran, Q. Han, J. Li, Y. Song, Piec, Zbigniew, Schaubel Kurt, "Development of 52 kA HTS Current Lead for the ITER CS Magnet Test Application Mon-Af-Po1.09–12[136], MT-25, August 27-Sept 3 2017.

21. R. Heller, W. H. Fietz, M. Heiduk, M. Hollik, A. Kienzler, C. Lange, R. Lietzow, I. Meyer, T. Richter, and T. Vogel," Overview of JT-60SA HTS current lead manufacture and testing," Mon-Af_Po1.09.

22. A. Ballarino. P. Bauer, B. Bourdini, A. Devred, K. Ding, E. Niu, M. Sitko, T. Taylor, Y. Yang, T. Zhou, "Qualification of Fin-Type Heat exchangers for the ITER Current Leads, 101, 2015.

23. J.A. Demko, I Sauers, D.R. James, M.J. Gouge, D. Lindsay, M. Roden, J. Tolbert, D. Willén, C. Træholt, "Triaxial HTS Cable for the AEP Bixby Project," IEEE Transactions on Applied Superconductivity, Vol. 17, Issue 2, Part 2, June 2007, pp. 2047–2050.

24. J. F. Maguire, J. Yuan, W. Romanosky, F. Schmidt, R. Soika, S. Bratt, F. Durand, C. King, J. McNamara, and T. E. Welsh, "Progress and Status of a 2G HTS Power Cable to Be Installed in the Long Island Power Authority (LIPA) Grid," IEEE Trans. Appl. Supercon., vol. 21, pp. 961–966, June 2011.

25. H-M Chang, Y.S. Kim, H. M. Kim, H. Lee, and T.T., Ko, "Current lead design for cryocooled HTS fault current limiters," IEEE Trans. Appl. Supercond. 17(2) (2007), p. 2244.

26. H.G. Lee, H. M. Kim, B.W. Lee, I.S. Oh, H.-R. Kim, O.B. Hyun, J. Sim, H.M. Chang, J. Bascunan, and Y. Iwasa, 'Conduction-cooled brass current leads for a resistive fault current limiter (SCFL) system," IEEE Trans. Appl. Supercond. 17(2) (2007), p. 2248.

27. F.K. Gehring, M.E. Hüttner, R. P. Huebener, "Peltier cooling of superconducting current leads,' Cryogenics, Vol. 41, pp. 521–528, 2001.

28. T. Kawahara, M. Emoto, H. Watanabe, M. Hamabe, J. Sun, Y. Ivanov, and S. Yamaguchi "Optimization of Peltier current lead for applied superconducting systems with optimum combination of cryo-stages," *AIP Conference Proceedings*, Vol. 1434, p. 1017, 2012.

29. Y. S. Choi, D. L. Kim, and M. S. Kim, "Electrical Contact Resistance of Multi-Contact Connector in Semi-Retractable Current Lead," *IEEE Trans. Appl. Supercond.*, Vol. 21, No. 3, 2011, p. 1050.

30. J. A. Demko, I. Sauers, D. Lindsay, "High Temperature Superconducting Cable", DOE Superconductivity Program for Electric Systems 2009 Annual Peer Review, Alexandria, VA. August 4–6, 2009.

31. J. Demko, I. Sauers, and D. Knoll, "High Temperature Superconducting Cable," 2010 Advanced Cables and Conductors Peer Review, U.S. Department of Energy, Alexandria, VA, June 29–July 1, 2010.

11 RF Power Input and HOM Couplers for Superconducting Cavities

Quan-Sheng Shu, Jonathan A. Demko, and James E. Fesmire

11.1 INTRODUCTION

Superconducting radio-frequency (SRF) cavities can deliver a quality value, Q, higher than that of copper cavities by ~105 and accelerating fields with *Eacc* higher by an order of magnitude in continuous wave (CW) mode. Huge cryogenic systems (multi-kW cooling capacities of refrigerators at both 2 K and 4 K) have been operated successfully associated with these applications. Cavities about several km in length have been in operation and development [1–5]. New projects of more than 20,000-m-long SRF systems have been proposed for applications in high-energy physics, free electron lasers (FEL), nuclear physics, material sciences, bio-sciences, and so on [1, 6].

There are irreplaceable functional insertion components (FICs), which are directly connected to a cold mass and the ambient environment and play crucial technical roles in these applications. Huge amounts of radio frequency (RF) power (megawatts) are transported to SRF cavities by RF input couplers (RFICs) from klystrons to generate strong electromagnetic accelerating fields. RFICs also provide tight vacuum breaks between ultra-high (UH) vacuum in accelerating space and ambient environments. High-order mode couplers (HOMCs) are employed to extract or dissipate unwanted RF energy present in the cavity to protect particle beams from instability.

However, such components also bring a huge heat leak to the cold mass. The heat leak is usually much greater than that through the entire evacuated multilayer insulation (MLI) insulation system and solid support structures combined. It is also a critical challenge to reduce the solid thermal heat leak through the RFIC into the cryogen (LHe) and also to minimize the large heat loads generated by high RF fields on the metal inner surface of the coupler.

The design and construction of RFICs and HOMCs are complicated task related to many highly technical challenges inherent in engineering and physics. The comprehensive trade-offs and optimizations in the designs, particularly the thermal aspects, are discussed. The methodology to balance between technical functional demands and minimization of the heat load are introduced. Selected applications as demonstrations from around the world are also summarized.

11.2 HIGH RF POWER INPUT COUPLERS

Challenges of RFIC. The primary crucial role of the high RF power input coupler is to transfer huge amounts of radio frequency power (up to megawatts) from klystrons at room temperature to cavity-generating strong electromagnetic accelerating fields to drive charged particle beams, as sketched in Figure 11.1. RFICs also provide tight vacuum breaks between UH vacuum in accelerating space and ambient pressure of high RF power passes. In the book, we are interested in RFIC only for SRF cavities cooled at 2 K–4 K. Therefore, RFICs also function as a solid bridge between RF sources at room T and SRF cavities in cryogenic T. From the point of view of cryogenic heat management, discussions are focused on how to reduce the heat flow through RFIC to the cold mass as much as

DOI: 10.1201/9781003098188-11

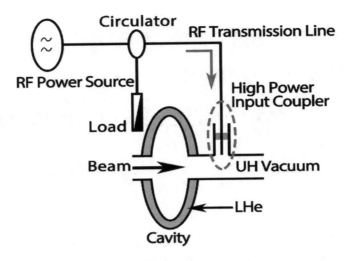

FIGURE 11.1 Function of the RF input coupler: transferring high RF power from a RF generator to charged beams through the accelerating cavity.

possible while efficiently transferring significant RF power to the SRF cavity and providing solid support and tight vacuum barriers.

The heat leak through RFICs is usually greater than that through the entire evacuated MLI system and solid support structures combined, as an example presented in Table 11.1. Challenges include not only the refrigeration loads to be met but also critical aspects of the RFIC design in satisfaction of highly restrictive, even contradictory, technical functions. Comprehensive trade-offs and optimizations are required in the designs.

Design Methodology: The first issue is to examine the balance between heat load and RF power delivery. To reduce the static conductive heat leak, a structure with thin walls of stainless steel for low electrical conductance is favorable, but to reduce the great amount of heat generated on the inner surfaces of RFIC by RF electromagnetic fields, it is better to use copper walls. Due to anomalous skin effects, RF fields can only penetrate a thin layer on inner surfaces. Therefore, a solution is to coat a thin layer of copper on the stainless inner surfaces of the RFIC. The thickness and residual resistance ratio (RRR) value of the copper layer are optimized to have minimal solid conductive heat leak and also sufficient thickness to carry the RF power. The vacuum-exposed window surface is coated with titanium nitride (TiN) to mostly eliminate multi-pact heating. Cooling and thermal anchors are employed for reducing the heat leak through the RFIC walls.

Many RFICs have successfully been developed with these methodologies, and they can be categorized into two types: coaxial RFICs and waveguide RFICs. The merits of waveguide RFICs and coaxial RFICs are summarized as in Table 11.2.

11.3 COAXIAL HIGH RF POWER INPUT COUPLERS

11.3.1 General Design Considerations

The RF input coupler (RFIC) connects the RF transmission line and the SRF cavity and provides crucial RF, thermal, and vacuum functions. RFICs transfer very high power levels in pulses or CW to to the SRF cavity and beam [7–9]. This has to match the impedance of the klystron and RF distribution system to the beam loaded cavity. There is a strong mismatch in the absence of the beam, which leads to full reflection in SRF cavities. The RFIC allows the change of the match for different beam loading [10–12].

TABLE 11.1
Static Heat Load, W of a ILC Cryomodule

	Full Set of 5 K Shield			Without 5 K Lower Shield		
	2 K	**5 K**	**40 K**	**2 K**	**5 K**	**40 K**
Thermal radiation	<0.001	1.14	54.4	0.10	0.18	54.6
Supports	0.32	2.06	16.6	0.23	1.06	19.0
Input coupler	0.26	1.29	17.6	0.26	1.60	16.8
HOM coupler (cables)	0.01	0.22	1.81	0.01	0.27	2.03
HOM absorber	0.14	3.13	−3.27	0.14	3.13	−3.27
Current leads	0.28	0.47	4.13	0.28	0.47	4.13
Cables	0.12	1.39	2.48	0.12	1.39	2.48
Sum	1.13	9.70	93.8	1.14	8.10	95.8

*The dynamic heat load by RF coupler is much greater than the coupler's static load.

Source: Of ILC Cryomodule* [7]

TABLE 11.2
The Merits of Waveguide and Coaxial RFICs

	Advantages	Disadvantages
Waveguide	Simple design	Larger size
	Better power handling, lower surface electric field	Greater heat leak, high thermal radiation
	Easier to cool	No easy tuning of the match
	Higher pumping speed	Rectangular parts—extensive
Coaxial	More compact	More complicated design
	Smaller heat leak	Worse power handling
	Easy RF match, change penetration of antenna	More difficult to cool
	Circular parts are easy to machine, assemble, seal	Lower pumping speed

The RF power coupler (PC) provides a vacuum barrier for the UH beam vacuum. It has to be easy cleaning, have clean assembly, and not contaminate the accelerator vacuum. In the case of SRF cavities, clean must meet ISO 4 standards ("dust free"). Low static and dynamic thermal losses to the cold mass (2 K–4 K) and mechanical flexibility for temperature cycles and thermal expansions are restrictive design requirements, while costs must always be kept in mind. Arcing, multipacting, or window heating will impact the accelerator performance. The damage to the coupler can make the acceleration system inoperable. Particulaly, a vacuum leak of the ceramic will cause serious contamination of the accelerator vacuum system, and recovery of the SRF cavity surface is very time consuming and expensive. As a successful example, an AMAC/DEASY [AMAC international collaborated with Deutsches Elektronen-Synchrotron (DESY)] design of RFIC for the TeV-Energy Superconducting Linear Accelerator/ International Linear Collider (TESLA/ILC) project by Shu et al. is illustrated in Figure 11.2 (Upper). For cryomodule operation with high RF powers, Cornell, KEK (High Energy Accelerator Research Organization, Japan), and others have separately developed coaxial RFICs with stronger cryogenic cooling of the inner and outer conductors. In addition to anchor cooling at flanges and tubes, "continuous" cooling with heat exchangers is utilized by Putselyk et al., as shown in Figure 11.2 (Lower) [10].

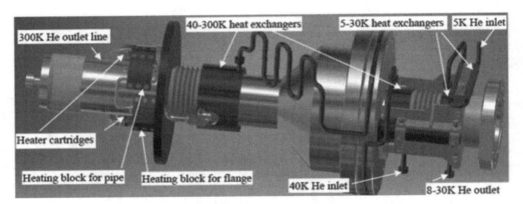

FIGURE 11.2 Upper: AMAC-DESY design of a TESLA/ILC high RF power input coupler [12]. Lower: Simplified view of RFIC coupler with continuous cryogenic cooling [10].

11.3.2 KEY ELEMENTS OF COAXIAL RFIC

Transition Element. The transition element connects the rectangular waveguide from the RF power source to the coaxial coupler, as shown in Figure 11.2 (Upper). In coaxial RFICs, the proper matching between coupler and cavity can be obtained by adjustment of the inner conductor. The doorknob shape functions as the RF match between the coaxial line and the rectangular waveguide.

Windows are designed to separate the vacuum of the cavity from the ambient pressure of the transmission line. As electromagnetic interfaces, they must satisfy strict matching requirements since dielectric materials (high-purity ceramics) are used for the transmission of RF power and involve complicated interfaces of conductors, dielectrics, and brazing metals. Electrons multipacting at the windows can be particularly dangerous, as large amounts of power can be deposited in small areas of the dielectric, potentially leading to failure. A careful choice of geometry and coating with low secondary electron emission coefficient materials can mitigate this phenomenon. The window can be designed as disk and cylindrical (or conical).

One Window vs Two Windows. For normal conducting cavities, the one-window structure is common. For SC cavities and low-gradient applications (<15 MV/m), one-window at room *T* is also

FIGURE 11.3   A: Window assemblies with ceramic elements indicated. B: Window with 5 K and 80 K anchors [13].

common. For high-gradient SRF cavities (≥20 MV/m), the two-window design is preferred and safe. As one window is at cryogenic T, the second window is at room T with an intermediate vacuum to avoid gas condensation on the cold window.

Assembly of Inner-Outer Conductors with Windows. To meet the RF, thermal, and ultra-clean requirements, windows are often designed and constructed together with inner/outer conductors and cooling systems (flange, anchor, and optional heat exchangers) as one window assembly. The assembly surfaces facing vacuum are normally coated by TiN to suppress electronic multipacting; Figure 11.3A shows warm and cold window assemblies, both using desk ceramic windows in the design [12–13]. Figure 11.3B shows a window assembly with 5 K and 80 K cooling anchors. The inner conductors can be cooled by conduction, water, or cryogens, depending on heating levels.

11.3.3   Design and Thermal Optimization

Systematic design of coaxial RFIC with two windows for the SRF cavity was initially conducted at DESY for the TESLA project, and successfully continuous efforts have since been achieved around the world. Shu et al. have reported several RFIC designs and test results for TESLA-ILC, SNS, and RARE projects [12–15]. As an example, the design and thermal analyses of a TESLA-ILC coupler by AMAC with DESY and CPI are shown in Figure 11.4.

11.3.3.1   Design Specifications and Procedurals

Key Tech Specifications: Frequency 1.3 GHz 1,110 kW; pulse length 1.3 ms; repetition rate 5 Hz; cryogenic losses 12 W at 70 K, 1 W at 4 K, 0.12 W at 2 K; movement 1.5 mm in cavity axis direction; double window coupler design DC bias up to 4 kV; coaxial line impedance between windows 50 Ω; coaxial line impedance on cavity side of window 70 Ω.

General Arrangement. The coupler is constructed in three separate sections, as illustrated in Figure 11.5 [15]. The cold window assembly is mounted on the cavity in an ultra-clean room and will provide the main vacuum seal between the coupler and cavity. The warm window assembly is mounted onto the cold window after the installation of the cavity in the cryostat. The WG is a WR650 element with a "door knob" for the transformation to the coaxial coupler line.

Ceramic Windows. These are made of aluminum oxide 99.5% for low RF dissipation and good thermal conductivity and are metalized at the edges for brazing to the copper parts.

Two General Designs. One warm window assembly is under vacuum, and another warm window assembly is filled with clean dry nitrogen at atmospheric pressure.

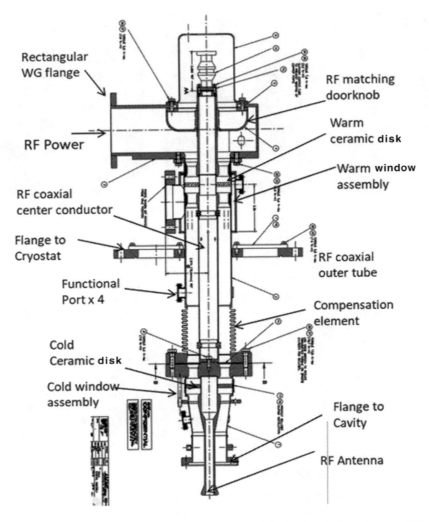

FIGURE 11.4 Cross-section of a TESLA-ILC RFIC by AMAC, showing all key elements [15].

Four Diagnostic Ports. Two for electron activity in the warm and cold sections, one for the temperature at the warm window, and one for detecting eventual arcing in the waveguide.

Cooling. The 7.2 kW average power level specified results in moderate T gradients in the ceramic, as shown in the thermal calculations. It allows the use of conductive cooling to ambient T and to the cold intercepts at 4 K and 70 K, thus eliminating the need for water cooling.

RF Calculations. Geometry calculations: HFSS finite analysis code was used to optimize the RF characteristics of the coupler. Tolerance analysis: To determine the sensitivity of RF performance to the engineering errors. Breakdown voltage: The HFSS code calculations for TESLA style design give a maximum voltage gradient (an electric field of 30.3 kV for 1.11 MW peak power), which is only about 1/3 of a standard coaxial line with the same impedance as the TESLA coaxial coupler.

Multipacting, Window Coating, and Tests. The multipacting properties of the coupler window sections were calculated at DESY [16]. The results show that this is not a problem for ideal surfaces with the designed geometry, but irregularities near the brazing region of the ceramic window could become potential electron emitters. The vacuum-exposed surfaces of the windows are coated with titanium nitride (TiN). A high DC voltage bias of up to 4.5 kV can be applied to the inner conductor to suppress eventual multipacting activities, and the inner conductor is electrically insulated from the waveguide assembly and the inner conductor pumping connection.

Thermal Calculations. The thermal calculations were performed with ANSYS. For material data in algorithms and thermal conductivity and electrical losses at each mesh point, refer to [17–19].

11.3.3.2 Key Small Model Calculation

The key small model geometry and thermal loads are illustrated in Figure 11.5. Please note that only the outer conductor portion from the 4 K flange to 70 K flange was under analysis. Thermal isolation boundary conditions are applied so that the T on the top flange is fixed to be 70 K, and the T on the bottom flange is set to be 4 K. RF heating of the outer conductor is applied as heat fluxes that are normal to the inner surface of the outer conductor. The 10.0 μm copper is plated on the inner surface of stainless steel (wall thickness 0.5 mm). RF dissipation is evaluated by Equation 11.1.

$$P_{out} = \frac{P}{\eta . \ln \frac{b}{a}} \cdot R_s \cdot \frac{1}{b} \cdot \frac{1}{2\pi b} \tag{11.1}$$

where P is the RF power transported in the coaxial line; $p = 7000.0$ W; a, b are the ratio of the inner conductor and outer conductor, respectively; η is the impedance = 376.7 Ω in this calculation; $R_s = \rho/\delta$ gives the surface resistance; ρ is the resistivity, which is temperature dependent; $\delta = \sqrt{\frac{2\rho}{\mu\omega}}$ is the skin depth; $\mu = 4.0 \times 10^{-7}$ H/m is the permeability of free space;

and ω is the circular frequency of the RF wave.

The heat intercepted by the bottom surface of the 4 K flange is calculated by ANSYS. The influence of RRR values on the heat transfer to the 4 K flange is investigated. The RRR values corresponding to the copper plating and solid copper after brazing were used in these algorithms. The

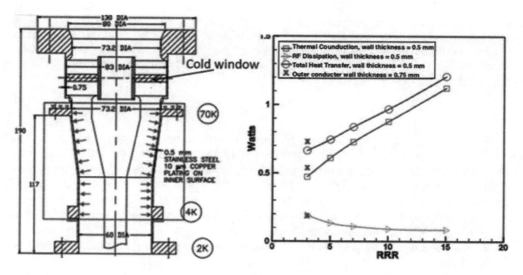

FIGURE 11.5 Left: Geometry and thermal loads for key small model analysis. Only the outer conductor components enclosed in the rectangular frame are considered in the small model; arrows stand for RF dissipation. Right: Heat transferred from 70 K to 4 K (static and RF dynamic) as a function of the RRR of the Cu coated on the inner surfaces of the outer conductor of the coaxial RFIC [15].

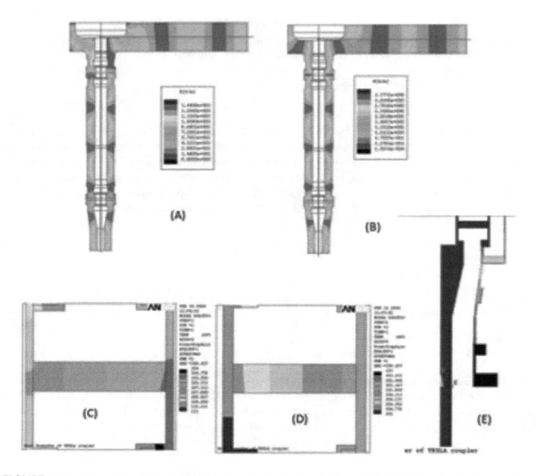

FIGURE 11.6 Part of the RF and thermal simulation results. A: Electric field distribution in the coupler; peak electric field occurs at the inner conductor taper near the bottom of the waveguide. B: Magnetic field distribution in the coupler; peak magnetic field occurs at (1) the inner conductor portion under the cold window and (2) the antenna portion below its taper. C: Temperature distribution in the cold window, RRR = 3. D: Temperature distribution in the warm window, RRR = 3. E: Temperature distribution in the cold part with RRR = 3. [15]

thermal conductivity values of the ceramic windows, material properties corresponding to the different RRR values, and expected RRR changes after brazing were taken from [20–21]. A series of Ni strike and copper plating thickness and several heat treatment T values were taken into account. The materials, part dimensions, and material combinations used for the AMAC-TESLA coupler design were varied until the required thermal performance was obtained. The most difficult specification heat loss value to meet was found to be in the section between the cold window and the flange attaching to the 2 K cavity because of the limited length available. Figure 11.6 (Right) shows part of the simulation results: static heat load, RF dynamic heat load, and total heat load as functions of copper RRR based on a wall thickness of 0.5 mm and 0.75 mm.

11.3.3.3 Heat Transfer Analysis of the Complete RFIC

This model includes all major components of the RFIC, including the antenna, cold ceramic window, adjustment bellows, warm ceramic window, and so on. The methodology is similar to the small model, and the detailed simulation/calculation was introduced in a report by Shu et al. [15]. There are several special considerations to be taken into account in calculations.

Anomalous Skin Effect on Surface Resistance. The anomalous skin effect is characterized by a dimensionless parameter α_s. α_s depends on ρ, L (the mean free path of electrons), and ω. when $\alpha_s \leq 0.016$, the classical expression of surface resistance applies. When $\alpha_s \geq 3.0$, it is suggested that the surface resistance should be re-evaluated [19]. The inner surface of the outer conductor is fully copper plated. If, after plating, the copper layer has RRR = 3, then classical surface resistance can be used for any temperature. However, if the RRR is greater than 5, there is no theoretical expression of surface resistance for $0.016 \leq \alpha_s \leq 3.0$. For the case of RRR = 200, when the temperature falls between 58 K and 136 K, $0.016 \leq \alpha_s \leq 3.0$. Thus, considering anomalous skin effects is more conservative when α_s is of intermediate value, that is, $0.016 \leq \alpha_s \leq 3.0$.

Dielectric Loss. In ceramic windows (loss tangent $\tan \delta = 0.0004$), dielectric losses are calculated on the basis of the electric fields through HFSS simulation at the center planes of warm and cold ceramic windows.

Nitrogen Gas. Design A: The space between the warm and cold windows is a vacuum. Design B: The space is filled with nitrogen. The thermal analysis considers both designs.

Results of Case Studies. The distributions of E and H fields are first simulated; accordingly, the dielectric loss and RF heating are calculated. The temperature maps for the entire RFIC are drawn, and the heat transfers to the 70-K, 4-K, and 2-K flanges are obtained, respectively, for different copper RRR values, thicknesses of stainless steel, and vacuum vs nitrogen between the windows.

Figure 11.6 demonstrates part of the RF and thermal simulation results.

Table 11.3 presents the comparison of heat transfer to 2 K, 4 K, and 70 K in the coupler as the RRR value changes. The minimum heat loads are 3 < RRR < 15, but for all cases, the resulting heat loads are lower than specifications. Table 11.4 gives the comparison between thermal results from vacuum and nitrogen gas models for RRR = 3, and both meet the specifications.

TABLE 11.3
Heat Load (W) in TESLA Coupler with Copper Plating of Various RRR Values

		RRR Values				
	Specifications	3	5	7	10	15
Heat transferred to bottom of 2-K flange	0.12	0.0697	0.0535	0.0418	0.0412	0.0464
Heat transferred to inner surface of 4-K flange	1.0	0.999	1.182	0.9499	1.513	1.878
Heat transferred to bottom of 70-K flange	12.0	7.694	7.873	7.213	8.151	8.490

Source: With Copper Plating of Various RRR Values [15]

TABLE 11.4
Thermal Data (W) with/without Nitrogen Gas, Cu-Plating, RRR=3

Heat transferred to bottom of 2-K flange	0.12	0.0697	0.0696
Heat transferred to inner surface of 4-K flange	1.0	0.999	0.9977
Heat transferred to bottom of 70 K flange	12.0	7.694	5.534

Source: For Model with/without Nitrogen Gas, Cu-Plating, RRR = 3 [15]

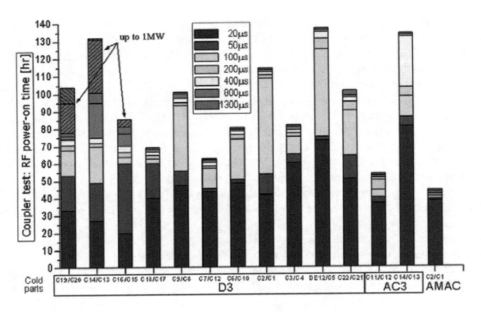

FIGURE 11.7   Coupler conditioning times: TTF3 and AMAC [12].

11.3.4 FRIEF TEST RESULTS

The final conditioning results for the AMAC-DESY TESLA input coupler show that the total RF power rise time was 44 h, total *RF* on conditioning time was 56 h; after conditioning, a standard power sweep done for 40 h showed almost no multipacting activity. The average RF on conditioning time for TTF3-type couplers at DESY is 60–130 h (Figure 11.7). The two AMAC-DESY input couplers meet the TESLA technical requirements.

11.4 COAXIAL RFICs WITH SRF CAVITIES IN CRYOMODULES

The Rare Isotope Accelerator (RIA) high RF power coupler was designed by AMAC to meet the specification requirements for the RIA accelerator cryomodule at Michigan State University (MSU) [22]. Communications & Power Industries (CPI) produced the two prototype couplers. The prototypes have been high RF power tested by Jefferson Lab and met or exceeded the requirements of the RIA technical specifications. The couplers were installed with superconducting cavities, as shown in Figure 11.8 (Upper). The couplers in the cryomodule were tested at a cryogenic temperature of 2 K. The performance of the couplers with cavities met the RIA specifications: 10 kW (CW) with a Voltage Standing Wave Ratio less than 1.05. The design load to the 2 K liquid helium is less than 2 W. The external Q of the coupler is about 2×10^7.

A high RF power coaxial input coupler vacuum tightly connects to the SRF cavity's beam tube (the cavity is inside the LHe vessel), as illustrated in Figure 11.8 (Lower) [10]. The outer cylinder of the RFIC is thermally intercepted to minimize the heat leak from ambient temperature to LHe. The connecting pipes of the fundamental power couplers to the beam tube are also cooled with LHe. Peterson et al. also summarize the features of TESLA-TTF input couplers with rich testing data for reference [23].

11.5 WAVEGUIDE HIGH RF POWER INPUT COUPLERS

The waveguide RFIC is also attractive for higher-frequency (size can be smaller than in low frequency) and very high-power applications (larger than 100 kW in CW). Its cooling is simpler, as

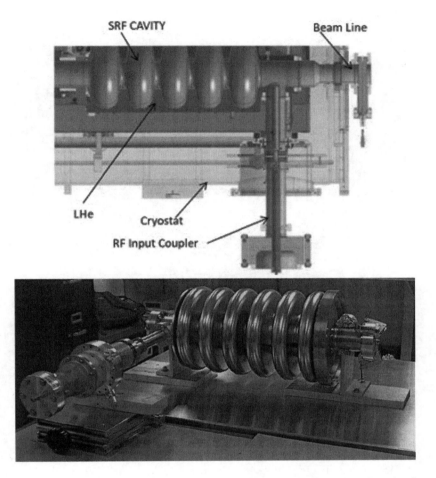

FIGURE 11.8   Lower: The AMAC's RIA high RF power coupler with an MSU RF cavity for external Q measurement [22]. Upper: The Euro Spallation Neutron Source (SNS) high power RF power coupler (RFIC) with SC cavity in a cryostat [23].

only the outer WG wall needs cooling. Also, the RF power density at the WG wall is much smaller than at the center conductor of a coaxial coupler, particularly in low frequency. However, the integration of a WG coupler with a cryostat is more complicate in mechanical structures than coaxial RFICs. In general, there seems to be no strong technical advantage of one approach over the other, but it appears there is a preference.

11.5.1   GENERAL FEATURES OF WAVEGUIDE RFICS

Because of the WG geometry, windows for WG couplers are generally more difficult to manufacture, and multiple circular windows within the WG cross-section were originally used. The coupling strength can be adjusted in three basic ways: (1) by the size of the coupling iris, (2) by the longitudinal location of the WG with respect to the cavity's end cell, and (3) by the location of the terminating short of the WG itself. Multipacting that occurs in WG can be moderated by the use of magnetic field biasing.

As a classic example of the WG RFIC, The RFIC for a Cornell Electron-positron Storage Ring (CESR-B) cavity is shown in Figure 11.9 [24]. It has a fixed coupling at $Q_{ext} = 2 \times 10^5$ via a coupling

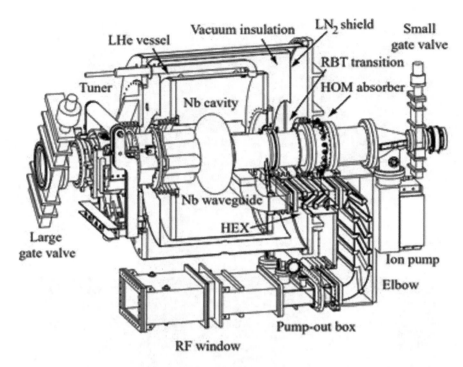

FIGURE 11.9   Cornell B waveguide coupler, layout of B-cell cryomodule SRF system for CESR luminosity upgrade [24].

slot in the beam pipe. The rectangular WG section immediately exterior to the helium vessel is cooled by cold helium gas flowing through the tracing welded to the WG walls. Next is the WG double E-bend elbow, which is cooled by liquid nitrogen. Following this is a short thermal transition to room temperature, a WG vacuum pumping section, and finally the WG vacuum RF window. This WG path was dictated by tight space considerations in the CESR tunnel. The RF window must transmit more than 300 kW CW at 500 MHz traveling wave and experience a >125 kW CW standing wave with the electric field maximum at the window ceramics when operating with a beam current of 550 mA. All waveguide parts are made of copper-plated stainless steel. Hence, RF windows for WG RFICs are usually planar inserts with one or more ceramics of different shapes and variable thicknesses [25]. If a cold window is used, a second window is needed in the warm part of the coupler to preserve vacuum on both sides of the cold window. When a warm primary window is used, a second cold window will normally not be demanded, even if the reliability issue is important.

11.5.2   HEAT FLOW INTERCEPT

As an important heat management issue, one has to decide on the best location to mount the 50 K intercept. To balance RF dissipation against heat conduction down the guide to meet the required minimum heat load for 2 K less than 1 W, a series of studies were carried out to optimize the distances between a 50-K thermal intercept and 2-K flange. The heat load to 2 K as a function of waveguide length (2 K to 50 K) and properties of material (stainless steel [SS] and Cu coating) is presented in Figure 11.10 [26]. The minimum length to optimize the heat load will depend on the total heat conductance of the SS waveguide wall and of the copper plating in addition to the RF loss in the copper. Although the Cu surface impedance is predominated by the anomalous limit between

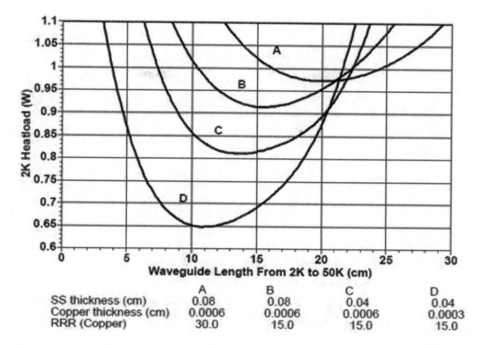

	A	B	C	D
SS thickness (cm)	0.08	0.08	0.04	0.04
Copper thickness (cm)	0.0006	0.0006	0.0006	0.0003
RRR (Copper)	30.0	15.0	15.0	15.0

FIGURE 11.10 The 2 K heat load from 50 K induced by the waveguide RFIC as function of the intercept location and properties of the material (SS and Cu coating) [26].

50 K and 2 K, the RRR and thickness of the copper layer have a significant influence and must be chosen to balance performance risk to against the manufactured tolerance.

11.5.3 WAVEGUIDE RFICS WITH SRF CAVITIES IN CRYOSTATS

Besides at Cornell University, various waveguide RFICs have been developed in several FEL and accelerator projects at Jefferson Laboratory. A cold ceramic window close to the beam line inside the cryostat and a polyethylene warm window (later changed for ceramic) were originally utilized. Currently in a 12 GeV upgrade project with 13 kW CW power, the seven-cell SRF cavity and double warm window RFIC were developed (each cryomodule has eight cavities) [27]. The WG RFIC is made of copper-plated stainless steel with thermal interceptions between 2 K and the outer cryostat envelope at 300 K. Each waveguide has two 90° elbows both to reduce radiation penetration and avoid the window charging to impact the 2 K flange. JLAB also developed the WG RFIC for high current cavity (Goal: >100 mA at 1.5 GHz and >1A at 750 MHz), which is integrated with the SRF cavity in the cryostat.

11.6 HIGH-ORDER MODE COUPLERS

Unwanted high order mode (HOM) of electromagnetic fields (so-called wakefields) is generated by a charged particle beam through the accelerating cavity. If the HOM is not sufficiently damped, it can lead to beam instabilities or loss of brightness. HOM also increases cryogenic losses due to the additional power dissipation in the cavity wall at LHe temperature. Cryogenic losses can be very serious when the beam currents become greater in many modern applications, as shown in Table 11.5.

TABLE 11.5

HOM Power Loss vs Beam Current

Project	CEBAF 12 GeV	XFEL	SPL	BERLIN —PRO	KEK-CERL	Cornell ERL	e-RHIC
Beam current (ma)	0.1	5	40	100	100	100	300
HOM power (per cavity, W)	0.05	1	22	150	185	185	

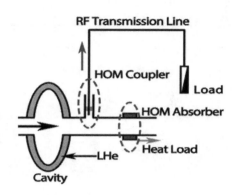

FIGURE 11.11   A HOM coupler extracts unwanted RF power from the cavity/beam and damps to load.

HOM couplers are developed and positioned in SRF structures to subtract and remove HOM waves from the cavity space and adjacent beam tube and damp them to absorbers at higher temperatures [28]. Simultaneously, a HOM coupler is required so that the coupling with the fundamental mode must be kept as small as possible. The HOM coupler must have effective cooling, compact structure, and minimal size to fit in the cryostat. Besides, low cost and being effectively dismountable are also important considerations. To efficiently damp HOM energy, broad-band absorbing materials with a low outgassing property in vacuum, such as Ferrite SiC, AlN, glassy carbon, and so on, are choices. A simplified sketch showing the function of a HOM coupler extracting unwanted RF power is shown in Figure 11.11.

Based on the different methodologies of extracting the HOM from the cavity/beam-tube at low temperatures and damping the HOM energy at high temperatures, HOM couplers can be categorized into three types: coaxial, waveguide, and HOM beam tube damper. One of the most important design parameters of a HOM coupler is the external quality factor, Q_{ext}, which describes the coupling to the modes or their damping by the coupler.

11.7   COAXIAL HOM COUPLERS

11.7.1   DESIGN CONSIDERATIONS

One of the first HOM couplers based on the coaxial line technique was applied for the 500-MHz four-cell Hadron-Electron Ring Accelerator (HERA) cavities at DESY and 352 MHz five-cell Large Electron–Positron Collider (LEP) cavities at CERN. Then a coaxial HOM coupler was successfully developed for the 1,300-MHz TESLA cavity [29–31], which was modified and innovated in many projects. The two types of coaxial HOM couplers are illustrated in Figure 11.12 [32]: one is L-type and another is I-type. Both HOM couplers are made of superconductor (Nb) to reduce RF heating.

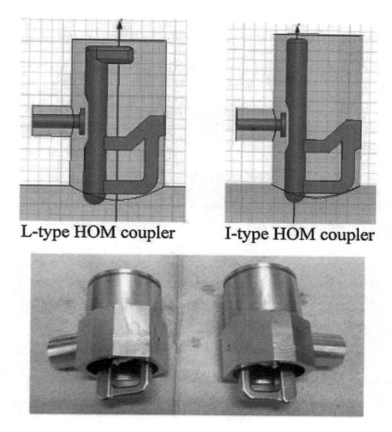

L-type HOM coupler   I-type HOM coupler

FIGURE 11.12   Examples of HOM coupler design approaches [32].

These couplers consist of resonant antennas shaped as loops or probes, which are designed to couple with potentially dangerous RF modes while sufficiently rejecting the fundamental mode. The HOM filter is to block the transmission of the fundamental mode while transmitting potentially dangerous HOMs to the RF cable. The design goals are optimizing Q_{ext}, low RF heating, depressing multipacting, efficient cooling, and less complexity. The thermal behavior of all three couplers is very similar and dominated by the heat transfer through the RF cable rather than by the electromagnetic field inside the cavity. A thermalization at 2 K on the coupler tube and further heat intercepts on the RF cable are required to sufficiently reduce thermal load.

Once the coupler quenches and Nb becomes normal conducting, the dissipated power in the surface rises by several orders of magnitude, which could potentially burn and melt parts of the coupler. There are two different heat sources for the HOM coupler, which act on the input and output side: (1) power dissipation caused by the electromagnetic fields in the coupler (extracted from the cavity) and (2) heat transfer via the RF cable, which extracts the HOM power.

11.7.2   GENERAL THERMAL ANALYSES

In order to define cooling requirements to ensure a stable superconducting state of the coupler surface, coupled RF and thermal simulations have been carried out. There are three heat sources in the coupler: RF heating, RF cable (extract HOM power out of coupler) conducting, and multipacting. As the coaxial HOM coupler is made of a superconductor to reduce the RF heating, peak temperatures rather than the average heat loss are more critical. A certain cable length inside the cryomodule and the use of heat intercepts may reduce the heat transfer to the coupler significantly. The evaluations of the temperature ranges in correlation to the surface resistance are performed to ensure that the coupler can operate stably. In the simulations for CERN's HOM coupler with HFSS and ANSYS, the surface resistance for bulk Nb is ~50 nΩ at 2 K, and RRR = 300 (worst case of 200 nΩ is used) is introduced.

RF Heating. The surface current and dissipated power are connected via the surface resistance composed by R_{BCS} according to the BCS theory and R_{res}, the temperature-independent residual resistance. R_{res} can be estimated to a few nano ohms; hence, it is a minor contribution to the total surface resistance. The thermal conductivity of niobium significantly depends on its RRR. When the pure Nb RRR is about 300, the heating of the HOM coupler by the electromagnetic fields appears to be minor unless the coupler is not well surface treated.

Heating from the Outside. Another heat source is the RF cable, which is thermally anchored to the 50 K shield and connected to the output port of the HOM coupler. The heat flows from the 50-K shield and sinks to the cavity wall at 2 K (coupler tube). The Nb critical temperature is 9.2 K. In order to keep the temperature-dependent surface resistance as low as possible, a heat load ≤1 W is required in the design. A cable length of 2 m with two heat intercepts for cooling the cable outer conductor can bring the heat transfer to less than 500 mW (enough of a margin). Subsequently, couplers were modified to suppress multipacting heat.

High-Current Beams. In the case of high-current beams passing the SRF cavity, the RF heating generated on the HOM coupler surfaces will greatly increase. Therefore, the RF heating becomes dominant and additional cooling is required in the HOM coupler design.

11.7.3 EXAMPLES OF COAXIAL HOM COUPLERS

TESLA-Type HOM Couplers [30]. As shown in Figure 11.12, the TESLA-style coaxial HOM coupler includes: (1) a loop antenna for coupling the HOM power, (2) a notch filter for rejection of the accelerating mode power, and (3) a HOM pickup probe for extracting HOMs to an external load. The HOM pick-up probe is made of SC (Nb) to reduce the RF heating on the tip surface. The notch filter is tuned to suppress fundamental mode coupling to the HOM pickup probe.

Jefferson Lab and Fermilab HOM Coupler[33]. If the heat is not sufficiently conducted away through the probe feedthrough, the tip may become normal conducting, leading to a thermal breakdown that dramatically lowers the cavity quality factor (*Q*). However, typical cryogenic RF feedthroughs provide a poor thermal conduction path for the probes. Reece et al. have developed a

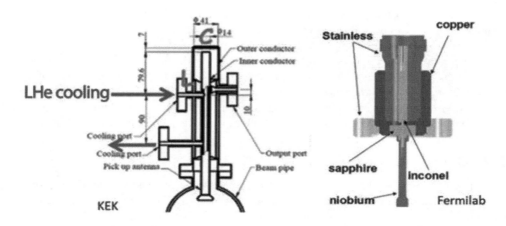

FIGURE 11.13   Fermilab 3D model of HOM coupler [33] and KEK design with active cooling [34].

solution, which is to directly braze the niobium probe to a single-crystal sapphire dielectric and the sapphire to a substantial copper sleeve captured in the stainless steel mounting flange. The copper sleeve is available externally for thermal strapping, if needed. A 3D model of this kind of HOM coupler at Fermilab is shown Figure 11.13. When the RF power goes to CW, the feedthrough enhanced with thermal strap still manages to remain stable at 15 MV/m. Within the range of geometries modeled, one may summarize the BCS heat generated on a niobium probe at 20 MV/m as 3–6 mW at 6 K and 16–30 mW at 8 K. This power must be sunk to 2 K through the feedthrough dielectric to the flange and cavity beam pipe or via supplemental strapping to the feedthrough body. Analysis indicates that with the external copper collar stabilized at 3 K, a heat flux of more than 100 mW could keep the probe below the critical T of niobium.

KEK HOM Coupler [34]. When CW beam current reaches higher levels, the RF heat generated on the HOM pick-up surface must be removed by active cooling. Figure 11.13 introduces the KEK design, which employs the active cooling of LHe through the center conductor.

11.8 WAVEGUIDE HOM COUPLERS

11.8.1 ADVANTAGES OF WG HOM COUPLERS

WG HOM couplers are preferred for many SRF applications because of their advantages both in physics and technical promise: (1) the high-pass filtering characteristics of the waveguide are used to bring the operating frequency of the cavity under the cut-off frequency of the waveguide. The higher-order modes can propagate from the waveguide. (2) They are able to handle high-power HOM, which is critical for high current and high power accelerators. The HOM loads can be damped at higher temperature regions, and static loss is tolerable. (3) They don't add much to the length of the cryomodule, which is beneficial for limited-size tunnels. Particularly, in applications of 100 mA, "Ampere class," such as ERL for the Navy, compact FEL for shipboard, machines of high electrical gradients, and high packing sizes, WG HOM couplers are the preferred choices. However, a rectangular WG HOM damper leads to a higher heat load for the cryomodule due to its big size.

11.8.2 EARLY WG HOM COUPLERS

HOM couplers for SRF cavities based on the waveguide technique were first designed at Cornell University in 1982 for 1.5-GHz cavities (Figure 11.14). About 160 of these cavities and HOM couplers were operated very successfully in the Continuous Electron Beam Accelerator Facility at Jefferson Lab (CEBAF) accelerator. The low nominal beam current of CEBAF and a small amount of HOM power allow termination of the HOM power with the WG absorber cooled by LHe inside the cryomodule. The original CEBAF absorber (glassy carbon, then synthesis of aluminium nitride (AIN)-based composites) mounted onto the HOM end flange and the modified design with extended copper cooling block for higher beam current application are also presented.

11.8.3 WG HOM COUPLERS FOR HIGH BEAM CURRENT

KEK, JLAB, Cornell, HZB, and others have developed several high-current CW beam machines (several hundred mA). One common feature is the use of WG HOM couplers with RF absorber materials at room temperature cooled by water. As an example, the BESSY-voltage sensitive relay (VSR) HOM coupler is for a future upgrade of the third-generation BESSY II light source. These 1.5-GHz cavities will operate in CW at high field levels ($Eacc$ = 20 MV/m) with high beam current 300 mA. Each cavity has four cells. The development of the HOM damping is a really challenging project due to the high current, high bunch charge, and multi-pass CW operation. The main HOM damping of the cryomodule is performed by 20 warm water-cooled WG HOM loads. These absorbers are made of silicon-carbide (SiC)-based composites with a maximum power specification of 460 W per load at room T. A detailed view of the WG HOM couplers and the tapered absorbers is shown in Figure 11.15 [35].

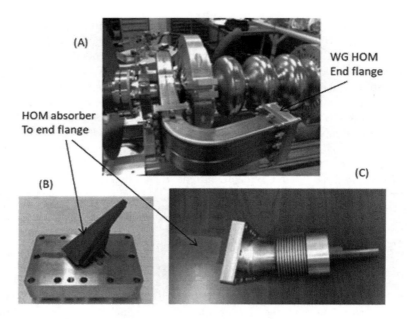

FIGURE 11.14   A: CEBAF cavity with WG HOM coupler. B: Original CEBAF absorber on end flange. C: End flange with extended copper cooling block for higher current. Source: *JLAB*

The absorbers are tapered on the height side of the WG and divided into 25.4 × 25.4 mm blocks in three columns and six rows so each block can be brazed onto a 4×4 Cu pegboard. The ceramics columns/rows are separated by 2.54 mm for brazing clearance. Each row of ceramics block has a different slope, forming a curved taper. Between 2 K and room T, there are two heat reception stations at 5 K and 80 K to reduce the heat reaching the 2 K cold mass. Calculations indicate about ~0.1 W in 2 K, ~0.6 W in 5 K, and 3.4 W in 80 K with the Cu-plated WG section between the bellow and the Nb flange. Without the Cu plating in that section, the Nb flange will quench.

11.9   HOM BEAM TUBE DAMPERS

11.9.1   GENERAL CONSIDERATIONS AND ABSORBER MATERIALS

Why a BT Home Damper. Utilizing a beam tube (BT) damper to substitute coaxial or waveguide dampers has been successfully demonstrated for storage ring cavities to handle average beam currents as high as one the order of 1 Ampere. Beam tubes are also enlarged (up to 200–300 mm ID) with BT dampers so that the lowest-frequency HOM can escape the cavity to be absorbed. It was first realized for the CESR and KEKB single-cell cavities, and then various other storage ring-based light sources adopted the technique. In general, the advantages of BT dampers can be briefly listed as the follows:

* BT is a natural high-pass filter with very broadband performance
* High power-handling capability
* Radial symmetry helps avoid beam kicks and helps all HOM be damped
* Can incorporate bellows sections between cavities
* Relatively simple design and efficient cooling and cost reduction

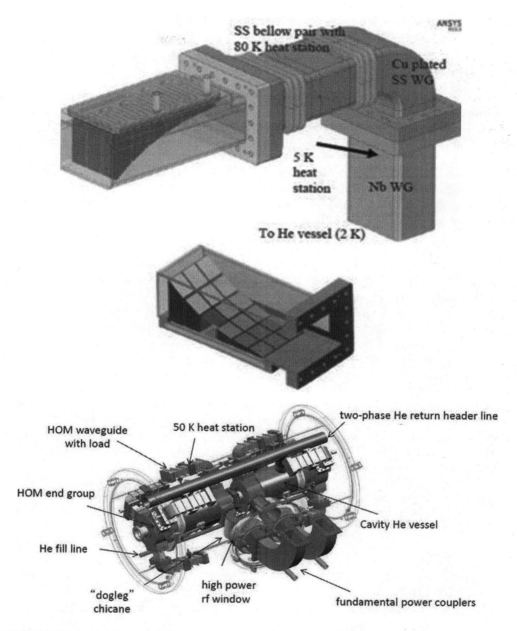

FIGURE 11.15 Lower: Four-cell-high power SRF cavity cut with waveguide input and HOM couplers [25]. Upper: Water-cooled HOM WG loads [35].

Cooling. When the HOM power level reaches several hundred W, it is better to place the BT damper at room *T* cooled by water rather than at cryogenic *T*. Therefore, a longer BT for *T* transmission is needed that will increase the estate size and cost.

Materials for TB Dampers. For BT dampers, various high-performance absorber materials have been developed, such as ferrite, glassy carbon, SiC (silicon nitride), and AIN. These materials must have the following properties: high thermal conductivity, low outgassing in vacuum, reliable cleaning for a dust-free environment, HOM broadband adsorbance, and good mechanical performance for shaping and brazing.

11.9.2   HOM BT DAMPERS AT ROOM TEMPERATURE

For single-cell superconducting structures in B-factories, synchrotrons, or light sources that operate with high currents, a common idea is to make all HOMs propagate into beam tubes and dissipate their whole energy at room T outside the cryomodule. KEK-B, CESR-B, and others utilize ferrite tiles as RT absorbers to dissipate several kilowatts of HOM power and water-cooling outside the cryostat.

The designed WG HOM cavity of BESSY is highly efficient but still has high HOM energy traveling outward from the module, about 1.5 kW. In order to avoid undesired load transfer to the ring or any backward reflection into the VSR module, a set of two new warm BT dampers are utilized. These warm loads use the SiC Coorstek.

11.9.3   HOM BT DAMPERS AT CRYOGENIC TEMPERATURE

ERL HOM BT Damper. The ERL at Cornell will operate with nominal beam current up to 100 mA. Stable operation requires suppression of about 200 W of the HOM power per seven-cell cavity of 2 K. The BT damper is installed between cavities inside the cryomodule [5]. Figure 11.15 shows a 3D view of the Cornell HOM BT damper and a picture taken during string assembly inside the clean room. The center assembly consists of the absorbing cylinder, which is shrink-fit into a titanium cooling jacket and flange. The cooling jacket and flange locate, support, and provide cooling at 80 K to the absorbing cylinder using a cooling channel inside the titanium. The absorbing material is silicon carbide, SC-35 from Coorstek. After a careful cleaning, no particle generation during the mounting procedure was observed. In addition, no Q degradation of the cavity mounted next to the absorber in a horizontal test cryo-module was observed. For some applications, more careful analysis was required for the 80 K circuit, where the coolant first cools the intercept of the power coupler and then the HOM absorber. The heat load of the HOM absorber is in the range of 0 to 400 W.

BT Damper for European XFEL. A high-frequency part of the HOM spectrum of the European XFEL facility is dissipated in the BT damper installed between eight-cavity cryomodules. The propagating HOM power is 5.4 W/cryomodule for operation with 40,000 bunches/s. CERADYNE ceramic rings are used for the absorption [30]. Dissipated power is transferred to the LN_2 line by

FIGURE 11.16   Left: 3D BT damper for the ERL at Cornell. Right: Mounting procedure inside the clean room [6].

the copper stub brazed directly to the ceramic. The stub holds the ring in the stainless steel vacuum chamber. The heat capacity of the absorber is more than 100 W. This extra margin is introduced for future upgrades of the facility.

REFERENCES

1. Omet M, Ayvazyan V, Branlard J et al., 2019 *Operation of the European XFEL towards the maximum energy,* SRF2019, Dresden, Germany, mathieu.omet@desy.de
2. Shu QS, Susta J, Cheng GF, 2003 *Five projects of high RF power input couplers & windows for SRF accelerators,* Superconducting RF Workshop, DESY, Germany.
3. Miller S, 2019 *FRIB cavity and cryomodule performance, comparison with the design and lessons learned,* SRF2019, Dresden, Germany.
4. Shu QS, 1998 Large applications & challenges of state-of-the-art superconducting RF (SRF) technologies, *Advances in Cryogenic Engineering,* 43.
5. Shu QS, Demko J, Fesmire J, 2019 Thermal optimization of functional insertion components (FIC) for cryogenic applications, *IOP Conf. Ser.: Mater. Sci. Eng.,* 755, 012139, 2020.
6. Eichhorn R, 2015 First results from the Cornell high Q CW full Linac cryomodule, CEC 2015, *IOP Conf. Series: Materials Science and Engineering,* 101.
7. Ohuchi N, Pagani C et al., 2012 Study of thermal radiation shields for the ILC cryomodule, *AIP Conference Proceedings,* 1434, 929.
8. Moeller W, 2011 *Design and technology of high-power couplers, with a special view on SRF,* CERN Yellow Report CERN-2011–007, pp. 209–222.
9. Kostin D, Moeller W-D, Kaabi W, 2013 *Update on the European XFEL RF power input coupler,* SRF2013, Paris, France.
10. Putselyk S, 2019 *Application of sub-cooled superfluid helium for cavity cooling at linac-based free electron lasers, energy recovery and proton linacs,* CEC 2019, IOP Conf. Series: 755 (2020) 012098.
11. Kako E, 2019 *High power input couplers and HOM couplers for superconducting cavities,* SRF2019, September 19.
12. Moeller WD, Shu QS et al., 2006 Development and testing of RF double window input couplers for TESLA, *PHYSICA C,* 441. ELSEVIER.
13. Shu Q-S, Susta J et al., *Five projects of high RF power input couplers & windows for SRF accelerators,* funded by a US Department of Energy DE-FG02–99–ER82739 & DE-FG02–00ER86102, a contribution from (CPI), a CRADA with Jefferson National Accelerator Facility, and a R&D agreement with DESY (Germany).
14. Moeller WD, 2004 for the TESLA collaboration, *Status and operating experience of the TTF coupler,* LINAC 2004, Luebeck, Germany.
15. Shu QS, 2004 *Novel, reliable, and cost effective input coupler for high RF power applications,* Final Report on DOE STTR, Date: June 30.
16. Proch D, Einfeld D, Onken R, Steinhauser N, *Measurement of multipacting currents of metal surfaces in RF fields,* DESY and Fachhochschule Ostfiesland, Emden, Germany, Particle Accelerator Conference, 1995.
17. Gall D, Lierl H, 2003 DESY, Private communication, April.
18. Cryodata (METALPAK).
19. V.Arp, private communication.
20. Nemoto T, Sasaki S, Hakuraku Y, *Thermal conductivity of alumina and silicon carbide ceramics at low temperatures,* Hitachi Limited, Ibaraken, Japan.
21. Singer W, et al., *Influence of heat treatment on thin electrodeposited Cu-layers,* RF superconductivity—7th Workshop, Gif-sur-Yvette, France, 1995.
22. Shu QS, Susta J, Cheng GF, Terry L, Grimm T et al., 2005 805 MHz high power input coupler for SRF cavity in a RIA cryomodule, *Applied Super Conductivity,* 15(2), June.
23. Darve C, 2014 The ESS elliptical cavity cryomodules, *CEC, AIP Conf. Proc.,* 1573, 639–646.
24. Belomestnykh S, 1999 *Operating experience with superconducting RF at CESR and overview of other SRF related activities at Cornell University,* SRF-1999.
25. Rimmer B, 2013 *JLAB waveguide couplers,* for JLab SRF Institute.
26. Delayen JR, *An RF input coupler system for the CEBAF energy upgrade cryomodule,* Jlab-ACT-99–03, 1999.
27. Reece C, et al., *Design and construction of the prototype cryomodule renascence for the CEBAF 12 GeV upgrade,* U.S. DOE under contract DE-AC05–84ER40150, 2005.

28. Mosnier A, 1989 *Developments of HOM couplers for superconducting cavities*, 4th Workshop on RF Superconductivity, August 14–18, Tsukuba, Japan.
29. Geng RL, Li YM, 2014 Comparative simulation studies of multipacting in higher-order-mode couplers of superconducting RF cavities, *Physical Review Special Topics—Accelerators and Beams,* 17, 022002.
30. Sekutowicz J, 2007 Multi-cell superconducting structures for high energy e+ e-colliders and free electron laser linacs, IES-WUT, CERN, EuCARD 2008.
31. Reece CE, 2005 High thermal conductivity cryogenic RF feedthroughs for higher order mode couplers, *IEEE, Proceedings of 2005 Particle Accelerator Conference*, Knoxville, TN.
32. Kako E et al., 2010 Cryomodule tests of four Tesla-like cavities in the Superconducting RF Test Facility at KEK, *Phys. Rev. ST Accel. Beams*, 13, 041002.
33. Wu G, Harms E, Khabiboulline T, 2008 *Evaluation of HOM coupler probe heating for 3.9 GHz cavities*, FNAL-TD-08–019, 2008.
34. Kako E, 2019 *Tutorial lecture of high power couplers and HOM couplers*, SRF2019.
35. Guo J, Fors F, Henry J, Rimmer RA, Wang H, Development of waveguide HOM loads for berlinpro and BESSY-VSR SRF cavities, *Proceedings of IPAC2017*, Copenhagen, Denmark, 2017.

12 Special Cryostats for Laboratory and Space Exploration

James E. Fesmire, Quan-Sheng Shu, and Jonathan A. Demko

12.1 INTRODUCTION

A cryostat is a device designed to hold a sample and apparatus at cryogenic temperature while providing all the interfaces (cryogen feeding, powering, diagnostics instruments, safety devices, etc.) for reliable testing or performing special tasks. The design, configuration, and operation of the cryostats are strong demonstrations of various technologies in cryogenic heat management. The T of the samples and apparatus in the cryostats is maintained by cryogenic liquids, cryocoolers, and mixed cooling methods. Normally, the sample and apparatus with cryogen or cryocooler are assembled in the vacuum vessel of the cryostat to minimize unexpected heat flows from the ambient environment. From a broad perspective, cryostats have numerous applications within science, engineering, and biomedicine. Therefore, cryostats have a great number of designs, structures, and functions for various unique applications. The contents of cryostats in this chapter are selectively not involved in the areas of calorimeters, SC accelerators, SC fusion, and so on because these subjects are discussed in detail in other chapters of this book. The focus of this chapter is cryostats developed for laboratories, industry, and space exploration. Commercially available cryostats will be only briefly mentioned.

From a practical point of view, the advanced design and performance of cryostats are also introduced as follows, while the main focus is on the thermal aspects:

- What cooling methods are preferred in specific applications
- Heat transfer enhancement between the sample/device and cryogen/cryocooler
- Reduction of heat flow from ambient environment to the sample/device and cooling structures
- Mechanical structure for stabilized support
- Unique high vacuum techniques for cryostats
- Easy loading of the sample/device
- Room temperature bore for samples and variable temperature sample probe
- Optical window for light/laser studies of samples
- Comprehensive implementations of these technologies for space exploration
- System integration

In Chapters 2–6 of the book, heat transfer (gas convention, solid conduction, and radiation), thermal insulation, and thermal anchors are discussed in detail. So, the method of derivation of thermal perspectives is not a focus in this chapter, but practical resolutions of these methods and principles for cryostats will be introduced. After a brief discussion of design considerations and methodologies, a series of representative examples of advanced cryostats are presented with their design schematics, operational techniques, and performance data.

DOI: 10.1201/9781003098188-12

12.2   METHODS OF COOLING SAMPLES/APPARATUS IN CRYOSTATS

The most important design issue of cryostats is to determine the cooling method, as illustrated in Figure 12.1.

The criteria to be followed when choosing a proper cooling method are as follows:

- The temperature to be reached and hold time
- The mass and velum of the sample/apparatus to be cooled
- Geometry of the sample/apparatus
- Assembly practicality
- Cryostat working environment
- Cost and other special requirements

Cryostat with Cryogen Bath. A common, simple cryostat cooled by an LHe bath is shown in Figure 12.1A, and the testing sample/apparatus can be placed in LHe or thermal contact with the LHe. The LHe may be refilled as often as every few hours or as long as months. The boil-off rate is minimized by efficiently reducing heat leaks through low thermal conductance support and multi-layer insulation (MLI). The additional LN_2 bath and concentric thermal shields are common practice. The cryogens in the cryostat can be LHe, LH_2, and LN_2 (or solid H_2) depending on the T needed: 1 K–4.2 K with LHe and 63 K–78 K with LN_2. The advantages of these cryostats include but are not limited to: no vibration, stable T, operation for several days to months (boil-off ~1% per hr). Disadvantages include the storage and handling of liquid cryogens.

Dry Cryostat (Cryogen Free) with Cryocooler. The cooling source of the cryogen-free cryostat is a cryocooler without the use of any cryogenic liquids, as shown in Figure 12.1B. The heat load of the sample/device is removed by a commercial two-stage cryocooler. The heat extraction is done by two thermal anchors made of high-conductivity copper. One of them is connected between the cryocooler's first stage and the active cooled thermal shield (40 K~80 K), and the other is thermally linked to the second stage and the sample/device (about 2 K–4 K). These cryostats are compact and have a low operation cost, high autonomy, and flexible orientation but require a high investment cost and could experience possible vibration.

Cryostat with Mixed Cooling Resources. As presented in Figure 12.1C, the cryostat is refrigerated by both cryogen and cryocooler. Normally, the first stage of the cooler is thermally anchored to the intermediate shields. The second stage of the cryocooler is either thermally anchored to the re-condenser on the top of the LHe bath for recondensation of the vaporized He gas or thermally connected to the LHe bath to greatly reduce the LHe boil-off. The arrangement of cryogen bath plus a cryocooler can be specially designed as a closed cycle cryostat, which can theoretically be operated for an unlimited time without additional liquid cryogen.

Continuous-Flow Cryostats. Continuous-flow cryostats are normally cooled by LHe or LN_2 from a storage dewar. The steady cryogen flow from the dewar is continuously replenished by the cryogen boils within the cryostat, as depicted in Figure 12.1D. The T of the sample is controlled by the cryogen flow rate into the cryostat together with a heating wire attached in the loop. The length of operating time is dependent on the volume of cryogens available. These cryostats are costly to operate but easily used to cool small samples in a restricted space, such as in microscopes.

Cryostat with Dilution Refrigerator (DR). To achieve T of a sample/device lower than evacuated liquid ^{3}He (0.3 K) and ^{4}He by a vacuum pump, such as lower than 1 K, dilution refrigerator or dry dilution refrigerator is used to reach as low as 2 mK. The cooling power of ^{3}He/^{4}He dilution refrigerator is provided by the mixing of the ^{3}He and ^{4}He isotopes, as shown in Figure 12.1E.

Cryostat with ADR. When a T of the sample/device lower than 1 mK is needed, an adiabatic demagnetization refrigerator will be employed in the cryostat. The basic operating principle of an ADR is to use a strong magnetic field to control the entropy of a material, often called the refrigerant. If the refrigerant is kept at a constant temperature through thermal contact with a heat sink

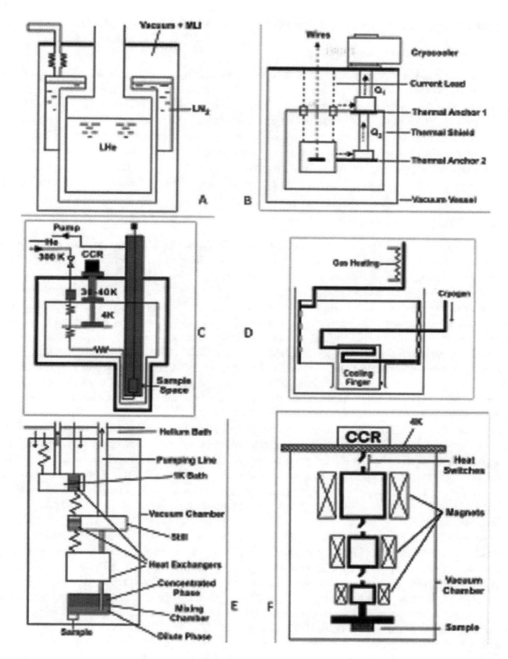

FIGURE 12.1 The main methods of cooling samples/apparatus at cryogenic temperature in cryostats: A: Cryogen bath. B: Cryocooler. C: Mix-approach with cryogen and cryocooler. D: Cryogen continuing flow. E: Cooled by dilution refrigerator. and F: Cooled by the adiabatic dilution refrigeration (ADR) method [1].

(LHe or cryocooler) while the magnetic field is switched on, the refrigerant must lose some energy because it is equilibrated with the heat sink. When the magnetic field is subsequently switched off adiabatically, the heat capacity of the refrigerant rises again, thereby lowering the overall temperature of a system with decreased energy.

12.3 CONFIGURATIONS OF CRYOSTATS FOR SAMPLES/APPARATUS

12.3.1 VERTICAL TOP-LOAD CRYOSTATS

A vertical top-load cryostat (VTLC) for testing samples/apparatus is the most common configuration and traditional design, regardless of which cooling method is chosen, as presented in Figure 12.1A–F. In general, the inner vessel of the vacuum enclosure is cooled to around 77 K by LN_2 or around 20 K by the first stage of the cryocooler. The LHe bath is vertically supported by its neck to top flange with/without a thermal anchor to intercept conducting heat leak. The sample/device with instrument wires and/or a magnet is immersed in LHe (or thermally connected to the second stage of the cryostat) and hung firmly from the top cover flange. If the neck has a wide opening, a group of thermal shields that act as buffers are assembled along the neck to block radiation.

12.3.2 OTHER SPECIAL CONFIGURATIONS OF CRYOSTATS

Due to various special requirements for testing samples/apparatus, many cryostats with sophisticated configurations have been developed and successfully employed around the world. These configurations can be generally categorized as in Figure 12.2.

Cryostat with Optical Windows for Laser and Particle Beam. The optical cryostat has become a foundational instrument in condensed matter physics research. For the study of material properties under irradiation of laser and the like at cryogenic *T*, the cryostat must have an optical tail assembly, four quartz/sapphire optical window sets (two warm and two cold), a sample holder, and an adjustable dewar mounting flange, as sketched in Figure 12.2A. Cryostats with four windows can also be cooled by either LHe bath or other cooling sources such as a cryocooler and refrigerator below 1 K. The optical viewports (windows) are tightly vacuum-integrated with the walls of the vacuum vessels. The assembly mounted on the cold stage can hold different optical samples with variable dimensions.

Cryostat with Variable Temperature Sample Probe. Beyond a critical base temperature, many experiments require sweeping through wide temperature ranges or stepping across transition temperatures. At each new temperature setpoint, the entire chamber must reach thermal equilibrium before taking a measurement, and waiting for thermal equilibrium can take minutes to hour. As shown in Figure 12.2B, the cryostat offers quick sample exchange and an operational range of 2 K to 300 K. The key component of the cryostat is the sample exchangeable and temperature variable probe (insertion).

Cryostat with Warm Bore. In many applications, such as MRI for medical applications, magnetic separators for water and mining, high-energy physics, and so on, experiments and processing take place at room temperature under very strong magnetic fields, which can only be provided by superconducting (SC) magnets in a cryostat. Therefore, a cryostat with warm bore (either vertical or horizontal) is essential for these applications, as illustrated in Figure 12.2C. The vertical warm-bore cryostat is cooled by a two-stage cryocooler for operating a SC magnet and the radiation shields and high-temperature superconductor (HTS) current lead. The warm bore assembly is the key component, which must meet both conflicting requirements simultaneously: (1) the space between the warm wall and the magnet inner hall should be as thin as possible to provide the largest testing space with strong magnetic fields, and (2) the thermal insulation system must block the heat flows between the warm wall and the cold magnet as much as possible.

Cryostat with Horizontally Openable Ends. When the testing subject is a device instead of simple samples, a cryostat with openable ends for functionally housing the device and providing cryogenic supports is crucial. These cryogenic test stands can be oriented horizontally or vertically. A cryostat with openable horizontal ends is illustrated in Figure 12.2D. The cryostat has a supply box for power, cryogens, pumping ports, and instruments in the center, while the test object (a superconducting radio frequency [SRF] cavity or SC magnet) is in its own cryogen vessel and slides into the cryostat horizontally from the end opening. Thermal shields are preinstalled, and

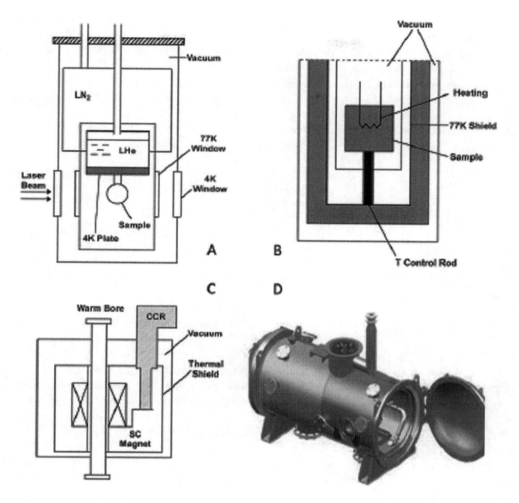

FIGURE 12.2 Simplified configuration sketches, A: Cryostat with laser window. B: Cryostat with variable *T* of sample hold. C: Cryostat with warm bore insertion. D: Horizontal cryostat with end openings for testing device [1].

MLI is wrapped on the test device to block the radiation heat. The main container is the outermost component under a high vacuum.

12.4 GENERAL CONSIDERATIONS OF CRYOSTAT THERMAL DESIGN

12.4.1 REDUCTION OF SOLID THERMAL CONDUCTION

Solid thermal conduction is the dominating heat flow contribution of cryostats in supporting systems and all feedthroughs (neck, sample holders, current leads, SRF couplers, instrumentation wires, etc.). There are several preferable methods in terms of minimizing the solid thermal conduction in cryostat design:

- Utilize materials with low thermal conductivity: stainless steel, special alloys (nickel alloys), G-10, and so on.

- Use of intermediate thermal interception at higher temperatures to block heat flow from ambient environment.
- Use of pipe, hollow cylinder post, and rope as support instead of solid beam to reduce heat flux while maintaining required mechanical strength.
- Re-entry structures of support to lengthen the heat flow pass within the limited space.
- Using thermal contact structure instead of solid connection (solder, weld, bolt) if possible.

Based on Fourier's law, for a beam of cross-section A and length l with the two ends of the beam at temperatures T_1 and T_2, conducting heat flow is presented as Equation 12.1.

$$\dot{Q} = \frac{A}{l} \cdot \int_{T_1}^{T_2} k(T) dT \qquad\qquad 2.1$$

The thermal conductivity integrals (W/m) for selected technical materials between indicated temperatures are presented in Table 4.1 of Chapter 4. To determine the best location and T of the heat interception, the Carnot power for varying intermediate temperatures at the optimum locations along the solid support is illustrated in Figure 4.6 of Chapter 4. With this information, one can perform general calculations and design cryostat support systems.

12.4.2 Minimization of Radiation Heat

MLI. Thermal radiation is the most important contribution to the thermal budgets for cryostats, and MLI is the dominant approach to eliminate radiation heat in cryostats. MLI functions as a multiple radiation reflection structure, which is inserted both between the warmer radiating surface (normally the outer vacuum vessel) and the intermediate thermal shield and between the thermal shield and cold surface of the internal device.

MLI materials and thermal performance between various pairs of T_h (warm) and T_c (cold) are introduced in detail in Chapter 3. Although MLI performance varies with structure and workmanship, several reference data points are briefly cited here for convenience: 300 K–77 K, 77 K–4 K, and 77 K–20 K. In general, from 30 to 45 layers of MLI are applied on the shield surface for 50 K–80 K and from 10 to 20 layers would be on the cold mass of the cold device being kept at approximately 4 K. When the shield T is about 20 K (as cooled by first stage of a cryocooler), the use of MLI layers on the testing device will be optional and depend on the conditions.

Multi-Conductive Shields. For a real cryostat structure, several practical engineering issues are crucial. An intermediate floating shield in a cryostat significantly reduces heat load to the cold surface and extracts the heat at a higher T, resulting in a lower cryogenic cooling cost. Both clean and well-polished Al and Cu have good thermal conductivity, small emissivity, and mechanical stress, but Al is lightweight and low cost. Gaps and slots in thermal shields and MLI blankets should be avoided since these gaps and slots will cause unexpectedly great heat loads on cold surfaces. When gaps and slots are unavoidable, typically due to thermal contraction and sophisticated geometries, MLI pads covering the gaps and slots are practical solutions to avoid radiation heat-through.

The wide opening neck of some cryostats can be a serious additional heat load due to the radiation heat transfer, as shown in Figure 12.3. Radiation heat transfer through the neck with a solid angle is presented as Equation 12.2. The Q_{neck} will increase with R enlarging and L reducing, which is true in some applications. One solution is to allocate the multi-low emissivity thermal shields along the neck, and the heat down the shields will be removed by the vapors. In practice, from three to five shields (layers) are enough; one shield is better anchored with LN_2 temperature and is used in combination with light insulation foams between shields in some cases.

$$Q_{Rneck} = S\varepsilon_e \sigma T^4 \quad \text{and} \quad \Omega = \text{solid angle} = \frac{R^2}{R^2 + L^2} \qquad\qquad 12.2$$

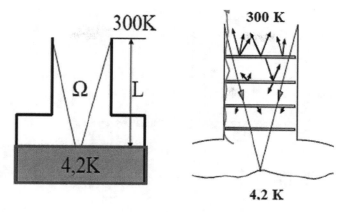

FIGURE 12.3 Radiation heat through wide neck and one solution.

12.4.3 Eliminating Gas Convection and Conduction

Generally, heat transfer by convection between surfaces at different T inside cryostats is negligible at moderate vacuum levels such as pressures below about 1 torr. However, residual gas conduction remains a considerable contribution that depends on the vacuum level, gas species, geometry, and temperatures.

Condensation starts to form if the gas pressure is higher than its vapor pressure at the temperature of the condensation surface. If the surfaces of a cold device in cryostat are around LHe temperature, the vacuum level can be very high due to condensation and freeze-out of the residual gases within the vacuum space. When $\lambda \ll L$, the viscous regime applies, and heat transmission is described in terms of thermal conduction, k, and does not depend on pressure. The molecular regime is reached as $\lambda \gg L$. The molecules travel from the warm to the cold surface, and heat transfer becomes proportional to residual gas pressure independent of wall distance, L. Of course, cryostats must keep actively pumping or a vacuum tight structure with high-absorptivity materials to absorb the gases released from solid surface. Although cryostats work under low pressure, a safety relief valve is needed to prevent over-pressurization due to leakage of cryogenic liquid to the warm region or sudden vaporization of any cryogenic liquid.

12.5 CRYOSTATS WITH CRYOGEN BATH FOR LAB TESTS

12.5.1 Classical Cryostats with Cryogen Bath

A classic LHe bath-cooled cryostat for laboratory tests is shown in Figure 12.4. The cryostat is made of stainless steel (SS) with LN_2-cooled guard vessel and radiation shields surrounding the LHe bath to reduce boil-off [1–2]. The cryostat can reach as low as 1.8 K by pumping in liquid ^{4}He. The radiation baffles in the cryostat wide neck prevent radiation from the top to LHe directly. Two groups of MLI layers are separately assembled on the surfaces of the LN_2 vessel and LHe vessel. The sample chamber submerged in LHe can be evacuated or filled with exchange gas (He ~1 mbar) to vary the sample's temperature (associated with a heater plus a heat switch in some cases). Normally, the sample cell and SC magnet (if needed) are supported by SS tubes or G-10 rods from the top flange with thermal anchors to minimize the heat leak. All instrument wires are anchored at a point at 4.2 K or other preferable T to reduce the heat leak. These cryostats are about from 1 m to 1.5 m height with a bursting disc pressure relief valve on the outer vacuum chamber. There are many configurations of these cryostats and sample chambers, which are commercially available worldwide. Users can also design and construct their own unique sample chambers and instruments to fit into an existing cryostat. Customers may also let a company produce a cryostat based on their tech specifications.

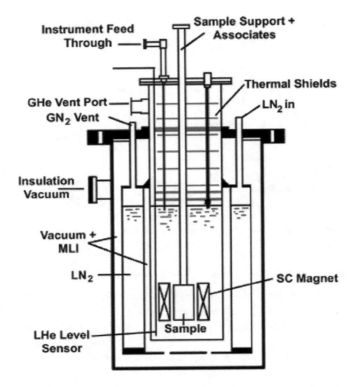

FIGURE 12.4   Sketch of a classical LHe bath-cooled cryostat for laboratory testing [1].

12.5.2   VERTICAL LHE II CRYOSTATS FOR MAGNET TESTS

A large vertical test cryostat used to test high-field SC magnets cooled by an LHe bath was developed at CERN and reported by Graen et al. [3], as illustrated in Figure 12.5. The design principles and methodology can also be adopted by similar types of cryostats. The T of the 3-m³ LHe bath is controlled between 4.2 K and 1.9 K. The dimensions of the cryostat are capable of housing magnets of up to 2.5 m in length with a maximum diameter of 1.5 m and a mass of 15 tons. All cryogenic supply lines are permanently connected to the cryostat to allow for faster insertion and removal.

Process Principles

There are two LHe baths in the cryostat, and these are separated by the lambda plate. Above the lambda plate is the saturated helium bath at 1.3 bar. Under the lambda plate is the sub-cooled liquid helium bath at 1.3 bar and 1.9 K in which the magnet is immersed. The lower bath heat exchanger is filled with saturated liquid helium at 1.8 K. Before entering the heat exchanger, the incoming liquid helium is subcooled in a gas/liquid heat exchanger by the outgoing pumped gaseous helium and finally expanded through a Joule Thomson (JT) valve. The power dissipated into the 1.9 K pressurized helium bath is extracted to the saturated bath at 1.8 K through the heat exchanger wall and is extracted by pumping on this bath. The magnet is precooled from 300 K to 90 K and then is cooled with liquid helium inserted directly in the helium vessel.

Key Functional Components

1. Cryostat. The cryostat is permanently connected to cryogenic lines and houses the heat exchanger, which saves precious installation space. The magnet is supported by two 50-mm-thick SS plates linked by three stainless steel rods with five evenly spaced copper

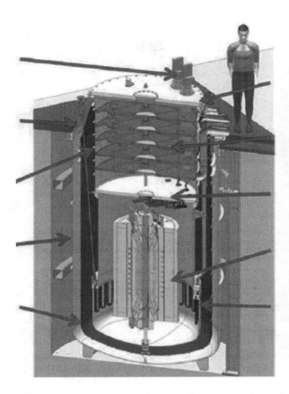

FIGURE 12.5   Section view of the cryostat and thermal shield assembly. *[3]*

baffles. The lower plate is the lambda plate, which ensures the separation of the helium vessel in the two bath temperatures, 4.2 K on the top and 1.9 K on the bottom.

2. LHe Vessel. The neck of the cryostat (from lambda plate to top) is a 1.6-m-long cylinder (3 mm thick) with a heat intercept placed at an optimized distance to limit the conduction heat leak to LHe. The temperature of the lower helium bath and the magnet is controlled by a copper heat exchanger, as shown in Figure 12.6.
3. Neck with Thermal Baffle Radiation Shields.
4. Thermal Shield. A 50 K–75 K thermal shield is composed of two main copper parts reinforced by a SS structure. The lower part is actively cooled with helium gas, and the top part is cooled by copper braids thermally anchored to the lower part.
5. Vacuum Vessel. The top conical cover is assembled via a flanged connection with a leak-tight O-ring sealing to the main cylindrical vessel. All the weight of the inner vessel and the insert is suspended on this flange. The thickness of the cone wall can be as thin as 10 mm. A 200-mm burst disk protects the vacuum vessel from over-pressurization (1.5 bar).

Heat Loads

Conduction heat load to the thermal shield is mainly due to the heat intercept installed on the neck of the helium vessel. Most of the conduction load dissipated in the 4.2 K bath is due to the four resistive current leads required to power the magnets. The static heat loads to the 1.9 K helium bath are mainly due to imperfections. The dynamic heat loads are due to power dissipation of magnet inductance. The summary of heat loads is given in Table 12.1.

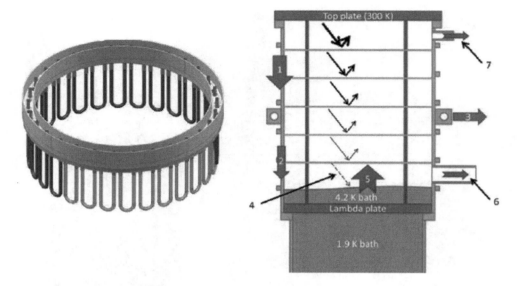

FIGURE 12.6 Heat exchanger tube welded to He vessel and thermal processing of the neck. *[3]*

TABLE 12.1
Heat Load

Source of Heat Load	20 K–75 K (Thermal Shield)	Temperature Level 4.2 K (Top Helium Bath)	1.9 K (Lower Helium Bath)
Conduction	25 W	75 W	6 W
Radiation	35 W	10 W	2 W
Imperfections (splice, tightness. . .)	/	/	17 W
Total static	**60 W**	**85 W**	**25 W**
Total dynamic	/	/	**15 W**

Source: Different Levels of Temperature [3]

Cryostats with Similar Design but without Lambda Plate

There are several similar configurations of other cryostats, which also have LHe II baths reaching ~1.8 K by pumping directly on the bath top, but some designs do not use the lambda plate. These cryostats have been designed and constructed successfully for the testing of SC magnets and SRF cavities at many leading laboratories [4–6].

12.5.3 Horizontal LHe Test Cryostats

Horizontal LHe test cryostats are developed primarily for horizontally loading and testing of SRF cavities, SC magnets, and other devices, which are equipped with their own LHe tanks, DC/RF power components, and instruments. These multipurpose cryostats generally consist of the horizontal vacuum vessel and top vertical feedcan (valve box) and accommodate different devices for testing at 2 K, 4 K, and extended *T* regions, reported respectively by Umemori et al. [7–9].

Figure 12.7 [9] shows the horizontal test cryostat, which has a vacuum chamber of 1 m diameter and 3 m length. An SRF cavity is in the cryostat. There are several flanges on both sides of the cryostat to access inside. Components, such as input couplers and frequency tuners, can be installed and tested. There are 2 K He lines, 5 K He lines, and 80 K N2 lines. A 4 K He pod, prepared inside

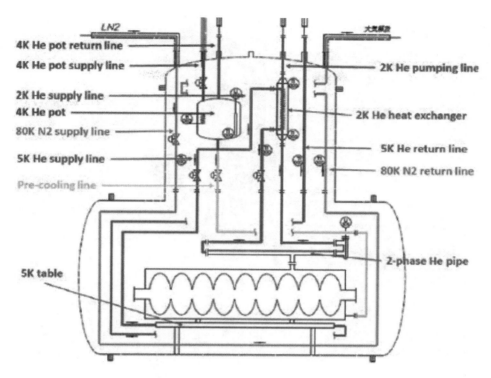

FIGURE 12.7 Conceptual view of the horizontal test cryostat [9].

the cold box, feeds 4 K He to the precooling line and 5 K He supply line and 2 K He to the 2 K He supply line through a JT valve. A He pumping system evacuates the 2 K line. Inside the vacuum chamber and cold box is covered with 80 K shields, which are cooled by liquid N_2.

Roger et al. presented another compact cryostat, which is used as a spoke cavity testing station, as shown in Figure 12.8.

12.5.4 CRYOGEN BATH CRYOSTATS WITH WARM BORE

The cryostat has a SC magnet with the warm bore (WB) that presented a unique challenge to the design [10, 11]. Racine et al. reported a successful modification of an existing horizontal magnet cryostat into warm bore-associated equipment. The magnet is mounted in a small helium vessel surrounded by a LN_2-cooled 80 K shield and a room T vacuum vessel. There are three paths for LHe from a 500-L dewar that enters the cryostat: One path is for cool down, entering the bottom of the He vessel. A second entry is used to fill the magnet once it is cold and maintain the helium level. A third line is used to force LHe into the space between the magnet coils and the warm bore in case there are He gas bubbles developing in the space and affecting magnet performance.

Helium Vessel and Warm Bore. Since the development of cryostats, space limitations on the bore coupled with the minimum diameter of the bore to provide for the test fixture limited the options available with very low vibration and minimal heat leak. G-10 rings attach the magnet to the helium vessel with a calculated heat load of 1.5 W to 4.2 K and 7 W at 80 K. The supports are two separate concentric rings with a copper liquid nitrogen-cooled shield between the rings. The copper LN_2 shield extends around, enclosing all the helium space. A 76.2-mm bore tube was picked for the inner magnet vessel, and a 50.8-mm bore tube was selected for the warm bore. A copper shield is placed between the two tubes and is cooled by liquid nitrogen. Calculated heat flow from

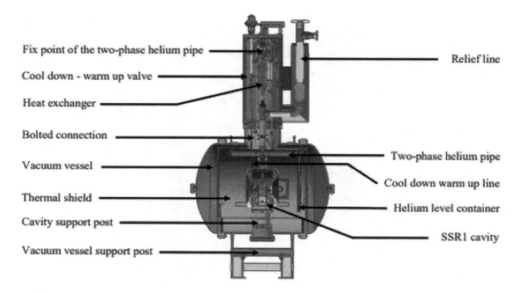

FIGURE 12.8   Cross-section of the LHe horizontal test cryostat with warm bore [10].

the warm bore to the copper shield is 4 W at 80 K, with a further loss of 1 W to 4.2 K. The total heat leak to the 4.2 K and 80 K levels in the cryostat is estimated to be 5 W and 16 W, respectively.

12.5.5   COMPACT LHe BATH TEST CRYOSTATS

LHe bath test cryostats usually have a cylindrical vessel with a vacuum jacket containing LHe and a vertical insertion (test object) emerging in the LHe bath supported from the top flange. A compact LHe Bath test cryostat normally lets the test insertion have its own LHe bath, which is surrounded by a room-temperature vacuum vessel. This arrangement is compact and needs less LHe to quickly cool down the test object.

Bruce et al. introduced a compact cryostat at CEA Paris-Saclay in which the magnets are cooled with LHe at 4.2 K and the liquid bath can be pressurized up to 2 bars to regulate the magnet temperature, as illustrated in Figure 12.9 [12].

To test SC solenoids, a specific insert made of non-magnetic SS 316L in the cryostat 1.7 m high and 410 mm inner diameter (ID). A LN_2 tank surrounding the cryostat act as a thermal shield. The helium circuit of the insert is composed of a current lead (CL) tank, a phase separator, and several connections with liquid helium at 4.45 K and regulates its pressure at 1.25 bar. Figure 12.9 shows the main elements surrounding the mechanical parts with the CL cluster composed of six current leads, the He transfer line, and the SS rods. The CL cluster and the phase separator tank (about 5 liters) are connected to the Y-shaped LHe circuit on the top of the solenoid. Five horizontal thermal shields are installed between the top of the phase separator and the main flange. The closest thermal shield to the main flange is made of copper, welded to the CL cluster tank and to the outlet helium tube. The Saraf solenoid package is shown in Figure 12.10. The system is controlled with LabVIEW.

12.6   CRYOGEN-FREE CRYOSTATS FOR LAB TESTS

12.6.1   CRYOCOOLER-COOLED CRYOSTATS WITH WARM BORE

Cryogen-free SC magnet systems (CFMSs) for measurement of physics properties and material processes within a warm bore at high magnetic fields are popular in many applications since CFMSs

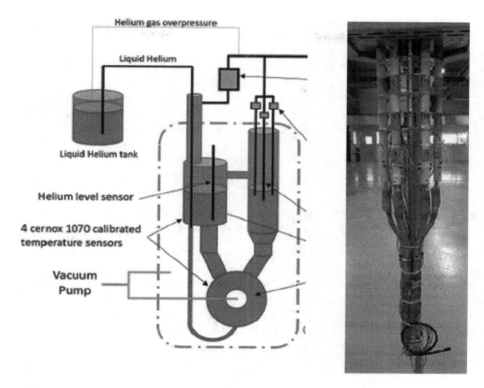

FIGURE 12.9 Left: Sketch of the compact LHe cryostat. Right: Current leads. *[12]*

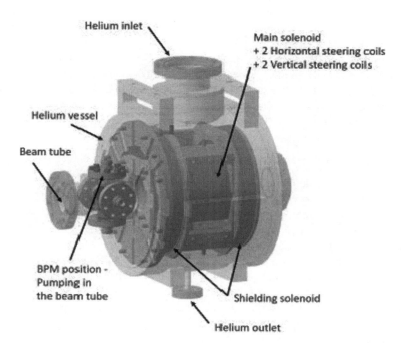

FIGURE 12.10 Saraf solenoid package. Source: *[12]*

are much easier to operate than the use of LHe [13, 14]. SC magnets are cooled to the desired T by conduction from a cryocooler (CCR). The cooling of the conventional SC magnet is as easy as immersing the magnet in a LHe bath. The operating T of a CFMS is dependent on several factors: the cooling capacity of the CCR, thermal conduction link between the second stage of the CCR and the magnet, heat loads caused by current ac losses in the SC wire, eddy current loss, thermal gradient between the cold head of the CCR, and winding of the magnet. In a cryogen-free environment, one of the difficulties presented is how to secure the stability of the SC magnet against any thermal disturbance. The thermal conductance of the link between the cold head and magnet greatly influences the charging rate and the stability of the magnet. Kar introduced a 6-T CFMS with a vertical 50 mm, room T working bore, cooled by a two-stage CCR (GM-cryocooler of 35 W at 50 K and 1.5 W at 4 K) for cooling the thermal radiation shield, hybrid current leads, and the NbTi magnet [13].

12.6.2 PULSE TUBE-COOLED CRYOSTATS FOR LASER/NEUTRON EXPERIMENTS

Despite the obvious advantage of operational simplicity, the use of CCRs (cryogen-free systems) is limited by the mechanical vibrations produced by conventional cryo-coolers. The pulse tube refrigerator (PTR) offers an effective solution: lower maintenance costs, less disruption, and service inspection [15–17]. Chapman et al. reported a cryostat for neutron scattering, which combines a Cryomech PTR (PT410–1 W at 4.2 K) and top-loading insertion with a He condensation loop, as presented in Figure 12.11 [17]. The ~50 K and 4 K thermal radiation shields are mechanically connected to the first and second stages of the PTR. The 50-mm variable temperature insert (VTI) (3) is thermally linked by copper braids to the flanges of the thermal shields. The sample stick (4), inserted into the VTI, has a set of baffles to reduce infrared (IR) radiation heat load on the sample.

In the cryostat, the helium gas first enters the first-stage heat exchanger (5), which consists of a copper capillary (1.5 m long, 3 mm diameter) hard-soldered to a copper form that is thermally connected to the first stage of the PTR. The helium gas then enters the second-stage regenerator heat exchanger (6). This heat exchanger is designed to sit on the regenerator tube of the PTR and is manufactured from a capillary tube silver-soldered to a high-purity copper jacket. On leaving the regenerator heat exchanger, the helium flows through the second-stage heat exchanger (7). Thermal contact between the VTI heat exchanger (10) and the sample (12) is achieved by cold exchange gas. A temperature range from 1.35 K to 300 K can be achieved by the appropriate use of exchange gas and sample heating. One of the main advantages of the VTI is the absence of liquid helium in the sample horizontal plane, which makes it ideal for neutron or X-ray scattering experiments.

The sample can be held continuously at a stable temperature in the range from 1.45 K to 300 K. The cooling power of 55 mW at 1.8 K is sufficient to run a standard dilution refrigerator insert in the continuous regime in the future.

12.6.3 CRYOSTATS WITH CRYOCOOLERS FOR ONLINE-OPERATING SC DEVICES

SC devices, such the SC undulator (SCU), wind generator, and so on, under online operation have a very long continuous work time. If cryocoolers can used in cryostats to replace continuing LHe supply from a liquefier, it would be preferable in the cost and operation [18–19].

Anliker et al. presented a new superconducting undulator cryostat for the Advanced Photon Source undulator (APSU). This cryostat is a cryocooler-based design providing refrigeration at 4.2 K for undulator magnets and helium reservoir (in zero boil-off), 14 K–17 K for the beam chamber, and about 40 K for the thermal shield reported [19]. The cryostat has three main components: the 508-mm-diameter SS vacuum vessel is 4.8 m. The thermal shield (3.2-mm Cu sheet) with 40 MLI layers consists of a main cylinder, two smaller turret cylinders, and multiple cover plates for access ports.

A cutaway model showing the different components inside of the cryostat is illustrated in Figure 12.12. The magnets are cooled via LHe through two parallel internal channels per magnet.

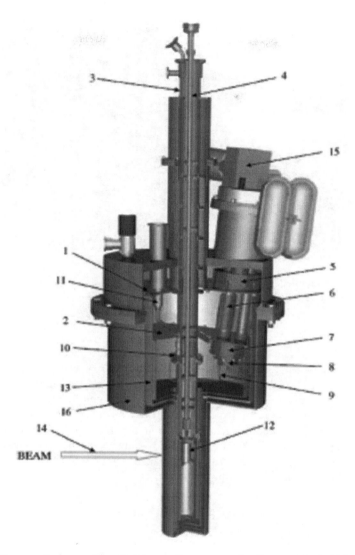

FIGURE 12.11 Cryogen-free cryostat with 50-mm-diameter top-loading sample facilities for neutron scattering experiments [17]:

(1)   50 K thermal radiation shield;
(2)   4 K thermal radiation shield;
(3)   50 mm variable temperature insert;
(4)   sample stick;
(5)   first-stage heat exchanger;
(6)   second-stage regenerator heat exchanger;
(7)   second-stage heat exchanger;
(8)   liquid helium chamber;
(9)   impedance;
(10) VTI heat exchanger;
(11) pumping line;
(12) sample;
(13) outer vacuum can (OVC);
(14) neutron beam.

Source: *[17]*

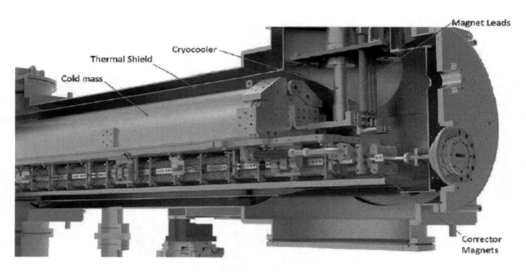

FIGURE 12.12   A cutaway model showing the interior of the APSU SCU showing the thermal shield and cold mass installed inside of the vacuum vessel. *[19]*

Magnet supports allow for adjustment in the vertical and axial directions. The beam chamber is notched periodically along the outer edges to provide space for the gap spacers. The aluminum beam chamber is supported by brackets connected to the bottom of the helium reservoir and thermally isolated from the reservoir with Torlon standoffs. The aluminum chamber is cooled via a distributed copper bus connected to a dedicated 20 K cryocooler. The SCU has four turrets that act as mounting points for the cryocoolers and instrumentation needed for operation. Two main end turrets house two 4 K cryocoolers. The 4 K stages are connected to the helium reservoir by copper foil links, while the first stages connect to a plate that distributes cooling power to current leads and to the thermal shield via braided copper links.

12.7   CRYOSTATS WITH COMBINED COOLING FOR LAB TESTS

12.7.1   Cryostats with LHe Bath/Cryocooler Re-Condensers

Conventional cryostats need periodic refilling of LHe. It results in frequent disruption in operation and heavy loss of LHe. These cryostats require a LHe supplier and a GHe recovery system, which add infrastructure and cost. Therefore, a He recondensing cryostat with various cryocoolers (resulting in a zero-loss system) is a logical and important development [21–24].

Wang et al. have developed a cryogen recycler using a 4 K pulse tube cryocooler for recondensing helium and nitrogen in an Nuclear magnetic resonance (NMR) magnet [21]. The liquid helium cooled NMR magnet has a liquid nitrogen cooled radiation shield. The magnet boils off 0.84 L/day of liquid helium and 6 L/day of liquid nitrogen. The recycler forms closed loops for helium and nitrogen. A two-stage 4 K pulse tube cryocooler, Cryomech model PT407 (0.7W at 4.2 K), is selected for the recycler as shown in Figure 12.13.

12.7.2   LHe II Bath Cryostats with Cryocooler Closed Loop

The closed-loop cooling system has great advantages of efficiency and compactness since no refill of helium is necessary. However, the thermal analysis to accurately predict the temperature distribution of the heat exchanger attached to the cold head for helium liquefaction is crucial to confirm

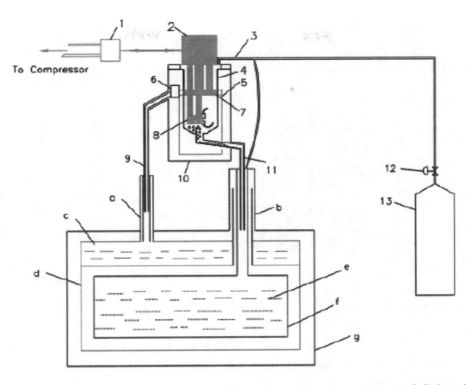

FIGURE 12.13 Schematic of the recycler with an NMR magnet. 1. Remote rotary valve. 2. Pulse tube cold head. 3. Vapor line. 4. Sleeve. 5. Radiation shield. 6. Nitrogen condenser. 7. Pulse tube first stage. 8. Pulse tube second stage/helium condenser. 9. Liquid nitrogen return tube. 10. Main assembly of recycler. 11. Liquid helium return tube. 12. Pressure regulator. 13. Helium gas cylinder. a. Liquid nitrogen fill port. b. Liquid helium fill port. c. Liquid nitrogen bath. d. Radiation shield. e. Liquid helium bath. f. Liquid helium container. g. Cryostat. *[21]*

the feasibility of the proposed cooling design. There are several successful efforts in design and development of the new closed-loop cooling LHe II cryostat systems [25–27].

As shown in Figure 12.14, the He II liquefaction of the closed-loop cooling system for the 21-T, Fourier Transform Ion Cyclotron Resonance SC magnet is introduced by Choi et al., Korea Basic Science Institute—National High Magnetic Field Laboratory, with an emphasis on the heat and mass transfer in the cryocooler heat exchanger [26]. The LTS magnets are immersed in a subcooled superfluid helium (LHe II) bath connected to the 4.2 K reservoir through a narrow channel. Saturated helium is cooled via a Joule-Thomson heat exchanger and flows through a JT valve and then enthalpically drops its pressure to approximately 1.6 kPa, corresponding to a temperature of 1.8 K as it enters the helium II heat exchanger. Helium leaves the He II heat exchanger as vapor, passing through the JT heat exchanger. The helium vapor leaves the cryostat and goes into a vacuum pump located outside of the magnet system. The helium vapor is then purified and liquefied by the cryocooler and is stored in a helium reservoir maintained at a certain liquid level. The helium liquefaction system using a two-stage pulse tube cryocooler in the closed loop without any replenishment of cryogen. Since the cold surface areas of a cryocooler are very limited, a cylindrical copper fin is thermally anchored to the first and second stage cold heads to extend the available heat transfer areas, as shown in Figure 12.14 (right). Since the temperature of second-stage cold head is below the saturation temperature of liquid helium, condensation could occur at the second stage.

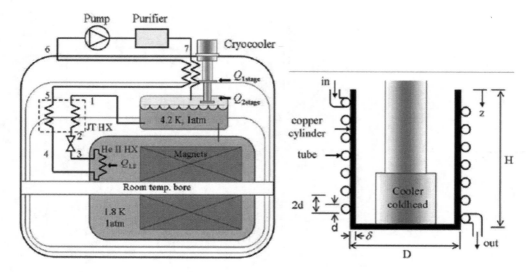

FIGURE 12.14 Schematic of cryostat with closed-loop cooling system for 21-T magnet and heat exchanger on extended surface of cryocooler. *[26]*

12.7.3 Special Inserting Cryostats for Applications with Another Background Cryostat

Low temperature combined with a high magnetic field has now become a basic requirement to study the physical properties of material due to the freezing of atomic motion. Variable temperature inserting cryostat (VTICs) integrated with another background cryostat with a high field SC magnet are drawing high interest [28–29].

Cryocooler-Based Variable Temperature Inserting Cryostats

A cryocooler-based inserting cryostat functioning as a VTI makes the cryogenic system compact and efficient with less power consumption. Nadaf et al. developed a VTIC covered from 5 K to 325 K for measurement of physical properties at low T and H (6 T) [29]. This VTIC needs to be integrated with an existing 6 T SC cryogen-free magnet system, which has a warm bore of 0.5 m deep and 50 mm ID. It poses a complexity in the VTIC design to achieve the desired requirements in the existing sample space.

A schematic of a VTIC system is shown in Figure 12.15 (Left), in which a GM cryocooler is fixed on the vacuum jacket and thermally linked with the thermal shield. The top-loading sample probe is inserted into its sample bore (25 mm), whose warmer section is made of SS-304, and the colder section is Cu. The warmer section of the sample bore is thermally anchored to the GM first stage. The colder copper section of the sample bore is thermally anchored at 120 mm below the ConFlat (CF) flange thermally anchored by copper braids to the GM second stage.

The sample probe has a sample holder at bottom of a 1.27-m SS-304 sample tube, and an instrumentation port is attached at the top of the tube. A set of radiation baffles were silver-brazed onto the sample probe tube. The sample probe can be positioned at the center of the magnetic field. A Cernox T sensor has been fixed on the sample probe. A 25-Ω non-inductive resistive heater of 100 W capacity has also been fixed on the sample probe to vary the sample temperature. The helium buffer vessel will provide necessary exchange gas to the sample bore space for thermal linking to the sample holder. The sample bore is also connected with a small vacuum pump, which can be used to evacuate, if necessary.

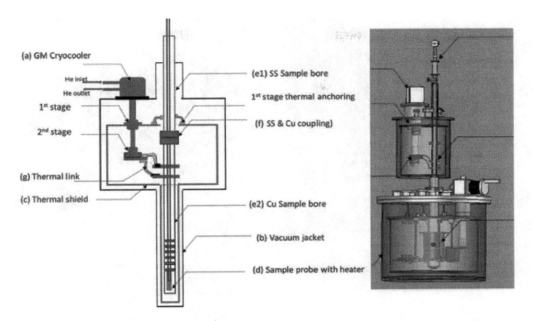

FIGURE 12.15   Left: Schematic of VIT system. Right: Integration of VTI and CFMS. *Nadaf [29]*

LHe Bath-Based Inserting Cryostats

Choi et al. at KBSI developed a 5-T HTS magnet-inserting cryostat for high field magnet applications, which should be inserted into a conduction-cooled 15-T background magnet cryostat with a 102-mm room-temperature bore [28]. The inserting cryostat of the HTS insert magnet is mainly composed of an outer vessel, radiation shield, and inner vessel (the small LHe vessel). The HTS insert magnet suspended by gravitational support is immersed in LHe. The LHe vessel is designed to be suspended from the outer vessel with axial suspension system. The radiation shield is heat-anchored to the LN_2 vessel above the background magnet. The thermal link (Cu) is connected between the LN_2 vessel and inner vessel to reduce heat leak along the wall.

Other Approaches to Varying the *T*

There are other very efficient approaches to varying the sample *T*, such as the use of heat switches and conductive bars with heaters, which are introduced in Chapters 8 and 9 in detail.

12.7.4   CRYOSTATS COOLED WITH CONTINUING FLOW CRYOGEN

Continuous-Flow Cryostats for Optical Microscopy (10 K to 350 K)

For applications such as the study of phase transitions in crystalline solids, a low-*T* stage should cover a wide range of temperature and not require any modifications to the microscope. The windows should be free from condensation and vacuum grease. A simple continuous-flow cryostat used with optical microscopes is introduced in Figure 12.16 by Cosier [30]. Its *T* working range is K to 350 K with stability +/–0.1 K. The specimen is mounted within an exchange-gas filled cell to ensure thermal equilibrium. Access to the specimen is via small, easily demountable, vacuum-tight low-*T* windows made using polytetrafluoroethylene (PTFE) gasket seals. To completely operate the apparatus requires the following components: a small high-vacuum system and a 25-liter LN_2 vessel or 50-liter LHe vessel coupled to the cryostat via a commercial flexible transfer line. A small oil-free diaphragm pump is used to produce the refrigerant flow.

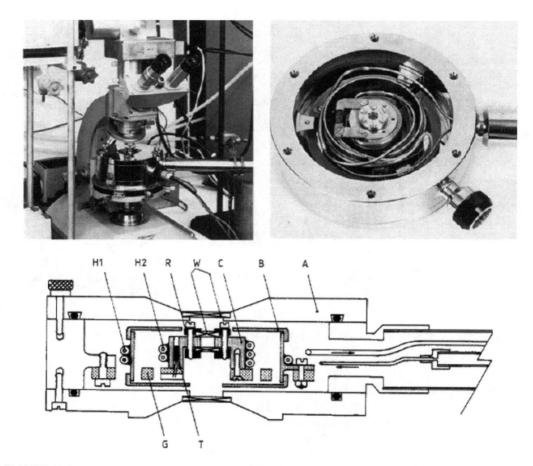

FIGURE 12.16   Upper: The low-T stage with a microscope and inside of the cryostat, shown top plates of vacuum case. Lower: Section of the cryostat. A: Outer vacuum case, B: radiation shield, C: copper cell, G: glass-fiber support, H: heat exchanger, W: Pyrex window, R: SS retaining plate, T: diode thermometer. *[30]*

Variable-T Continuous-Flow Cryostats Inside Scanning Electron Microscopes

Scanning transmission electron microscopy (STEM) can study microscopic structure at the atomic level. Börrnert et al. [31] introduced a concept for a dedicated in-situ STEM for flexible multi-stimuli experimental setups with the capabilities of holographic recording and scanning electron microscopy-type imaging of about 1 nm. A variable-T liquid-helium continuous-flow cryostat for nanometer-resolved imaging and diffraction works at controlled temperatures between 10 K and 300 K. The cryostat has two operation modes, one with two cooled radiation shields for temperatures below 10 K and one without the shields for free sample access from outside the cryostat at temperatures down to 20 K. It can map the phase structure of different electronic and magnetic phenomena, like charge density waves (CDWs).

Counter-Flow Cryostats for Solid Hydrogen Targets

Nixon et al. reported a special counter-flow cryostat, which will produce a solid hydrogen plane (target) perpendicular to the positron beam to perform studies [32]. The cryostat chamber will allow for annealing of the hydrogen crystal to remove any defects that may have been produced during the deposition process. Due to many restrictions, the cryostat utilizes a counter-flow heat exchanger

(HX) for cooling the hydrogen instead of a cryocooler. The hydrogen is cooled from 300 K to 77 K in the HX first stage by He entering at 45.7 K. In the second stage, the hydrogen is further cooled to 14 K through heat exchange of helium entering at 13.8 K.

12.8 CHALLENGES AND CONSIDERATIONS OF SPACE CRYOSTATS

Cryogenics for space is as old as the space era itself, such as the propellants used to launch first Sputnik using liquid oxygen (LOX) in 1957 and then the upper stages of the Saturn V Moon rocket using LO_2 and LH_2, the electrical power systems in orbit and on the Moon using LH_2 and LO_2, and the first dewar used in space on the Apollo missions in 1968. Since then, each important leap forward of space exploration has been accompanied by cryogenic engineering and science. Based on incomplete data, a total of ~110 flight coolers have been and will be worked in space worldwide between the early 90s and ~2025 [33]. The scope of cryogenic technology in space applications has kept up with time and continues to progress into science payloads, electrical power, life support, and propulsion systems.

On the other hand, the farther satellites travel, the weaker the signals that are received. Furthermore, the intrinsic noise (white noise) is proportional to the third power of the sensors' temperature. To receive and analyze useful signals, the T of the sensors of instruments must be reduced to such low temperature that the noise signal is less than the useful signals received. Liquid or solid cryogen dewars for instrument cooling in space were popular in the 1980s and the first half of the 1990s. Cryostats with instruments cooled by closed-cycle cryocoolers and mixed cooling methods are currently preferable. Finally, the durations of space missions become much longer as the missions go into the deep universe and stay in space with long lifetimes.

Wide Scopes of Application. Instruments in cryostats cooled by various cooling methods have numerous applications: Earth observation, imagery (civil or military), monitoring of lands (fires, etc.), meteorology, atmospheric chemistry, science, black hole studies, gamma ray studies, chemistry of exoplanets, and zero-boil-off for long cryogenic propulsion missions. To fulfill these missions, engineers and scientists face a great number of challenges in design and construction of these cryogenic systems, which are briefly summarized in the following.

Challenges of Cryostats in Space. First, there are very limited resources in space, such as power, with which to launch masses and volume of launched payloads. So, heat leaks in the design of cryogenic systems are limited to watts, mW, and microwatts (μW). The cryocoolers used in space must be very light with high thermal efficiency. Second, lifetime requirements are crucial to maintain performance during about ten years of mission times of continuous operation. Most instruments in cryostats with cryocoolers (particularly optical instruments) often have pointing requirements. So, the thermo-mechanical performance of the coolers and instruments, such as thermal link assembly, vibration reduction structure, and heat flow control, are important. Outgassing from cryogenic components must be taken into serious account, as it will affect the performance (contamination and pollution) of the instruments and cryogenic systems. Finally, thermal tests and control of cryogenic systems used in space must be strictly tested on the ground since there is almost no second chance to repair their failures in space.

12.9 SPACE CRYOSTATS WITH CRYOGEN BATHS

12.9.1 Solid H_2 Cryostat for Space Wide Field Infrared Survey Explorer Mission

The mission of the wide field infrared survey explorer (WISE) is to perform a high-resolution, all-sky survey in four IR wavelength bands. Its science payload is a 40-cm-aperture cryogenically cooled IR telescope with four IR focal plane arrays (2.8 to 26 μm). A two-stage solid H_2 cryostat provides cooling to temperatures less than 17 K and 8.3 K at the telescope and focal planes, respectively, for a minimum seven months effective orbit mission [34]. A cryogenic scan mirror freezes the field of view on the sky over the 9.9-second frame integration time.

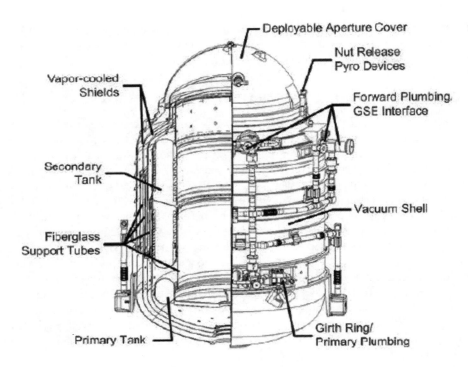

FIGURE 12.17   The WISE cryostat [34].

WISE Cryostat. Larsen et al. reported the two-stage WISE cryostat, which was provided by Lockheed-Martin and consists of two cryogen tanks housed within a vacuum shell, as shown in Figure 12.17 [34]. This design provides two separate cooling zones: The 7.3 K primary tank is mounted off the secondary tank and cools only the Si:As devices. The secondary tank operates at 10.2 K; cools the optical subassembly to less than 17 K; and absorbs the parasitic heat loads from the outer shell, environment, MCT arrays, and telescope. The vapor pressure and temperature of each cryogen tank are determined by the heat loads and the size of the vent lines to space. The aperture cover seals the vacuum space and protects the cryostat interior and optical subassembly during ground and launch operations. The cover is fastened to the cryostat with pyro-actuated separation nuts and will be abandoned on orbit. The cryogen tank vent valves are also pyro-actuated. Cryogens from the secondary tank will be vented through two low-thrust vents to reduce any torque caused by the exiting hydrogen. The primary tank vent rates are not high enough to impart significant torque on the flight system. To limit the parasitic loads into the cryogen, the vacuum shell will operate below 200 K on orbit by sun synchronal orbit and MLI blankets and leave the space-facing side open to radiate to space. Two vapor-cooled shields, mounted intermediately along the support structure, use the hydrogen effluent vapor to absorb a significant portion of the incoming parasitic heat. The heat dissipations for each FPAs are about 3–8 mW, including conduction heat through cables.

12.9.2   He II Bath Cryostats for HSO Space Missions

The He II Bath Cryostat for the Herschel Space Observatory (HSO) of the European Space Agency (ESA) science program, successfully launched in 2009, was designed for six days ground hold time and 3.5 years lifetime in orbit, as shown in Figure 12.18 by Langfermann et al. [35]. In the HSO cryostat, the helium is below the lambda point in the superfluid state, and its evaporation occurs far below atmospheric pressure on ground into vacuum pumps and into space when in orbit (Figure 12.18). The main part of the cryostat is the superfluid helium tank, which contains up to 337

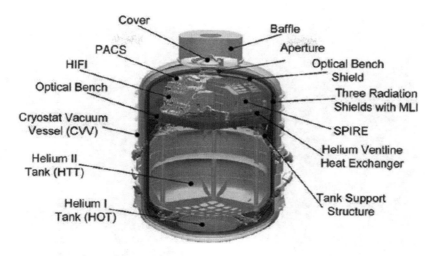

FIGURE 12.18 He II bath cryostat for HSO space mission [35].

kg of helium. To separate the fluid and gas phase under low gravity, a passive phase separator (PPS) is mounted on top of the tank. The PPS limits the evaporation and hence the cooling power in low gravity to <2.1 K and <20 mg/sec mass flow. For pre-launch and test operations, an auxiliary tank (helium one tank, HOT) with a volume of 80 l is mounted at the bottom side in the cryostat vacuum vessel (CVV). Both tanks can be filled and closed via the filling port and by switching of the electro-mechanical acting LHe valves. The scientific focal plane instruments (FPUs), optical bench (OBA), and thermal shields (TS1–3) will be cooled by the evaporating helium gas.

12.10 SPACE CRYOSTATS COOLED BY CRYOCOOLERS

12.10.1 CRYOCOOLER SUBSYSTEMS FOR MID-INFRARED INSTRUMENT MISSIONS

The cryocooler for the mid-infrared instrument (MIRI) on the James Webb Space Telescope (JWST) provides cooling at 6.2 K on the instrument interface. It has components that traverse three primary thermal regions on the JWST: (1) approximated by 40 K, (2) approximated by 100 K, and (3) at the allowable flight temperatures for the spacecraft bus [36–38]. One of the serious challenges is to verify and characterize the performance of the MIRI cryocooler subsystem.

MIRI Cooler Subsystem. Durand et al. introduced the cooler subsystem for the MIRI of the JWST. The JWST requires cooling at 6 K for the SiAs focal planes, provided by the active cooler instead of a cryogen tank. The cooler system consists of a three-stage pulse tube pre-cooler, Joule-Thomson circulator and upper-stage recuperators (located on the JWST spacecraft bus), and the final-stage recuperator and 6 K JT expander (located at the remote instrument module with a 12-m round-trip line at from 18 K to 22 K between the spacecraft and instrument). To improve the thermal efficiency, an actively cooled thermal shield surrounding the MIRI optical module to increase the overall thermal efficiency and thereby increase the margin between the cooler lift capability and the expected heat loads has been created. The current cooler requirements are 55 mW at 6.2 K and 232 mW total load on the shield heat exchanger and on the refrigerant lines.

MIRI Cooler Subsystem Test. For this test, the full capability of the chamber is utilized to provide the requisite thermal zones to represent a flight-like environment. The overall chamber configuration and flight and flight-like GSE (Ground Support Equipment) hardware for the test campaign are shown in Figure 12.19.

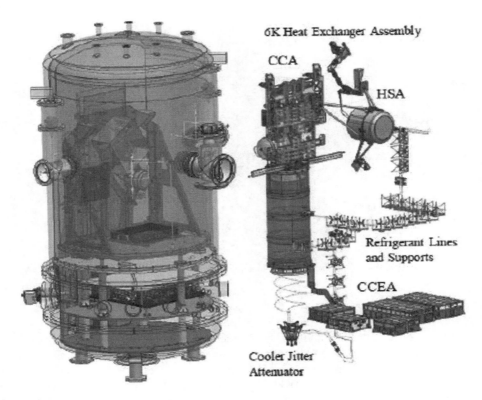

FIGURE 12.19   Left: Overall ATP3/4 and E2E test chamber configuration. Right: Flight and flight-like GSE under test [37].

12.10.2   Cooling and Heat Rejection on Planck Spacecraft

Planck was a space observatory operated by the European Space Agency from 2009 to 2013; then the mission substantially improved upon observations made by NASA. Planck has a comprehensive cryogenic system to fulfill missions with very challenging cryogenic heat management for cooling instruments and rejecting the heat load from the Planck spacecraft. The main objective of the Planck mission is to map the temperature fluctuations or anisotropy of the cosmic background radiation over the whole sky, with a sensitivity of $\Delta T/T < 2 \times 10^{-6}$ and an angular resolution of 10 arc minutes.

   Comprehensive Cryogenic System for Planck Mission. The Planck spacecraft relies upon a modular approach with the Planck payload module (P-PLM) housing the "cold" part of the satellite and the focal plane units as presented, while the service module (SVM) houses the equipment as well as the payload "warm" units, as shown in Figure 12.20 by Reix et al. [39]. The Planck cryogenic system is shared between the satellite and instruments. The performance of the spacecraft in flight is very similar to that predicted and measured during the ground test. Figure 12.21 gives the status of Planck cooling 44 days after launch. 0.1 K is expected to be reached about 50 days after launch.

12.11   CRYOSTATS FOR APPLICATIONS BELOW 1 K

Sub-Kelvin temperatures are produced by a combination of mechanical cryocoolers or LHe (less popular currently) at the upper temperature end and specialized sub-Kelvin coolers at the cold end. In terms of sub-Kelvin cooling, there are dilution refrigeration, adiabatic demagnetization refrigeration, and mixed refrigeration. Cryostats for these applications below 1 K have been introduced

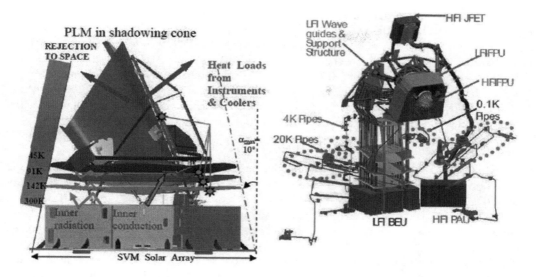

FIGURE 12.20 Planck coolers: Left: Passive cooler by payload module. Right: Active coolers together with instrument FPU. *[39]*

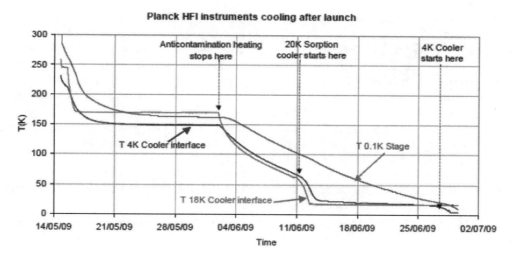

FIGURE 12.21 Planck HFI instruments after launch—flight data cooling of the Planck pay load module and FPU. *[39]*

and discussed in detail in books authored by Lounasma [40] and White [41]. We will very briefly introduce several examples herewith for readers' convenience.

12.11.1 Cryostats for Tests below 1 K with Dilution Refrigerators

He-3/He-4 dilution refrigerators are very common and crucial in sub-Kelvin temperature research, as it is the only method which provides temperatures between 0.3 K and 0.005 K for unlimited working periods. Cryogen-free dilution refrigerators are about to replace traditional cryostats with liquid helium precooling [BL-1–3,5]. Due to two-stage pulse tube cryocoolers having minimum vibration, the dilution circuit is always precooled by a two-stage PTR; therefore, refrigeration capacities are available to the cooling power at the T of the two stages of the PTR and furthermore at three temperatures of the dilution circuit (~0.7 K–still, 0.1 K–heat exchanger, ~0.01 K–mixing chamber). Based on the same cooling principle, there are many variations in how to design the details of the PTR circuit. For example, some PT precooled DRs use a separate He-4 circuit that condenses the He-3 of the dilution loop, but in others, the dilution circuit and the He-4 circuit are separate.

Cryostats of Dry Dilution Refrigerator with Separate 1 K Circuits

There are several applications (e.g. quantum information processing or astro-physics) where the cooling power of the still near ~1 K is not sufficient to cool amplifiers and electric lines. Uhlig presented a DR that added a ^{4}He cooling circuit in the cryostat to the dilution circuit [42]. This ^{4}He circuit provides up to 60 mW of refrigeration capacity in addition to the cooling capacity of ~30 mW of the still. The dilution circuit and the 1 K circuit can be operated together or separately, as shown in Figure 12.22.

A pulse tube refrigerator, the Cryomech PT405-RM (0.5 W at 4 K, second stage), is used to cool this DR. The cryostat has an inner vacuum can (214 mm OD) that can be evacuated and charged with exchange gas for fast cooldowns and serves as a radiation shield. The gas streams of the DR and of the 1 K stage are both cooled and purified in separate but identical charcoal traps that are thermally connected to the first stage. Finally, the gas streams of the DR are cooled in heat exchangers to the base T of the second stage of the PTR. Then, the helium flow of the 1 K cooler is expanded in a flow restriction, and the resulting liquid-gas mixture is run to a pot (V_{pot} = 53 cm^3) where the liquid can accumulate. The flow restriction was made from a capillary. Two rotary pumps in parallel (2 × Alcatel 2033H) are used, and the combined pumping speed is 66 m^3/h. The dilution unit is designed to circulate ^{3}He with two turbo pumps and two rotary pumps in parallel. The dilution unit is equipped with three heat exchangers, a continuous counterflow heat exchanger, and two discrete heat exchangers. The bottom of the mixing chamber is made from silver wires and sponges to establish good thermal contact between the liquid ^{3}He and the bottom of the mixing chamber. Finally, the cooling capacities reached were up to 60 mW at 1.35 K. Parallel to the 1 K stage, a powerful DR unit in the cryostat reached up to 700 µW at a mixing chamber T, 100 mK, and its base temperature at small ^{3}He flow was 10 mK. It was noted that with a greater cooling power, such as 1–1.5 W at 4 K, the DR could be used for big projects with SC magnets.

The 1 K stage in the left half of the cryostat consists of a heat exchanger at the second regenerator of the PTR, a heat exchanger at the second stage of the PTR, a flow restriction, and a pot. It has its own charcoal purifier attached to the first stage of the PTR.

The He-3/4 dilution unit on the right side of the refrigerator is a standard cryogen-free DR (for details, see text) [42].

12.11.2   Sub-Kelvin ^{3}He Sorption Cryostats for Large-Angle Optical Access

Hähnle et al. introduced lens-antenna-coupled microwave kinetic inductance detectors, which become an increasingly attractive option for large-scale imaging instruments [43]. A pulse tube-cooled cryostat provides 0.9 W cooling power at 4.2 K, where a ^{3}He sorption cooler (CRC-7B-002, Chase Research Cryogenics) is mounted that can reach temperatures down to 240 mK. Radio-transparent multilayer insulation was employed as a recent development in filter technology to efficiently block near-IR radiation. The sorption cooler is a two-stage, single-shot system consisting of a ^{3}He cold head as the mounting point for experiments and a ^{4}He buffer head. The windows are in a cone-like configuration, providing a half opening angle of θ = 37.8° to the center of a detector chip mounted on the cold stage of the sorption cooler. The measured load value of 6 µW is well within the expected accuracy of the model.

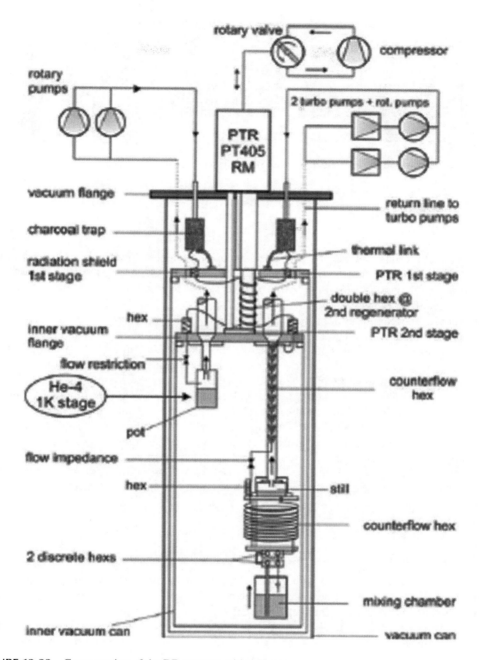

FIGURE 12.22 Cross-section of the DR cryostat with 1 K stage.

12.11.3 Cryostats for Tests below 1 K with ADRs

Adiabatic demagnetization refrigeration was the first method developed for cooling below 0.3 K (the lowest T reachable by pumping liquid ^{3}He vapors) and has become one of the most popular methods to obtain T as low as 10 mK with multiple stages of ADR. Newer generations of detector arrays need several µW of cooling at 50 mK or lower. Therefore, detectors of far IR space missions must have sufficient sensitivity to overcome the photon noise present in dark space. ADR has many

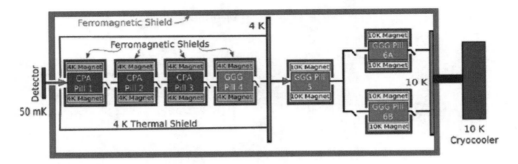

FIGURE 12.23   Upper: A block sketch of cryostat for the proposed CADR. Source: *[44]*

advantages that are uniquely suitable for space applications, such as gravity independence, milli-Kelvin *T* capability, high thermodynamic efficiency, and a lack of moving parts. Currently, there are many design variations of ADR with different AD stages, heat switches, precooling approaches, and cryostat configurations.

Tuttle et al. reported the development of a space-flight ADR, which can provide continuous cooling at 50 mK with more than ten times the current flight ADR cooling power and reject the heat at 10 K and will also continuously cool a 4 K stage for instruments and optics [44]. A block sketch of a cryostat for the proposed CADR and a preliminary model of the full CADR layout are illustrated in Figure 12.23. The proposed CADR consists of seven ADR stages, each being a paramagnetic "pill" suspended inside a SC magnet by tensioned Kevlar ropes. These stages are grouped into two different CADRs, which will be tested separately before integration into the overall system. The left side of the image is the four-stage 0.05 K to 4 K CADR. Its magnets are wound with niobium-titanium wire and are thermally and structurally attached to the continuous 4 K stage. The switch between the first two stages is a SC rod surrounded by a Helmholtz coil. All other heat switches in the CADR are passive gas gap switches. The 4 K to 0.05 K CADR is surrounded by a 4 K thermal shield to eliminate the heat load from 10 K thermal radiation. The 10 K to 4 K CADR is on the right side of the image and absorbs heat at 4 Kelvin, while the other dumps heat at just above 10 K to the cryocooler.

12.12   CRYOSTATS FOR BIO-MEDICAL APPLICATIONS

Over the past several decades, many types of cryostats for biomedical applications have been developed by research institutes and industries. Their fields range from cryopreservation and transmission of blood, organs, tissues, sperm, human eggs, and so on to whole human-body cryopreservation. The design principles and methodologies of biomedical cryostats are generally consistent with the technologies introduced in Chapters 2 through 5. In this section, only two unique developments of special applications are cited as examples.

12.12.1   BIOLOGICAL CRYOSTAT FOR CONTAMINATION-FREE LONG-DISTANCE TRANSFER

Cryo-electromagnetic imaging techniques have been intensively developed for resolving structural information of biological samples at the subcellular or even atomic level. Cheng et al. [45] introduced a passive transfer system for long-distance cryogenic transfer of samples, which are kept at stable cryo-*T* for long time periods and contamination free, as shown in Figure 12.24. With 125 mL of liquid nitrogen stored, one cryo-sectioned sample was maintained around 120 ± 1 K and a pressure of about 3×10^{-7} mbar for at least two hours. With a total transfer weight of 5 kg, this system

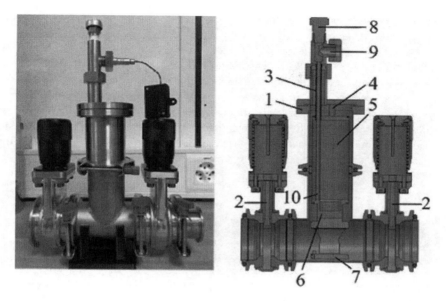

FIGURE 12.24 Left: A sample carrier cryostat (left) with details inside. Right: 1) top flange, 2) mini gate valves, 3) filling tube, 4) supporting tubes, 5) LN$_2$ reservoir, 6) reservoir bottom, 7) sample cooling stage, 8) relief valve, 9) connection for electronics, 10) level meter [45].

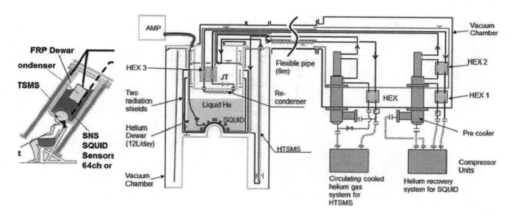

FIGURE 12.25 Diagram of the zero boil-off cooling system, responsible for cooling the HTSMS and the dewar shield and re-condensing the evaporated helium gas. *[46]*

can be easily handled and carried by any transportation means so that the same sample can be used for different imaging centers located remotely, permitting correlative studies.

12.12.2 ZERO BOIL-OFF CRYOSTATS FOR SC MAGNETO-ENCEPHALOGRAPHY

Mobile magneto-encephalography (MEG) developed by Sumitomo Heavy Industries, Ltd. (SHI) employs an HTS magnetic shield (HTSMS) and SNS type (SQUID) sensors. The HTSMS and SQUID are cooled by a zero boil-off cooling system, which consists of a circulating cooled helium gas system for cooling the HTSMS below a temperature of 90 K and a helium recovery system for cooling the SNS-type SQUID sensors to LHe temperature as shown in Figure 12.25. Narasaki et al. [46]

introduced the achievement of the first measurement of neuron current in the brain by using SHI's MEG with a helium zero-boil-off cooling system in April 2018.

The zero boil-off cooling system consist of two closed-cycle cooling subsystems. One is a helium recovery system for cooling the SQUID sensors to LHe temperature, and the other is a circulating cooled helium gas system for cooling the HTSMS below 90 K. To minimize the impact of GM coolers' moving parts, a flexible connecting pipe of ~8 m length and 100 mm OD is used to connect the coolers to the vacuum chamber. The dewars are made of a fiber-reinforced plastic (FRP). Two-stage radiation shields of copper meshed type are installed between the inner shell and outer shell. Ten layers of MLI are covered on the outside of the outer radiation shield. The He loss rate of the dewar is 12 liters per day without the re-condensing system.

REFERENCES

1. QS Shu, *Methodology of cryostat design*, Tech Note, AMAC—TN 09M14, 2007.
2. QS Shu et al., 2015 *Developments in advanced and energy saving thermal isolations for cryogenic applications*, CEC/ICMC-2015.
3. A Craena et al., 2014 New vertical cryostat for the high field superconducting magnet test station at CERN, *AIP Conference Proceedings,* 1573, 229—[LL-10].
4. S Yang et al., 2017 Mechanical design and analysis of LCLS II 2 K cold box, *IOP Conf. Series: Materials Science and Engineering,* 278, 012091.
5. L Sun et al., 2019 *2 K superfluid helium cryogenic vertical test stand of PAPS*, SRF2019, Dresden, Germany.
6. J Polinski et al., 2014 Design and commissioning of vertical test cryostats For XFEL superconducting cavities measurements, *AIP Conference Proceedings*, 1573, 1214.
7. NR Chevalier et al., 2014 Design of a horizontal test cryostat for superconducting RF cavities for the FREIA facility at Uppsala University, *AIP Conference Proceedings*, 1573, 1277.
8. J Knobloch et al., 2003 *The HoBiCaT test facility for SRF cavities*, 11th Workshop on RF Superconductivity, Lübeck, German.
9. K Umemori, K Hara, E Kako, Y Kobayashi, 2016 Construction and 2K cooling test of horizontal test cryostat at KEK, *Proceedings of IPAC2016*, Busan, Korea.
10. TB Weber et al., 2006 *A warm bore cryostat for the testing of a potential ILC superconducting quadrupole magnet at SLAC*, SLAC-PUB-12007 July.
11. V Roger, T Nicol, 2017 Upgrade of the spoke test cavity station, *IOP Conf. Series: Materials Science and Engineering*, 278, 012111. doi:10.1088/1757-899X/278/1/01211.
12. R Bruce et al., 2020 Compact cryogenic test stand for superconducting magnets characterization, *IOP Conf. Series: Materials Science and Engineering,* 755, 012147.
13. S Kar et al., 2012 Experimental studies on thermal behavior of 6 Tesla cryogen-free superconducting magnet system, *AIP Conference Proceedings*, 1434, 909.
14. YS Choi et al., 2010 Development of cryocooled binary current lead in low yemperature superconducting magnet system, *AIP Conference Proceedings,* 1218, 553.
15. P Micke et al., 2019 Closed-cycle, low-vibration 4 K cryostat for ion traps and other applications, *Rev. Sci. Instrum.*, 90, 065104.
16. T Trollier et al., 2014 Optical cryostat realizations at absolut system, *AIP Conference Proceedings*, 1573, 99.
17. CR Chapman et al., 2011 Cryogen-free cryostat for neutron scattering sample environment, *Cryogenics*, 51, 146–149.
18. Y Shiroyanagi et al., *Thermal modeling and cryogenic design of a helical superconducting undulator cryostat*, DE-AC02–06CH11357, yshiroyanagi@aps.anl.gov
19. E Anliker et al., 2020 A new superconducting undulator cryostat for the APS upgrade, *IOP Conf. Series: Materials Science and Engineering*, 755.
20. JD Fuerst et al., 2014 Cryogenic performance of a cryocoolercooled superconducting undulator, *AIP Conference Proceedings* 1573, 1527; https://doi.org/10.1063/1.4860888
21. C Wang, B Lichtenwalter, 2015 A cryogen recycler with pulse tube cryocooler for recondensing helium and nitrogen, *2015 IOP Conf. Series: Mater. Sci. Eng.*, 101, 012189.
22. PK Muley et al., 2020 Development of a helium recondensing cryostat, *IOP Conf. Series: Materials Science and Engineering*, 755.

23. YS Choi et al., 2012 Development of cryostat for 5 T HTS insert magnet, *AIP Conference Proceedings*, 1434, 1680.
24. S Pattalwar et al., 2011 Development of recondensing cryostat for PAMELA, *Cryocooler*, 16.
25. C Wang et al., 2014 A vibration free closed-cycle 1 K cryostat with a 4 K pulse tube cryocooler, *AIP Conference Proceedings*, 1573, 1387.
26. YS Choi et al., 2008 Helium-liquefaction by a cryocooler in closed-loop cooling system for 21 T FT-ICR magnets, *AIP Conference Proceedings*, 985, 367.
27. M Takahashi et al., 2008 Compact He II cooling system for superconducting cavities, *AIP Conference Proceedings*, 985, 119.
28. YS Choi et al., 2012 Development of cryostat for 5 T HTS insert magnet, *AIP Proceedings*, 1434, 1680.
29. A Nadaf et al., 2017 Helium exchange gas based variable temperature insert for cryogen-free magnet system, *IOP Conf. Series: Materials Science and Engineering*, 171.
30. J Cosier et al., 1983 A simple continuous-flow cryostat for optical microscopy in the range 10–350 k, *J. Phys. E: Sci. Instrum.*, 16.
31. F Börrnert et al., 2016 A variable-temperature continuous-flow liquid-helium cryostat inside a (scanning) transmission electron microscope, *Microsc. Microanal.*, 22 (Suppl 3).
32. CA Nixon et al., 2012 A variable-temperature continuous-flow liquid-helium cryostat inside a (scanning) transmission electron microscope, *AIP Conference Proceedings*, 1434, 384.
33. Thierry Tirolien, 2019 *Cryogenics in space missions,* European Space Agency ESTEC, Thermal Division, (presentation), October 3.
34. MF Larsen, J Drake et al., 2008 Wide-field infrared survey explorer science payload update, *Proceedings of SPIE—The International Society for Optical Engineering*, August.
35. M Langfermann et al., 2010 Launch preparation of the Herschel cryostat, *AIP Conference Proceedings*, 1218, 1530.
36. D Durand et al., 2010 Launch preparation of the Herschel cryostat, *AIP Conference Proceedings*, 1218, 1530.
37. B Moore et al., 2017 Mid infrared instrument cooler subsystem test facility overview, *IOP Conf. Series: Materials Science and Engineering*, 278.
38. J Cha et al., 2020 Thermal design and on-orbit performance of the ECOSTRESS instrument, *IOP Conf. Series: Materials Science and Engineering*, 755.
39. J-M Reix et al., 2010 The Herschel/Planck programme Planck PFM testing campaign, *AIP Conference Proceedings*, 1218, 1520.
40. OV Lounasma, 1974 *Experimental principles and methods below 1 K*, Academic Press, London.
41. K Guy, 2002 *White experimental techniques in low-temperature physics,* Oxford University Press; 4th edition.
42. K Uhlig, 2012 Cryogen-free dilution refrigerator with separate 1K cooling circuit, *AIP Conference Proceedings*, 1434, 1823.
43. S Hähnle et al., 2013 Large angle optical access in a sub-Kelvin cryostat, *J Low Temp Phys*, 193, 833–840.
44. J Tuttle, 2017 Development of a space-flight ADR providing continuous cooling at 50 mK with heat rejection at 10 K, *IOP Conf. Series: Materials Science and Engineering*, 278, 012009.
45. T Cheng et al., 2017 A new passive system for contamination-free long-distance cryo-transfer of biological tissues, *IOP Conf. Series: Materials Science and Engineering*, 278.
46. K Narasaki et al., 2020 Development of zero boil-off cooling systems for superconducting self-shielded MEG, *IOP Conf. Series: Materials Science and Engineering*, 755.

13 Demonstration of Cryogenic Heat Management in Large Applications

Quan-Sheng Shu, James E. Fesmire, and Jonathan A. Demko

13.1 LIQUID HELIUM—BEST CRYOGEN FOR LARGE SC MACHINES

The most important tasks in cryogenic heat management are how to safely and cost-efficiently keep cryogenic devices cold and preserve and transfer cryogenic liquids with a zero or minimal evaporation rate. Due to the irreplaceable advantages of superconducting devices, more large machines have been successfully developed and operated for high-energy physics, nuclear physics, fusion reactor, and materials sciences. Sophisticated SC magnets and SRF cavities operated at temperatures close to 0 K are enclosed inside high-vacuum environments. Some machines have masses of thousands of tons and lengths of tens of kilometers. Hydrogen is the energy carrier of the future. LH_2 is the likely basis for the logistics of moving large amounts of renewable energy around the world. To store and move around large amounts of energy, LH_2 will be at the heart of the matter. LH_2 remains the signature fuel for space launch and exploration, both government and commercial. LH_2 has been successfully used since 1963 in the evolution of upper-stage Centaur rocket engines. LNG continues to be a major business throughout the world. Container ships with capacities of over 200,000 m^3 are used.

13.1.1 RAPID DEVELOPMENT OF LARGE LTS PROJECTS/MACHINES

Due to the unique and irreplaceable advantages of low temperature superconducting (LTS), more and more large LTS projects/machines have been successfully developed, operated, and constructed for high-energy physics, nuclear physics, fusion reactor, materials sciences, FEL, and other applications. These machines demonstrate successful applications of superconducting (SC) magnets [1–3] and superconducting radio frequency (SRF) cavities [4–6]. Sophisticated SC magnets and SRF cavities operate at a temperatures close to absolute zero-degree K and are enclosed inside high-vacuum environments. These SC devices in some machines have weights in the thousands of tons and lengths of tens of kilometers.

Development of SC Magnet-Based Machines

Superconductors have zero electric resistance and can carry large amounts of electric current without generating Joule heating below the critical temperature, T_c. Therefore, SC magnets can generate magnetic fields up to ten times stronger than ordinary ferromagnetic-core electromagnets (which are limited to around 2 T). The magnetic fields are more stable and less noisy, and the absence of an iron core provides significantly larger working space. Also, large magnets can consume much less power, and the power consumed is for the refrigeration to keep the cryogenic temperature.

One of the representative machines of SC magnets is the LHC accelerator. Niobium-titanium (NbTi) magnets operate at 1.9 K to allow them to run safely at 8.3 T. Each magnet stores 7 MJ. In total, the magnets store 10.4 gigajoules in a 27-km ring with the huge cold mass immersed in 400 m^3 of pressurized superfluid He. Another great machine that uses SC magnets is ITER, in which the

DOI: 10.1201/9781003098188-13

magnets designed for the fusion reactor use niobium-tin Nb_3Sn. The central solenoid coil can carry 46 kA and produce a field of 13.5 T and the 18 toroidal field coils are at maximum field of 11.8 T.

Development of SRF Technology-Based Machines

The technical advantages of superconducting RF (SRF) accelerating cavities over conventional RF approaches are: (1) the quality values, Q, of SRF cavities are about 10^9–10^{10} (surface RF resistance Rs is 10^5 times lower than that of Cu cavities), and (2) its accelerating gradients are around 15–30 MV/m in continuous wave (CW) mode, an order of magnitude higher than Cu cavities in CW. The benefits of SRF cavities in large-scale applications can be summarized as follows: Produce very high-quality, high-energy particle beams and reduce operational cost. Tremendous efforts have been contributed to the development of SRF technology for decades in many laboratories and by industrial partners around the world. These efforts result in the production of higher RRR (>250) of Nb materials and overcoming of multipacting, field emission (FE), and thermal breakdown (TB). At the present time, a total of about 2,000 m of SRF structures with associated large 2 K cryogenic systems have been developed and successfully operated, such as CEBAF, SNS, TTF, XFEL, LEP, ILC, KEKB, and others.

13.1.2   Liquid He—The Only Practical Cryogen for Large LTS Machines

LTSs, such as Nb-Ti (critical T_c ~ 9.2 K) and Nb3Sn (T_c ~ 18.36 K) are the only practical superconductors used in the previous large applications [7]. Although high-T superconductors (HTSs) can work at around 77 K, such as YBCO and Bi-2233, and around 30 K–40 K, such as MgB2, HTS devices can only carry moderate current density, which can't meet the strict technical requirements in these machines. The critical current, Ic, and critical magnetic field, Hc, of LTSs increase further with the decrease of the operational temperature. LHe is capable of providing a T region of 1.8 K–4.2 K, which is much lower than LTS critical temperature. Therefore, LHe is the only practical cryogen for large LTS applications, while LN_2 (boil-point 77 K) is very often used for precooling and thermal shielding of LHe-cooled machines. Of course, many modern cryocoolers are capable of cooling LTS devices down to 2 K–4 K, but their cooling powers are far less than the large machines need.

13.1.3   Optimized LHe Operational Points for Best SC Machine Cooling

Challenges of SC Machine Cooling

The challenges of cooling large SC machines may be attributed to the following aspects: (1) the huge masses to be cooled; (2) a great amount of static heat load to cold mass from ambient T through MLI and supports be removed; (3) significant dynamic heat load generated internally by SRF heating in cavities, high power current leads, beam synchrotron radiation heating in magnet beam tubes, and so on, which must be removed; (4) the sophisticated geometry of SC devices; and (5) how to pass LHe through a SC machine with a length of several km while avoiding vaporized two-phase flow and maintaining efficient heat exchange between LHe and cold masses.

Practical Operation Points for Best Cooling of LTS Machines

Lebrun et al. introduced summaries of LHe in LTC machines and the helium phase domains, which show operation points of LTS devices, as illustrated in Figure 13.1 [8].

Point A, saturated He I for SRF devices—HERA, LEP, KEKB (and testing cryostats): Straightforward using saturated helium I either in pool boiling or forced flow. An advantage is its simplicity, fixed T and high heat transfer by means of the latent heat up to 104 W·m⁻², but there is the drawback of two phases, resulting in local dry-out and boiling crisis.

Point B, pressurized He I for SC magnets—HERA, Tevatron: Monophase supercritical pressurized helium is often used in forced circulation for cooling large extended systems such as strings of

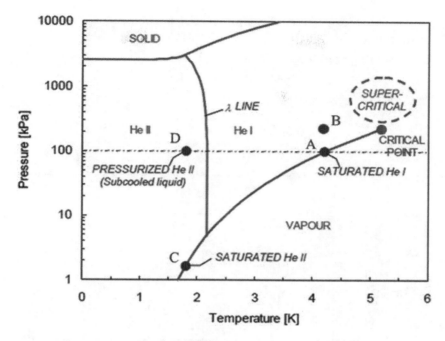

FIGURE 13.1 Operation point of representative large superconducting projects in He phase. Source: *Modified from Lebrun [8].*

magnets in accelerator tunnels. Heat loads are then absorbed by the sensible heat, at the cost of an increase in temperature of the circulating fluid and need high flow rate.

Point C, Saturated He II (superfluid), for SRF—CEBAF, TTF, SNS, EXFEL, ESS, ILC: Saturated helium II exists only at low pressure (below 5 kPa) at T below 2.2 K, in particular, the low viscosity and high apparent thermal conductivity and low viscosity exhibited are very useful for improving the stability of superconductors. Its drawbacks are low pressure, thermodynamic cost, and dielectric breakdown.

Point D, pressurized He II for magnets—LHC, Tore Supra: Overcome saturated He II drawback and subcooled below the lambda line, which is monophase and an excellent cooling medium in large magnet systems. Heat transport 10^3 times that of cryogenic-grade OFHC copper peaking at 1.9 K and very high specific heat stabilization10^5 times that of the conductor per unit mass and 2×10^3 times that of the conductor per unit volume.

13.1.4 CONTINUING IMPROVEMENT OF THERMAL EFFICIENCY

Table 13.1 shows how many liters of LHe or LN_2 will be required to cool down 1 kg iron in different cases. It also indicates that the thermal efficiency of using latent heat plus gas enthalpy is much higher than latent only, and using LN_2 as precooling will greatly help. Based on the second law of thermodynamics, refrigeration work is required to extract the heat load, Q, from the superconducting device at temperature, T, and to reject it at ambient temperature, Ta. Carnot efficiency is the efficiency of an ideal Carnot refrigerator, which reaches the maximum efficiency and requires minimum work. The advent of large superconducting projects has opened a market and thus stimulated industrial development of efficient, high-capacity cryogenic refrigerators, now available worldwide from a few specialized engineering firms. The continuing improvement of large LHe refrigerators is shown in Figure 13.2.

TABLE 13.1
Cryogen Required to Cool Down 1 kg Iron

	Latent Heat Only	Latent Heat and Gas Enthalpy
LHe 290–4.2 K	29.5 L	0.75 L
LHe 77–4.2 K	1.46 L	0.12 L
LN$_2$ 290–77 K	0.45 L	0.29 L

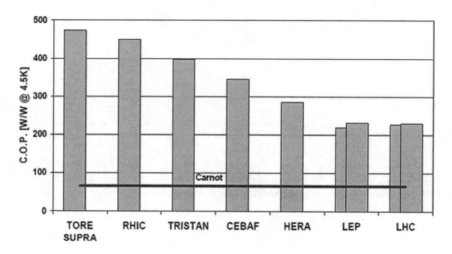

FIGURE 13.2   Coefficient of performance for lager He refrigerators. *Y* axis: inverse efficiency [9].

13.2   LARGE CRYOGENIC MACHINES BASED ON SC MAGNETS

13.2.1   COMMON FEATURES AND CHALLENGES

Long-Pass Distribution and Narrow Cooling Channels

Large superconducting circular accelerators, colliders, synchrotron light sources, and others based on SC magnets normally have quite large accelerating rings (kilometers), and their various magnets all have very narrow cooling channels. The straightforward practice of using helium I at saturation, either in pool boiling or forced flow, requires control of the level and flow-rate of a two-phase, boiling fluid, which may present instabilities and local dry-out by phase separation. Therefore, forced flows of monophase supercritical helium and pressurized He I or He II, is often used in forced circulation in large cryogenic machines based on SC magnets.

Careful Tradeoff between High Magnetic Field and Cost

As shown in Figure 13.3, the practical high magnetic fields of NbTi cable can reach from 7 Tesla at 4.5 K to 10 Tesla at 1.8 K [7,9]. The higher field of magnets results in reduction in sizes and construction cost. In contrast, the thermal efficiency of refrigeration is about 280 W/W at 4.5 K and about 900–1,100 W/W at 1.8 K. So, a trade-off decision must be made to balance the technical advantage and cost reality over particular cases.

Highly Restricted Requirements of Heat Load and Geometry Size

Due to the highly restricted requirements, the magnets and cryostats must be designed to be as compact as possible in length and diameter. The magnets must have continuing cooling channels

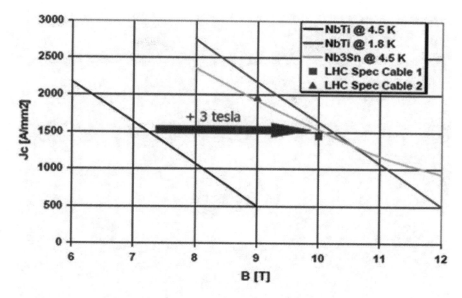

FIGURE 13.3 LHC cable specs and critical *B* vs *Jc* of NbTi and Nb3Sn at 4.5 K and 1.8 K [9].

for LHe flow, and the cryostat needs to contain the magnets and all possible pipes for forward and returning cryogens. For optimization of thermal design, focus is given to the following resources: external heat loads (radiation, residual gas, solid conduction), internal heat sources (SC magnet Joule heating, power lead heating), and beam-induced heating (synchrotron radiation, beam image current, and photoelectrons). So, the intermediate *T* shields with MLI, low heat leak support, beam screens, and high-vacuum associates are sophisticatedly designed to meet the requirements.

Sectional Design and Multi-*T* Output of Cryo-Plants

Large cryogenic machine heat loads are absorbed by cryogens at the cost of an increase in *T* of the circulating cryogenic fluids. At low *T*, even small temperature gradients will increase the work of refrigeration significantly as well as the capital and operational costs of the cryogenic plant. For example, if the *T* is raised by 0.5 K, the refrigeration work will increase by 1.10 at 4.5 K and 1.22 at 1.8 K. Cooling of large systems featuring high heat loads therefore requires large mass-flow rates, as well as periodic or continuous re-cooling of the supercritical stream by saturated helium, in order to contain its temperature excursions within allowed limits. Liquid produced can be used for cooling by its latent heat and for non-isothermal cooling up to room temperature by the large remaining cooling capacity of its vapors. Therefore, several sectional cryo-plants with multi-*T* output cryogens (1.8 K, 4.5 K, 50–75 K, etc.) and many re-cooling devices are allocated along the accelerator ring to cool the SC magnets and re-cool the pressurized LHe as well.

13.2.2 LHC—The Largest SC Accelerator in the World

General Features of LHC

The Large Hadron Collider (LHC) is the largest superconducting machine and highest-energy particle accelerator-collider in the world. It was built by the European Organization for Nuclear Research (CERN) between 1998 and 2008 in collaboration with hundreds of universities and laboratories from about 100 countries. It set the world record for highest energy of 6.5 TeV per beam (0.56 A) and 13 TeV total collision energy in 2018. The colliders and other devices, like ATLAS, CMS, ALICE, and so on, study the Higgs boson and look for clues to several physics mysteries, including the origins of mass and quark–gluon plasma shortly after the Big Bang and "antimatter" [9–11].

FIGURE 13.4 LHC SC magnets in the underground tunnel. Source: *CERN*

The greatest stars of the scientific and engineering achievements are the SC magnets and cryogenic systems, which lie in a tunnel 27 km in circumference and as deep as 175 m under the France–Switzerland border near Geneva, as shown in Figure 13.3. A total of about 1,800 SC magnets, including 1,232 bending dipole magnets, 392 focusing quadrupole magnets, and many other multipole magnets of correction, are used to guide the beams into four colliders. With the dipole magnet having a mass of over 27 tons, the total mass of the cold masses at 1.9 K is about 37,000 tons (with about 7,600 km of SC Rutherford cable and 1,250 tons of NbTi materials). Approximately 96 tons of LHe II is needed to keep the magnets at their operating temperature of 1.9 K (−271.25°C). To fill the huge cooling tasks, the LHC has the largest cryogenic complexes in the world at LHe *T*. There are eight refrigerators, each carrying 18 kW cooling power at 4.5 K and 20 kW at 1.8 K, as shown in Figure 13.4. The total LHe inventory is about 96 tons around the 24 km accelerator. During LHC operations, the CERN site draws roughly 200 MW of electrical power (the LHC accelerator and detectors draw about 120 MW).

LHC Magnet and Cryostat

The LHC dipole magnet cryostat has two beam tubes with respective paranal SC magnet coils inside a common yoke. The cold mass is contained inside a stainless-steel helium vessel with an external diameter of 0.6 m, length 15 m. The cold mass is 30,000 kg. The dipole inner coils have a Cu/Sc ratio of 1.7, and the outer layer is 1.9. The critical current density is about 10 T at 1.8 K. The force containment consists of coil clamping collars, the iron yoke, and the shrinking cylinder. The iron yoke is split vertically into two parts with a gap at room temperature, which is closed at 2 K. As the protons are accelerated from 450 GeV to 7 TeV, the field of the superconducting bending magnets will be increased from 0.54 to 8.3 T. The magnets are enclosed vacuum-tightly in the cold mass shell and cooled by the pressurized He II (superfluid He) through special narrow channels, as shown in Figure 13.5, the cross-section of LHC main dipole cryostat.

The LHC dipole cryostat (1 m diameter and z kg weight) provides well-designed thermal insulation for the magnet, associated cryogenic pipes, and beam vacuum system. Also, the cryostat

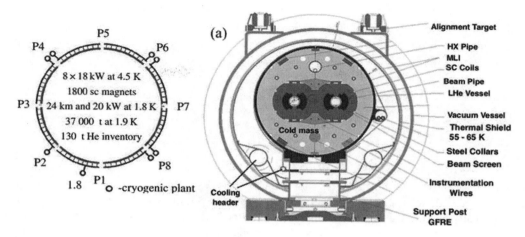

FIGURE 13.5   A: Cross-section view of the LHC dipole magnet and cryostat. B: Schematic of heat flows and heat intercept temperatures [9].

functions to rigidly support and precisely position the heavy dipole cold mass with respect to the external alignment fiducials. Its other functions include magnetic screening of the stray field or radiation shielding of the beam halo. It is therefore clear that cryostat design and optimization must integrate all boundary conditions and constraints set by the superconducting magnet and cryogenic and other relevant accelerator systems.

At a minimum, the radiation heat loads from ambient temperature to the 2 K temperature of the magnets, the intermediate T thermal shields with MLI are employed. In the initial design of the LHC cryostat, there are outer thermal shields with 30 MLI layers at 77 K (LN_2) and inner thermal shields with 10 MLI layers at around 5 K. After careful trade studies of thermal efficiency and cost, the inner intermediate 5 K shields are limited, and the 10 MLI layers are directly applied on the cold mass. Simultaneously, the outer intermediate T thermal shields with 30 MLI layers are cooled by returning cold He vapor at 50 K–65 K, as presented in Figure 13.5. To greatly reduce the solid thermal conduction, the magnets are supported on three support posts made of low thermal conductivity material (GFRE) and also thermal anchored at 50 K–65 K and 6 K, respectively. The entire cryostat is placed inside vacuum vessels in order to remove heat loads from gas convection and conduction.

When particle beams reach very high energy (multi TeV), high current in a circular accelerator ring will induce large synchrotron radiation heating, wake fields, and photo-desorbed gases. Beam screens (also called beam pipe liners) are unique and crucial devices, which minimize the presence of photo-desorbed gases in the particle beam line vacuum and remove the heat load at a temperature higher than beam pipe temperature (also improving the cryogenic thermal efficiency). Shu et al. have designed the first beam liners of 80 K inside along the SSC 4.2 K beam pipes in 1990s. Angerth et al. successfully designed, developed, and implemented an approximately 4 K beam screen system inside the LHC 1.9 K beam pipe.

Finally, the design and LHC dipole magnet and cryostat meet the LHC budgeted heat loads based on Evans's data [11]. Simplifies heat load budgets are listed in Table 13.2.

LHC Cooling and Distribution

Thermal Efficiency of LHe Refrigeration. Attributable to continuing effort in improvement of refrigeration efficiency in decades, the $\eta = W_{min}/W_{real}$ (W_{min} is the Carnot limit) of the large 4.5 K refrigerators reached 30% at LHC and were provided by both Linde and Air Liquid. This means to extract 1 W heat at 4.5 K to 300 K, the minimum refrigeration $W_{real} = W_{min}/\eta = 65.7/0.3 = 220$ W/W.

TABLE 13.2

A LHC magnet Budgeted Heat Loads, W/m

T	50–75 K	4.6–20 K	1.9 K He	4 K VLP
Static heat leak	7.7	0.23	0.21	0.11
Resistive heating	0.02	0.005	0.10	0
Beam-induced heat	0	1.58	0.09	0
Total nominal	7.7	1.82	0.40	0.11

The state-of-the-art large 1.8 K refrigeration plant at LHC has an efficiency of 15%. So, the W_{real} is about 990 W/W. Real efficiencies (COP W/W) at 4.5 K and 1.8 K are shown in Figure 13.6, in which the points are measured, and the lines are fitted. Note the highest efficiencies are at the design point of the cryogenic plant, which is usually not run continuously at full capacity (data from Gruehagen and Claudet) [12].

Cryogen Distribution and Re-Cooling. At low refrigeration temperature, a few minute temperature gradients due to various heat loads to cryogen will increase the work of refrigeration significantly, which adds to the capital and operational costs. Therefore, the accelerator is sectorized to reduce effective heat transport distances in distribution of the cooling power, and the T gradient is usually kept well below 1 K [9,13]. The heat transfer can be enhanced by increasing the wetted perimeter and surface finish between cooling fluid and saturated liquid re-coolers, where heat is transported across the solid wall of a heat exchanger. The principle of flow circulation between adjacent cryo-plants is presented in Figure 13.7.

13.2.3 FROM TEVATRON TO OTHER LARGE SC MACHINES

Since Tevatron (1980), the first large SC machine successfully constructed, SC magnets and cryogenics have now become key technologies in modem accelerators and colliders, such as the HERA, RHIC, UNK, SSC (closed), and LHC (2008).

Tevatron—The First Largest SC Accelerator in the World

The world's first highest-energy proton/antiproton collider at Fermilab, the Tevatron 6.5 km superconducting ring provided for 777 dipoles, 216 quads, and 204 correction elements located in a 2.1-m-diameter concrete tunnel buried 6.0 m below ground [14]. The cryogenic system consists of a hybrid system of a central helium liquefier feeding twenty-four 1 kW satellite refrigerators through a 6.5-km transfer line and supplying the liquid helium (LHe) for all the SC magnets and liquid nitrogen for the thermal shielding, as shown in Figure 13.7 (Upper). Tevatron upgrades were completed in 1996 and allowed a potential decrease of magnet operating temperatures from 4.9 K to 3.9 K (Tevatron energy from 900 to 990 GeV). The original CHL helium liquefaction capacity was 4,000 liters/hour. The helium inventory of the cold refrigeration system is 20,000 liters, plus an additional 10,000 liters in the Tevatron transfer line. The Tevatron was retired and the original facilities were modified into a Muon collider at Fermilab.

CMS and Others

The Hadron Electron Ring Accelerator (HERA) designed for providing collisions of 30 GeV electrons or positrons with 820 GeV protons in a ring tunnel of 6.3 km circumference has been operating at DESY since 1991 [15]. The proton ring consists of 422 SC dipoles, 224 quadrupoles, and correction magnets. In order to avoid bubbles of helium gas, the magnet coils are cooled by supercritical, one-phase helium at 4.4 K and 2.5 bar. A group of 53 dipoles and 26 quads is cooled in series within a 623-m-long magnet octant. The refrigerator, with a capacity of 25 kW at about 4 K, was originally

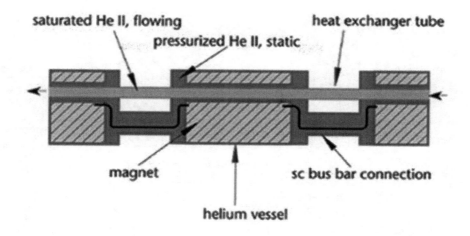

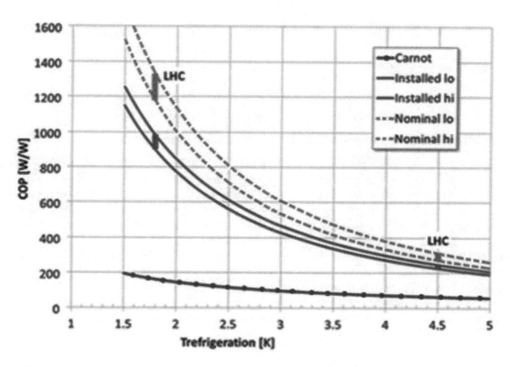

FIGURE 13.6   Upper: Measured thermal efficiency of LHC refrigeration at 4.5 K and 1.8 K. Lower: The principle of He II cooling of LHC magnets [9,12].

designed to match the load for the Colliding Beam Accelerator Project. The CMS detector at LHC contains one of the largest superconducting magnetic systems, as shown in Figure 13.8 (upper is a 3D design of the CMS, and lower is a photo of the real system).

Future Large Colliders

After the LHC era, CERN proposed a new worldwide international collaboration, which will design and develop an energy-frontier 100 TeV proton and heavy-ion collider of 80–100 km and a high-luminosity electron-positron with 16–20 T dipole models [16].

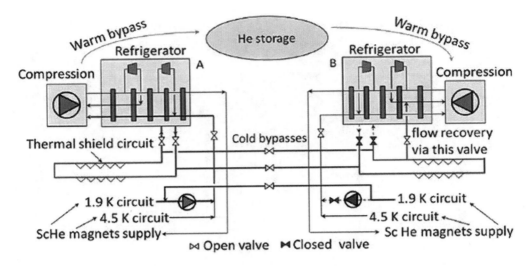

FIGURE 13.7   Principle of the flow circulation between adjacent cryo-plants during run 2. *[13]*

13.3   LARGE CRYOGENIC MACHINES BASED ON SRF TECHNOLOGY

13.3.1   General Considerations of SRF Technology-Based Machines

Advantages and Limitations. As shown in Figure 13.9, the 1.3-GHz SRF Nb accelerating cavity can generate high-frequency electromagnetic fields with an accelerating electric field *Eacc* along the cavity axial direction. An electrically charged beam that enters the cavity will be further accelerated to higher energy by the *Eacc*. The surface resistance of the Nb cavity is 10^4–10^5 times less than a copper cavity, so the quality factor Q(Nb) will be large than Q(Cu) by 4–5 orders of magnitude. It is for this reason that almost all high-accelerating fields and high CW beam current have employed SRF cavities instead of Cu cavities. However, there are still some fundamental and technical burdens that prevent reaching the highest *Eacc*, as the so-called Q vs *Eacc* curves will drop dramatically after certain *Eacc* values.

SRF Surface Resistance and Q. The quality factor, Q_0, is inverse to the surface resistance of the cavity, with a geometry conversion factor. This surface resistance $R_s(T) = R_{BCS}(T) + R_0$:

$$R_{BCS}(T) = \frac{A\omega^2}{T} \exp\left(-\frac{\Delta_0}{K_b T}\right)$$ (13.1)

where $R_{BCS}(T)$ is surface resistance by BCS theory, R_0 is residual resistance, Δ_0 is the energy gap, K_b is the Boltzmann constant, A is the material constant, T is the operating temperature, and ω is the operation frequency. The lower the cavity temperature, the lower the temperature-dependent BCS losses in the cavity wall, which results in direct savings in capital and operation. The ratio at 2 K between R_{BCS} and R_0 for the European XFEL cavity production for 1.3 GHz is about 2:1, and the Qo is about 2×10^{10}.

Push to Reach the Maximum Eacc. The higher the *Eacc* used in an accelerator, the less both the accelerator size and construction cost. Therefore, pursuing the maximum *Eacc* is one of the ultimate goals for SRF scientists and engineers. First, optimization of the cavity geometry design may help obtain higher ratios of both *Eacc* to maximum surface, E, and *Eacc* to maximum surface

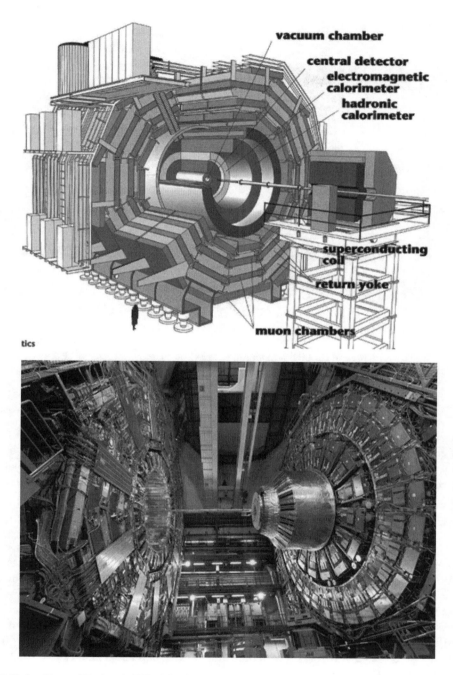

FIGURE 13.8 Upper: 3D sketch CERN CMS collider. Lower: Photo of CMS. Source: *CERN*

magnetic field, H [17]. Then after using high-RRR (>250 if possible) Nb materials to construct the cavity, a series of processing of the cavity (such as various cleaning processes, He processing, or heat processing) are employed to depress the field emission sources and thermal breakdown defects.

Cryomodule and Cryogenic System. Different from SC magnets with zero CD electrical resistance, the SRF cavity has RF surface resistance, and hence a strong electromagnetic field will

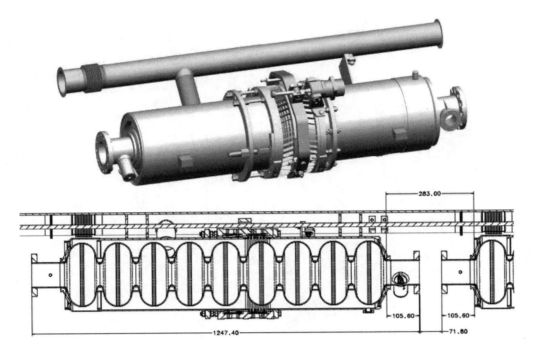

FIGURE 13.9   Lower: Nb 1.3 GHz SRF accelerating cavity for charged particle TESLA-ILC, and Upper: The TESLA cavity in a LHe vessel [20]

generate significant heat load in the cavity inner surface, which is greater than static heat from the ambient environment to the cavity in many practices. To extract the RF heat load from the cavity, the cavity is usually immersed in saturated HE I or He II liquid in a vacuum-tight container, which is installed in a cryomodule. Various thermally well-insulated cryomodules have been developed, and high cooling capacity refrigeration plants have been constructed [18–20].

13.3.2   XFEL—THE LARGEST CRYOGENIC MACHINE BASED ON SRF CAVITIES

From TESLA to XFEL

The TESLA (TeV Electron Superconducting Linear Accelerator) Collaboration led by DESY Germany is an international R&D effort toward the development of an e+e-linear collider of 20 km with a 500-GeV center of mass for multiple purposes. The TESLA Collaboration first successfully developed the highest CW accelerating SRF cavities of 25–30 MV/m with high RF input/HOM couplers and cryomodules and then constructed the prototype TESLA test facility (TTF) of a 500-MeV SC linear accelerator (~100 m). TTF was modified and expanded into a free-electron light source in Hamburg (FLASH).

The TESLA Collaboration stopped short mainly due to financial reasons, but the core technologies and project architectures have continued under a new collaboration name, the International Linear Collider (ILC), as shown in Figure 13.10 [18–20]. The main Linacs consist of many SRF cavities operating at 2 K over the entire length of about 11 km each, and ten cryo-plants are located along the main Linacs to cool down the cavities. Each cryo-plant provides 19 kW at 4.5 K equivalent and 3.6 kW at 2 K.

While the participants of ILC look for huge financial resources, various projects subsequently adopted the TESLA cryomodule design concept with some modifications depending on specific

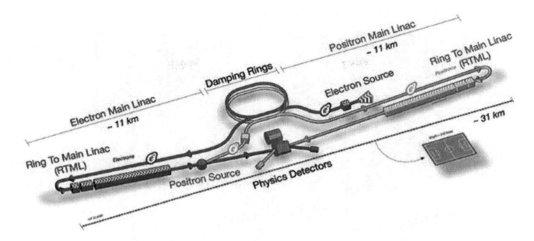

FIGURE 13.10 Schematic layout of ILC [19].

requirements and innovations. These projects include the X-ray Free Electron Laser (XFEL) at DESY, the Linac Coherent Light Source II (LCLS-II) at SLAC, and Shanghai FEL in China.

Linear Accelerator and Cryomodule of EXFEL

The €1.54-billion European XFEL is a 3.4-kilometer-long facility located mainly in underground tunnels, which starts at DESY Hamburg into Schleswig-Holstein [4,19–20]. It generates ultrashort X-ray flashes—27,000 times per second with a brilliance of a billion times higher than the best conventional X-ray sources. In September 2017, the XFEL started generating extremely intense X-ray flashes for mapping atomic details of viruses, film chemical reactions, studying the processes in the interior of planets, investigating materials, and medical applications. The European XFEL accelerator currently uses 768 SC Nb cavities at 2 K with *Eacc* about >28 MV/m over a 1.7 km length and energizes electrons up to 17.5 GeV. At temperatures below 2.2 K, the superfluid He II has a low viscosity and high apparent thermal conductivity. Saturated helium II about 2.2 K exists only at low pressure (below 5 kPa).

An XFEL cryomodule of about 12 m with eight SRF cavities and the partial section of a TTF 3-style cryomodule mock-up with the complete cavity and key elements are illustrated in Figure 13.11. A cryogenic unit consists of 12 cryomodule strings. To take out a large mass flow of 2 K He gas vaporized by RF power dissipated along the cryomodule string, a large He return pipe (HGRP, 300 mm) is needed to reduce the pressure drop. This pipe can be made large and stiff enough so that it can act as the main structural backbone for the module cold mass. The HGRP hangs from the vacuum vessel by three low thermal conduction composite posts (similar to LHC-style posts), and the entire cold mass (including eight cavities, one magnet, RF power couplers, shields, etc.) is connected to the HGRP. Each SRF cavity has its own helium tank. The center support post is rigidly fixed to the vacuum vessel of the cryomodule, while the other two can be movable longitudinally to adjust the shrinkage of the gas return pipe while cooling down from room temperature to 2 K. There are two types of cryomodules: type A has nine cavities, and type B has eight cavities plus one SC quadrupole magnet. The cryomodule has two thermal shields in it; helium-gas-cooled shields intercept thermal radiation and thermal conduction at 5 K–8 K and at 40 K–80 K.

Thermal Performance of the Cryomodule. The development of the XFEL cryomodule was a long evolution, and many types of cryomodules are undergoing continuous improvement at many prestigious institutions around the world. Table 13.3 represents the XFEL thermal budget heat load and the measured thermal loads of static and dynamic loads based on an ILC-style cryomodule.

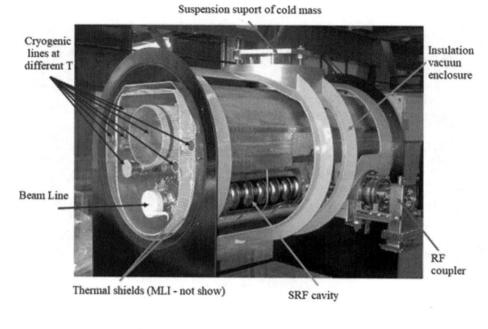

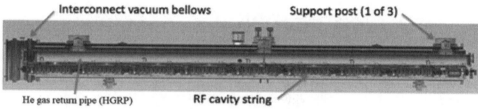

FIGURE 13.11 Upper: TTF 3-style cryomodule mock-up. Lower: Simplified layout of the XFEL cryomodule [20].

TABLE 13.3
Thermal Load Budgets and Measured Heat Loads of a LIC-Style

	XFEL Refrigerator Budget W			ILC Cryomodule Measured W*		
T	**Static**	**Dynamic**	**Total**	**Static**	**Dynamic**	**Total**
40/80 K	83	40	123	74	105	179
5/8 K	13	2.3	15.3	13.0	4.87	17.85
2 K	4.8	8.6	13.4	3.5	8.37	11.87

Source: Cryomodule [20]

Dynamic W* Measured from cavities, RF power input couplers, and HOM couplers, while *Eacc* = 31.5 MV/m, Q = 1 × 10[10] under ILC RF operation conditions.

European XFEL Cryogenic System

The existing cryogenic infrastructure at DESY servicing the HERA was designed for providing cryogenic cooling at 4.5 K and consequently had to be modified in order to meet the new requirements as presented in Table 13.4. Two of the three existing HERA helium refrigerators were

TABLE 13.4

Heat Load Specification for the XFEL Helium Refrigeration System

| Cooling Loop | Unit | Case 17.5 GeV | | Case 20 GeV |
		W/O Margin	With 50% Margin	With 50% Margin
2 K isothermal	kW	1.46	1.90	2.46
CC string	g/s	74.0	96.0	143
5 K–8 K shield	kW	2.40	3.59	4.15
40 K–80 K shield	kW	16.0	23.9	30.0

Source: Helium Refrigeration System [20]

required to cover the design heat load of the XFEL-linac. The existing refrigerator infrastructure was overhauled, modified, and extended by a 2 K cooling loop whose main component consists of a string of four cold compressors generating approximately 1.7 kW isothermal cooling capacity at 2 K. Details of the conceptual design have been reported previously [4,5,21].

The current status of this project, commissioning results, and particular challenges are presented in [21]. While most of the performance tests of the XFEL helium refrigeration system were successfully completed, the last function tests and the demonstration of long-term reliability over 800 consecutive hours have to be performed.

13.3.3 Other Advanced Machines Based on SRF Technologies

Due to their great performance advantages, many large cryogenic machines based on SRF technologies have successfully been developed and are under operation. These machines can be categorized as cooled by saturated He I and saturated He II or can be grouped into different usages. CESR, CEBAF, KEKB, TTF, and LEP are generally for nuclear and high-energy physics. XFEL, ESS, ERL, Shanghai FEL, and many light sources are for lasers with different frequencies, while SNS is for the Spallation Neutron Source.

The SNS is a 1 GeV negative Hydrogen ion accelerator with up to 2 MW power producing a source of neutrons for materials research [6]. The cryomodule (CM) is based on the CEBAF CM with improvements. The helium cryogenic system at Spallation Neutron Source (SNS) provides cooling to 81 SRF cavities. SNS refrigerator capability at preliminary design: for Linac cavities in saturated He II about 2 kW at 2.1 K, 0.041 atm, 120 g/s and for shield 8.3 kW at 35–52 K, 4.0 atm, 90 g/s. The diagram of central He liquefier is shown in Figure 13.12. At KEKB the saturated He I is employed: A cryogenic system consisting of a helium refrigerator (4 kW at 4.4 K) and a LHe distribution transfer system for TRISTAN 508 MHz 32 X 5-cell SRF cavities was designed and constructed. Cornell University's X-ray light source based on the Energy-Recovery-Linac (ERL) principle [22], which promises superior X-ray performance as compared to conventional third generation light sources. Its cryomodule faces the following challenges: (1) A high current beam with up to 500 kW of total power transfer, (2) significant HOM excitation in the SRF cavities by the high current beam, and (3) the preservation of the ultra-low emittance of the electron beam. At the radiation source ELBE a superconducting radio-frequency photoinjector (SRF gun) was developed and put into operation.

13.4 SUPERCONDUCTING FUSION MACHINES AND CRYOGENICS

13.4.1 Development of Superconducting Fusion Machines

Nuclear fusion on a large scale in an explosion was first carried out on November 1, 1952, in the Ivy Mike hydrogen bomb test. Fusion merges atomic nuclei to create massive amounts of energy, which

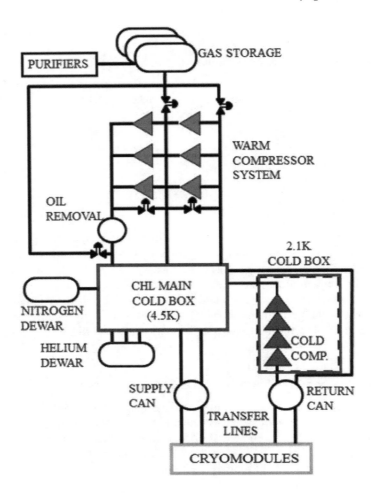

FIGURE 13.12   Diagram of SNS central helium liquefier (CHL) [23].

is the opposite of the fission process. Inertial confinement fusion (ICF) is a type of fusion energy research that attempts to initiate nuclear fusion reactions by heating and compressing a fuel target, typically in the form of a pellet that most often contains a mixture of deuterium and tritium. ICF is one of two major branches of fusion energy research, the other being magnetic confinement fusion. The design of newer and larger machines would finally reach the goal of clean energies.

The International Thermonuclear Experimental Reactor (ITER), with 35 partners and a total cost of about US$22.5 billion, is being built to test whether the long-sought dream of nuclear fusion—the atomic reaction that powers the sun—can be harnessed to generate near-limitless clean energy. ITER is considered the most expensive scientific endeavor in history. The Experimental Advanced Superconducting Tokamak (EAST) in Hefei, China, internal designation HT-7U, is another experimental magnetic fusion energy reactor. Both ITER and EAST have employed superconductors to construct their huge SC magnet complexes, which not only can reach the highest magnetic fields but also provide a large space of strong magnetic fields for confinement of hot plasmas. Consequently, advanced cryogenic technologies and sophisticated refrigerators have also played a crucial role in these projects.

The United Kingdom has embarked on a step toward building the world's first nuclear fusion power station, where it can be plugged into the electricity grid. The Spherical Tokamak for Energy Production (STEP) is hoped to begin with the plant operating as soon as 2040.

13.4.2 ITER—The World's Largest SC Fusion Machine

General Introduction to ITER

ITER is the world's largest nuclear fusion research project and is based in France, with the European Union having a 45% stake, and the United States, China, India, Japan, Russia, and South Korea each have a 9% stake. ITER is expected to be completed in 2025 and should demonstrate the scientific and technological feasibility of fusion power by achieving extended burn of D–T plasmas with steady state as the ultimate goal. It has been designed to create a plasma of 500 megawatts (thermal) for around 20 minutes while 50 megawatts of thermal power are injected into the tokamak, resulting in a ten-fold gain of plasma heating power. The other goals are to integrate and test all essential fusion power reactor technologies and components of fusion for future machines.

Serio et al. introduced the SC and cryogenic aspects of the design and construction of ITER with several papers [24–26]. The major radius of the magnets is 6.2 m, and the minor radius is 2 m. The magnetic field at the center of the plasma is 5.3 T. The ITER magnetic field is composed of four systems: the toroidal magnetic field system, the CS, the PF system, and the correction coils (CC). They all use NbTi- and Nb3Sn-based conductors. The total mass of the magnet system is about 10,000 tons. The conductors are CICC made up of SC and copper strands assembled into a multistage, rope-type cable inserted into a conduit of butt-welded austenitic steel tubes. The coils are cooled with supercritical helium at an inlet temperature of 4.5 K. All magnets are contained by a huge cryostat, and the high vacuum of the machine is maintained by a group of cryopump systems. Supercritical LHe and associated LN_2 are produced and delivered by the cryo-plant. The ITER 3D schematic is illustrated in Figure 13.13.

ITER SC Magnets

The International Thermonuclear Experimental Reactor (ITER) superconducting (SC) magnets are 10,150 t. The central solenoid coil will use SC Nb3SN (six modules) to carry 46 kA and produce a field of up to 13.5 T. The 18 toroidal field coils will also use Nb3Sn. At their maximum field strength of 11.8 T, they will be able to store 41 gigajoules. They have been tested at a record 80 kA. Other lower-field ITER magnets (PF and CC) will use NbTi for their SC elements. The magnet model computes the thermo-hydraulic behavior of a forced-flow supercritical helium (SHe), with specified thermal energy deposition along the conductor in the case of deuterium-tritium (DT) operation as well as plasma disruption followed by the fast energy discharge of CS and PF. One sector of the ITER Tokamak SC magnets is illustrated in Figure 13.14.

ITER Cryostat

The cryostat is a large 3,800-t stainless steel structure surrounding the vacuum vessel and the superconducting magnets in order to provide a super-cool vacuum environment. Its thickness, ranging from 50 to 250 mm, will allow it to withstand the atmospheric pressure on the area of a volume of 8,500 cubic meters. On June 9, 2020, Larsen and Toubro completed delivery and installation of the cryostat module. The cryostat is the major component of the tokamak complex, which sits on a seismically isolated base.

ITER Vacuum Vessel

The vacuum vessel is the central part of the ITER machine and provides a hermetically sealed plasma container. It is a double-walled steel container in which the plasma is contained by magnetic fields. Each of the nine torus-shaped sectors will weigh about 400 t. It would be a total of 5,116 t with all shielding and ports. The whole structure will be 11.3 m high.

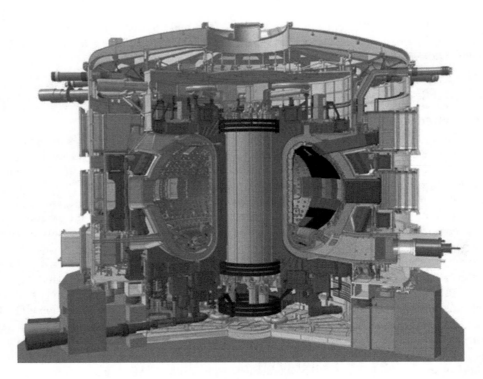

FIGURE 13.13   The 3D design schematic of ITER. Source: *ITER*

ITER Cryogenic Plant and Distribution for Magnet Cooling

Most of the ITER heat will be removed by a primary water-cooling loop through a heat exchanger within the tokamak building supplying water. This system will need to dissipate an average power of 450 MW during the tokamak's operation. The ITER cryogenic system will be the largest concentrated cryogenic system in the world, with an installed cooling power of 75 kW at 4.5 K (helium) and 1,300 kW at 80 K (nitrogen). Three identical helium refrigerators will work in parallel to provide LHe for cooling ITER's 10,000 t of SC magnets. Thermal shields and the cryo-sorption panels of the cryopump ensure high-quality vacuum to the large volumes of the cryostat (8,500 m³) and vacuum vessel (1,400 m³). The three units will provide a total average cooling capacity of a maximum cumulated liquefaction rate of 12,300 L/hr. The layout of cryogenic distribution consists of more than 50 cold boxes, 3 km of cryolines, and 4,500 components, as shown in Figure 13.15. The cryoplant termination cold box (CTCB) is the largest one and plays an essential role in distributing the cold helium fluid with the highest mass flow rate of 4 kg/s.

13.4.3   EXPERIMENTAL ADVANCED SUPERCONDUCTING TOKAMAK

The EAST mainly consisted of 17 D-shaped TF SC magnets, including one spare coil; 13 PF SC in Hefei magnets, including one spare central solenoid (CS); vacuum vessel (VV); thermal shields (TSs); cryostat vessel (CV); and support systems, as illustrated in Figure 13.16. Wan et al. stated that its total weight is 414 tons, the toroidal field at the plasma center is 3.5 T, the peak field on the TF coil is 5.8 T (NbTi/Cu Cable in Conduit Conductor), and the operating current is 14.3 kA with a total stored energy about 300 MJ. The 3 kW/4.5 k (1 kW/3 K) helium cryogenic system has been developed and is successfully in duty. The subcooled helium cryogenic testing platform is composed of a 2.5 kW@4.5 K refrigerator and a cryogenic distribution system to provide LHe by a helium circulating pump and subcooled helium by cold compressors. The 3 K subcooled helium is for the

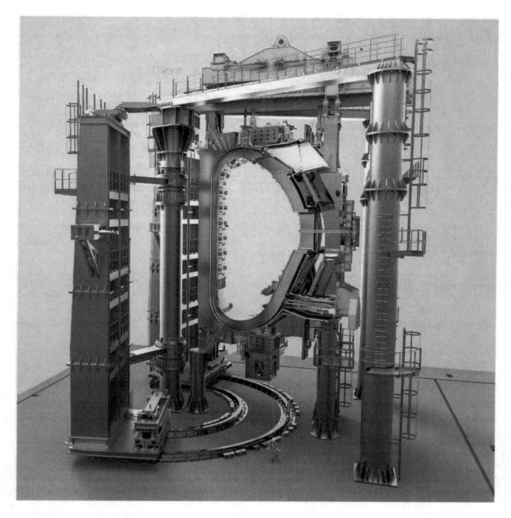

FIGURE 13.14 One sector of the ITER Tokamak SC magnets. Source: *ITER*

cooling of the SC magnets (the total cold mass of the SC magnet system is 194 tons). In 2018, EAST reported that the tokamak reached a milestone of 100 million °C electron temperature (higher than the temperature in the sun's core) in its core plasma.

13.5 ADVANCED APPLICATIONS OF H₂

Top Candidate for Energy Carrier. Hydrogen is no longer the top candidate to be the energy carrier of the future. Hydrogen is the energy carrier of the future. To be more specific, liquid hydrogen is the basis for the logistics of moving large amounts of energy around the world or from any geographical region to another. Batteries of lithium ion and gas bottles of hydrogen will always have their place for small items, but to store and move around large amounts of energy, liquid hydrogen will be at the heart of the matter. To keep things simple, liquid hydrogen in this context is meant to cover any cryogenic state of hydrogen, whether it be liquefied hydrogen (LH_2), cryo-compressed (or supercritical hydrogen), or cryo-adsorbed hydrogen, as the main point of cryogenics is the means of providing a practical, high-density method of energy storage.

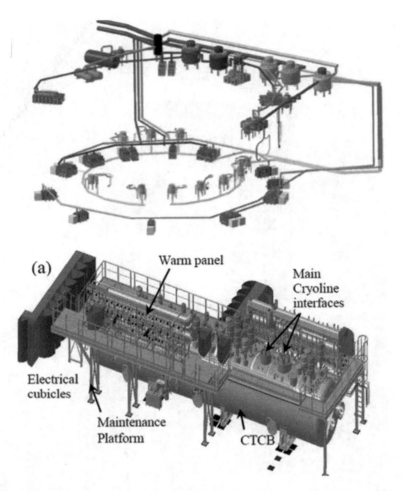

FIGURE 13.15 3D layout of ITER cryo-distribution boxes in the tokamak building and ITER cryogenic distribution lines [25].

Advanced Applications of Hydrogen. The use of LH_2 both as a rocket fuel and as a supply for electrical power began about the same time, around 1959, by NASA. These are two very different things, as one involves combustion and the other electrochemistry, but both applications made possible the space programs to come. Nearly everyone recognizes the use of big tanks of liquid hydrogen and liquid oxygen (a.k.a. space launch vehicles) for rocket propulsion. But the vital importance of electrical power in space, also made possible by long-duration storage tanks of LH_2, is obscured and sometimes overlooked. This dichotomy is finally beginning to change. For example, in 2019, Blue Origin announced plans for a lunar lander system that will have an LH_2 tank to service both the propulsion and electrical power needs on the Moon. The boil-off from the LH_2 tank will serve to cool the liquid oxygen tank and provide feedstock for the electrical power system.

There are certainly places and applications for all of the above in the energy world, but hydrogen is king because of its simplicity, versatility, and potential to be completely clean. As the most abundant element on Earth and in the known universe, it can be obtained from anywhere and just about anything. Hydrogen can be combusted as a fuel, exhausting water as the by-product, but the beautiful thing is that it is now used directly for electric power. There is no fuel burning and no moving parts, just pure electrical power produced by the hydrogen electrical cell (or "fuel cell"). These electrochemical cells have moved into the practical realms of cost and functionality for electricity

FIGURE 13.16   Photo of the EAST SC system. Source: *EAST*

supply for everything. Well, almost everything. What in the world today is going away from being run on electricity? Not much. More and more electricity, at scales from micro to mega, means the demand for more and more hydrogen. And that means liquefied hydrogen.

But first, a little history. Most do not realize that hydrogen production on an industrial scale also began in the early 1900s, followed by industrial-scale liquefaction starting in the 1950s. By the 1980s, hydrogen production was a mature business. For example, more than 50 tanker trailers of liquid hydrogen (LH_2) were required for every on-time launch of the Space Shuttle (135 flights over 30 years). In 2018, global hydrogen production was around 70 million tons per year. That much hydrogen, in liquid form, represents about 1 billion cubic meters (or, in gas form, about 30 trillion cubic feet). That hydrogen is piped and trucked around, and you never hear about it, a good thing from the safety perspective. One cup of water weighs about the same as one gallon of LH_2. When spilled or leaked, hydrogen is so lightweight and fast moving that it disperses into the atmosphere very rapidly. Hydrogen is powerful, which is why we want it, but it is hard to get it to explode with damaging effects.

From the 1930s, it was recognized that hydrogen and oxygen would be the ultimate combination for fuel and oxidizer to power rocket propulsion. Hydrogen is an extremely light and powerful rocket propellant. It has the lowest molecular weight of any known substance and, in combination with oxygen, burns with extreme intensity. Liquid hydrogen yields the highest specific impulse, or relative efficiency, of any known rocket propellant. Research and development of the handling of liquid hydrogen in the 1940s and 1950s paved the way for NASA space programs of the 1960s, which enabled the Apollo 11 Moon landing in 1969. Liquid hydrogen also enabled living in space through the hydrogen electric cell ("fuel cell"), which provided the necessary electrical power and water. Despite early setbacks and daunting engineering challenges, the taming of liquid hydrogen proved one of NASA's most significant technical accomplishments.

13.6   PROPULSION FUEL OF SPACE LAUNCH AND EXPLORATION

Today, liquid hydrogen is the signature fuel for space launch and exploration, both government and commercial. In addition to the Saturn V upper stages for the Apollo missions, LH_2 has been successfully used since 1963 in the evolution of upper-stage Centaur rocket engines. The modern Centaur is used today by United Launch Alliance with its Atlas V and Delta IV launch vehicles. The Space Shuttle, with its three main engines, was successfully powered by LH_2 for all 135 flights. The European Space Agency developed the highly successful liquid-hydrogen stage for the Ariane rocket in the 1970s, which continues today with the Ariane 5. Blue Origin has dedicated its rocket development to use LH_2 technology and plans the maiden flight of New Glenn from Florida in 2022. United Launch Alliance, with its plans for the new Vulcan Centaur launch vehicle, includes the next generation of LH_2-fueled upper stages.

13.6.1 New Space Launch System

NASA's new space launch system (SLS) heavy-lift rocket for the Artemis program is nearing completion with the recent hot fire test of the core stage, as shown in Figures 13.17 and 13.18. The SLS LH_2 tank makes up the bulk of the rocket, holding 2,033 m^3 (537,000 gallons) m^3 of LH_2 in its 8.4-m diameter by 40-m height. Construction of an additional LH_2 storage tank at Kennedy Space Center's Launch Complex 39B is also nearing completion to provide a total on-site storage capacity of 8,000 m^3 (2,100,000 gallons). Research and development are also underway by NASA for exploring the Moon and establishing long-term residence. As evidenced by the NASA Tipping Point selections in late 2020, dedicating nearly 70% of the total $370M funding to the topic of cryogenic fluid management technology demonstration, a key technological challenge is the first long-term storage and use of large amounts of liquid hydrogen in space.

13.6.2 Formation of Hydrogen Storage

Hydrogen is stored in different formats depending on the need. Cryogenic formats include liquid, cryo-compressed (supercritical), and cryo-adsorbed (molecular). Cryogenic formats provide the ultimate in density but also offer key ancillary benefits, such as purity and process cooling. For comparison, the density of liquid is 71 kg/m^3 at its 20 K normal boiling point, while a 700-bar gaseous storage tank is only about 41 kg/m^3. Moving hydrogen around is a challenge in any case,

FIGURE 13.17 The core stage for the first flight of NASA's Space Launch System rocket is seen in the B-2 Test Stand during a hot fire test January 16, 2021, at NASA's Stennis Space Center near Bay St. Louis, Mississippi. Source: *NASA (Television)*

FIGURE 13.18 Construction of the world's largest LH$_2$ storage tank with a liquid capacity of 4,700 m^3 of at NASA Kennedy Space Center in 2020. Source: *CB&I Storage Solutions, Photo Credit to McDermott International, Ltd.*

but the storage and transfer of LH$_2$ is now worked out by mature engineering standards and safety codes.

Even the storage tank boil-off losses and tanker offloading losses are now a thing of the past with the incorporation of active refrigeration technology. The new 4,700-m^3 storage tank at NASA Kennedy Space Center includes the provision for integrated refrigeration and storage (IRaS) technology to provide energy-efficient control and zero boil-off capability. The new tank also includes glass bubble insulation technology as part of its vacuum-jacketed thermal insulation system for a near 50% improvement in thermal performance compared to a perlite powder system. These technologies were proved through years of research and development by the CTL team in Florida (see Figure 13.19).

The big problem with LH$_2$ has always been the energy-intensive cost of liquefaction. With the advent of renewable energy from the sun, wind, and other sources, we are now beginning to see the dream of clean and abundant energy emerge into reality. The problem then becomes not that of the high cost of liquefaction but the inability to store all that free energy when it is not being generated. But . . . wait a minute . . . we can just liquefy and keep the hydrogen for when it is needed. The LH$_2$ is the battery for storing the excess energy, like electricity in liquid form. And in liquid format, we can move it around whenever and wherever it is needed.

Even the boil-off problem for cars, trucks, boats, and other vehicles is solved, if the vehicle is electric, then it probably also has a battery in addition to its hydrogen electric cell. The boil-off charges the battery. Hydrogen equals electricity (via the electrochemical cell). Electricity equals hydrogen (via the electrolyzer). What more perfect synergy can there be in the energy world?

FIGURE 13.19 Engineers complete a test of the Ground Operations Demo Unit for liquid hydrogen at NASA's Kennedy Space Center in Florida. The system includes a 125-m³ storage tank with an internal heat exchanger supplied from a cryogenic helium cycle refrigerator. Source: *NASA/Cory Huston*

LH₂ continues to lead the way in the aerospace world of propulsion. The ultimate chemical rocket is still the one that combines oxygen and hydrogen to make thrust and steam. Space exploration also relies on LH₂ for electrical power and drinking water. Why LH₂? The same reason that filling stations around the world are now moving away from gaseous storage to liquid storage: for the density, logistical practicality, economic effectiveness, and performance that the cryogenic systems provide. Cryogenics is all about energy and energy density, and energy is a lot about economics.

Research on energy storage abounds, and yet LH₂ makes a strong argument for being the king of all. Although one technology is rarely optimal for all situations, LH₂ systems are now being developed to work at scales from small drones to mega-scale carrier ships. In the middle are a lot of other things like semi-trucks, ferry boats, airplanes, and backup power generators. At the heart of it all is electrification. Hydrogen electric cells like those first pioneered by NASA for space travel (see Figure 13.20) are being developed for electrical power applications of all kinds and at all scales. Could it be that all other energy storage methods will fall away? Certainly not, as there are millions of different situations to cover, but look at it the other way around. Is LH₂ storage likely to fall away at either end of the spectrum? Not a chance.   .

The driver for hydrogen is renewable energy. The vector for putting that energy to use is hydrogen. The future is bright for cryogenic engineering, product development, and manufacturing around hydrogen liquefaction, storage, transfer, servicing, and global transport.

From propulsion to energy storage to renewables, as well as a host of industrial applications, hydrogen is versatile and valuable. To use it at scale and for its full potential, liquid hydrogen must be made, dealt with around the world, and put to work for end-use applications. After three-fourths of a century, the art of cryogenic engineering for LH₂ systems may be just beginning (Figure 13.20).

13.7 LIQUEFIED NATURAL GAS

Liquefied natural gas (LNG) continues to be a major business throughout the world. The first ocean-going tanker ships began in the late 1950s. The 1970s saw major growth and expansion as

FIGURE 13.20 Liquid natural gas tanker designed and built by AECOM employees to support NASA's Morpheus program. Source: *NASA, www.aecom.com/blogs/wp-content/uploads/sites/103/2015/07/LT-81-LNG_690x355.jpg*

FIGURE 13.21 Field demonstration of liquid hydrogen densification and zero boiloff storage with a 125-m3 storage tank and integrated helium refrigeration system at the NASA Kennedy Space Center's Cryogenics Test Laboratory. Source: *Fesmire*

mega-scale transport of LNG throughout key shipping lanes of the world became commonplace. Today, container ships with total capacities of over 200,000 m³ are used. A floating LNG (FLNG) production factory called Prelude was deployed by Shell off the Australian coast in 2018. Prelude, with an annual production capacity of 3.6 million metric tons of LNG, is the largest facility ever placed in water (Figures 13.21 and 13.22).

Shore-based storage tanks of similar size scales are used. These tanks are typically flat-bottom designs (and therefore not vacuum-jacketed) and insulated by specific combinations of perlite powder, polyisocyanurate foam, and cellular glass foam. Issues such as rollover were identified decades

FIGURE 13.22   Prelude FLNG exports second cargo. Source: *Shell*

ago and are now dealt with by combinations of internal design features and operational parameters to keep the thermo-fluid properties of the liquid under control.

LNG is used to generate electrical power on the grid, heat homes, fuel industrial processes, and produce hydrogen by steam methane reforming (SMR). Today, the bulk of the 75+ million tons of hydrogen produced in the world each year is done through SMR. As this process produces at least eight times the amount of carbon dioxide, by mass, future processes are being developed to improve this situation. However, the second leading method of hydrogen production, coal gasification, is drastically worse on the environment. In general, LNG-based power generation has led to dramatic improvements in clean air, less carbon, and overall energy efficiencies compared to the heavier fossil fuels such as diesel fuel and bunker oil.

Starting around 2000, the proliferation of natural gas for transportation vehicles began. The use of compressed natural gas (CNG)-powered vehicles was followed by LNG-powered long-haul trucks for significant improvements in both operational costs and environmental impact. A modern development in the use of LNG has been its use as a rocket fuel for space launch vehicles. Under flight test and development today are liquid oxygen- and LNG-fueled rockets for heavy lift vehicles such as the New Glenn by Blue Origin, the Vulcan by United Space Alliance, and the Starship by SpaceX. The use of LNG (or perhaps liquid methane) is also being developed for different Moon and Mars exploration vehicles.

13.8   HIGH-TEMPERATURE SUPERCONDUCTING POWER

There is a need for increased reliability and capacity in the supply of electric power. This is particularly important for large, densely populated urban areas that frequently host critical centers of finance, trade, and government. Should disruptions occur to the electric power grid, there can be severe impact to the regional and national economy and security. As a result, there are advances in the research, development, and application of high-temperature superconducting power system equipment.

High-temperature superconducting power cable systems have been demonstrated in several commercial utility grids [29–32]. These cables are cooled with circulating liquid nitrogen with an operating temperature ranging from 65 K to 80 K. The triaxial HTS cable design, shown in Figure 13.24, is a very compact design which has been used in several projects [30–33]. It consists of a stainless-steel form onto which alternating layers of dielectric and HTS conductor are wound. The cable has three high-current, high-voltage (13.8 kV) layers and a ground layer on the outside, as described in [29]. One such installation was performed by Ultera. They installed a single 200-meter-long high-temperature superconducting three-phase triaxial design cable at the American Electric Power (AEP) Bixby substation in Columbus, Ohio. This cable connected a 138/13.2-kV transformer to the distribution switchgear serving seven outgoing circuits. It carried 3,000 amperes root mean square (rms) amps under normal operation and demonstrated that it was capable of handling fault current of over 20 kA.

One advantage of the triaxial design is the compactness, such that all three electrical phases can be put into a single cryostat. The cryostat heat load is one of the larger components that depends on the length of the cable. If the phases were in separate cryostats, there would be as much as three times the thermal load from the cryostats. Other significant thermal loads come from the current leads in the termination, which provides for warm electrical connections that transition to the liquid nitrogen temperature HTS cable. The termination from the AEP Bixby project is shown in Figure 13.23. Another feature that has a thermal impact on a HTS cable system are joints. Cables lengths are limited to the amount which can be placed on a spool for transport. For long-length cable systems, field joints must be made which join the HTS cable and the cryostat containing the cable. A field joint that was assembled in a standard concrete manhole is shown in Figure 13.24. The trixial cable was also used in the German AmpaCity project, which started in 2011. This project consisted of a 10-kV, 40-MVA HTAS cable system that was 1 km long and installed and operated in Essen, Germany. This cable was installed with a HTS fault current limiter. This provided the capability to limit an available fault current of above 60 kA to just over 20 kA. The Electric Power Research Institute (EPRI) conducted an extensive review of the technology issues, manufacturing readiness, costs, and commercialization potential of HTS cables, which can be found in [35].

In addition to cables, electric utilities have encouraged the development of HTS transformers. Conventional transformers are filled with a dielectric oil, about which there are environmental and safety concerns. The use of HTS in the transformer windings offers several advantages, including a smaller size and lower weight, as well as the elimination of the transformer oil. The transformer can now be filled with liquid nitrogen and refrigerated with cryocoolers [35].

In addition to the development of technology for the utility power sector, HTS power is being applied in other areas, such as all-electric aircraft as proposed by the CHEETA project [31]. The concept uses cryogenic hydrogen in fuel cells to generate electricity, as shown in Figure 13.25. Liquid hydrogen is also used to cool a high-efficiency superconducting electrical system.

FIGURE 13.23 Cross-section of the triaxial HS cable and triaxial HS cable termination used in the AEP Bixby project [29].

FIGURE 13.24 Field splice of a triaxial HTS cable and cryostat for the AEP Bixby project [29].

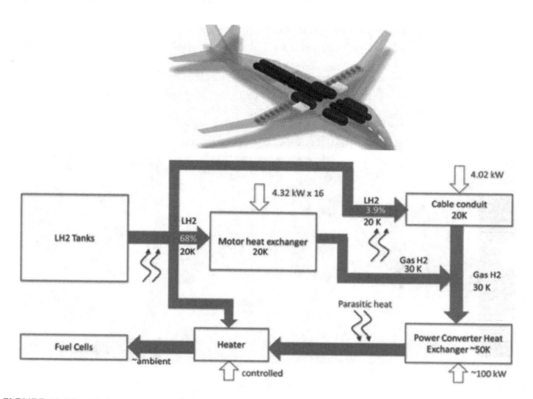

FIGURE 13.25 Hydrogen-powered electrical aircraft concept from the CHEETA project [36].

REFERENCES

1. Védrine P, 2014 Large superconducting magnet systems, *Proceedings of the CAS-CERN Accelerator School*, CERN—2014–005.
2. Lebruna Ph et al., 2014 Lessons from the LHC, *AIP Conference Proceedings* 1573, 245.
3. Maekawa R, et al., 2019 Dynamic simulation of ITER cryogenic system under D-T operation, *CEC 2019*, IOP Series: 755 (2020).
4. Omet M, et al., 2019 *Operation of the European XFEL Towards The Maximum Energy*, SRF2019, Dresden, Germany.
5. Shu QS, 1998 Large applications & challenges of state-of-the-art superconducting RF (SRF) technologies, *Advances in Cryogenic Engineering*, 43.
6. Henderson H, et al., 2014 The Spallation Neutron Source accelerator system design, *Nuclear Instruments, A*, 763, 610–673.
7. Godeke A et al., 2006 Limits of NbTi and Nb₃Sn, and Development of W&R Bi—2212 High Field Accelerator Magnets, ASC, Seattle, WA.
8. Lebrun P, 1995 Cryogenic Systems for Accelerators, CERN-AT-95–08 (CR), 1995.
9. Lebrun P, 2009 Cryogenics for particle accelerators, CAS Course, in General Accelerator Physics, Divonne-les-Bains.
10. Parma V, Cryostat Design, 2013 CERN School "Superconductors for Accelerators", May.
11. Evans L, and Bryant P (editors), 2008 LHC machine, *Journal of Instrumentation*, iopscience.iop.org, Institute of Physics Publishing and SISSA.
12. Lebrun, private communication with Shu, Dec. 2020.
13. Brodzinski K et al., 2019 Cryogenic operational experience from the LHC physics run2 (2015–2018 inclusive), *IOP Conf. Series: Materials Science and Engineering*, 755 (2020), 012096.
14. Rode C, 1983 Tevatron cryogenic system, Fermilab tech report, Batavia, IL.
15. Wolff S, 1986 Superconducting magnets for HERA, *13th International Conference on High-Energy Accelerators*, Novosibirsk, Soviet Union.
16. Philippe Lebrun, et al., 2015 Beyond the Large Hadron Collider: A first look at cryogenics for CERN future circular colliders, *Physics Procedia*, 67, 768–775.
17. Shu QS, 1989 A study of the influence of heat treatment on field emission in superconducting RF cavities, *Nuclear Instruments and Methods in Physics Research*, A278.
18. Kostin D et al., 2018 Progress towards continuous wave operation of the SRF LINAC at DESY, Superconductivity & Particle Accelerators 2018, Krakow, Poland, November.
19. H Nakai et al., 2017 Cryogenic system configuration for the International Linear Collider (ILC) at mountainous site, *IOP Conf. Series: Materials Science and Engineering*, 171, 012.
20. Carlo Pagani, *ILC Cryomodule and Cryogenics*, INFN, International Accelerator School for ILC, 2006 Japan.
21. Blum L et al., 2015 Status and commissioning results of the helium refrigerator plant for the *European XFEL*, *Advances in Cryogenic Engineering*, 61, 2015.
22. Eichhorn R., et al., 2015 The Cornell main linac cryomodule, *Physics Procedia,* 67, 785–790.
23. Howell M et al., 2015 Cryogenic system operational experience at SNS, *IOP Conf. Series: Materials Science and Engineering*, 101, 012127.
24. Serio L, 2009 *ITER Cryogenics,* MATEFU Spring School, ITER Organization Headquarter, Paris, France.
25. Patel P., et al., 2020 ITER Cryoplant termination cold box, *IOP Conf. Ser.: Mater. Sci. Eng.* 755 012088.
26. Maekawa R et al., 2019 Dynamic simulation of ITER cryogenic system under D-T operation, *Advances in Cryogenic Engineering,* 65.
27. Wan YX, et al., 2004 Progress of the EAST Project in China, Institute of Plasma Physics Report: FT/3–3, HeFei, China.
28. Wei J et al., 2010 The superconducting magnets for EAST tokamak, *IEEE Transactions on Applied Superconductivity*, 20(3), 556–559.
29. Demko JA, Sauers I, James DR, et al., 2007 Triaxial HTS cable for the AEP Bixby project, *IEEE Transactions on Applied Superconductivity*, 17(2), pp. 2047–2050.
30. Stemmle M, Merschel F, et al., 2013 Ampacity project—Worldwide first superconducting cable and fault current limiter installation in a German city center, *CIRED 22nd International Conference on Electricity Distribution Stockholm*, 10–13 June 2013.
31. Weber C, Lee R, Ringo S, 2007 Testing and demonstration results of the 350 m long HTS cable system installed in Albany, NY, *IEEE Transactions on Applied Superconductivity*, 17-(2), 10.1109/TASC.899828.

32. Maguire JF, Schmidt F, Bratt S et al., 2007 Development and demonstration of a HTS power cable to operate in the long island power authority transmission grid, *IEEE Transactions on Applied Superconductivity,* 17-(2), 10.1109/TASC.2007.898359.

33. Maguire J, Folts D, Yuan J, Henderson N., et al., 2010 Status and progress of a fault current limiting HTS cable to be installed in the Con Edison grid, *Advances in Cryogenic Engineering*, 55, 2010.

34. Weber C, Reis CT, Hazelton DW, et al., 2005 Design and operational testing of a 5/10-MVA HTS utility power transformer, *Applied Superconductivity*, 15(2), 2210–2213.

35. Summary Report: Technical Analysis and Assessment of Resilient Technologies for the Electric Grid: High Temperature Superconductivity. EPRI, Palo Alto, CA: 2017.

36. Balachandran T, Haran K., et al., 2022 Feasibility study on superconducting motor topologies for a hydrogen-powered all-electric commercial aircraft, *CEC Virtual Presentation* (on line) *Advances in Cryogenic Engineering*, 67, 2022.

Appendix A

Cryostat Test Data for Select Thermal Insulations

Benchmark data summary from Cryostat CS-100 testing with LN_2:

- Boiloff flow rate (m-dot)
- Heat (Q)
- Heat Flux (q)
- Effective thermal conductivity (k_e)

Materials/systems include:

- Vacuum Only (black sleeve)
- Three different MLI systems
- Layered Composite Insulation (LCI) system
- Aerogel blanket
- Fiberglass blanket
- Glass bubbles
- Perlite powder
- Aerogel particles
- Spray-on foam insulation

Test Specimen	CVP	WBT	Flow	Q	q	k_e	Specifications			
	millitorr	K	sccm	W	W/m²	mW/m-K				
A114 Vacuum Only	0.003	293.1	7447	30.8	88.4	10.44	*x*	*n*	A_e	*ρ*
	0.02	292.9	7620	31.5	90.5	10.69	mm	-	m²	kg/m³
	1	292.8	8917	36.9	105.9	12.52				
Black sleeve.	1	291.9	8912	36.9	105.8	12.57	-	-	0.304	-
	10	292.4	12907	53.4	153.3	18.16				
	100	292.5	15961	66.0	189.5	22.44				
	760000	242.9	34094	141	404.8	62.37				
A154 MLI Mylar-Net (4.9, 10, 42)	0.001	293.0	75	0.312	0.997	0.023	*x*	*n*	A_e	*ρ*
	0.02	290.7	83	0.341	1.09	0.025	mm	-	m²	kg/m³
	0.1	293.4	88	0.363	1.16	0.026				
Double-aluminized Mylar	0.3	293.3	126	0.519	1.66	0.038	4.9	10	0.313	42
and polyester net spacer.	1	291.5	271	1.12	3.58	0.082				
Layer by layer installation.	3	293.0	623	2.58	8.24	0.187				
	10	290.0	1688	6.98	22.3	0.512				
	100	285.3	8220	34.00	109	2.55				
A125 MLI Mylar-Net (16, 40, -)	0.01	293.8	32	0.132	0.398	0.028	*x*	*n*	A_e	*ρ*
	0.02	293.1	35	0.143	0.431	0.031	mm	-	m²	kg/m³
	0.1	293.1	44	0.182	0.549	0.040				
Double-aluminized Mylar	1	292.8	81	0.333	1.00	0.072	15.5	40	0.332	-
and polyester net spacer.	10	293.3	518	2.14	6.46	0.464				
Layer by layer installation.	10	292.9	521	2.16	6.51	0.468				
	100	292.5	3604	14.9	45.0	3.24				
	1000	292.8	8983	37.2	112	8.06				
	10000	292.4	11341	46.9	142	10.2				
	100000	293.0	15058	62.3	188	13.5				
	760000	292.2	19376	80.2	242	17.4				
A128 MLI Foil-Paper (21, 80, -)	0.003	292.0	42	0.176	0.516	0.051	*x*	*n*	A_e	*ρ*
	0.3	293.6	46	0.192	0.563	0.055	mm	-	m²	kg/m³
	1.1	293.1	54	0.221	0.648	0.064				
Aluminum foilContinuous rolled	10	294.0	188	0.780	2.29	0.223	21.1	80	0.341	-
and microfiberglass paper spacer.	100	293.0	1214	5.03	14.7	1.44				
Continuous-rolled installation.	1000	292.7	5293	21.9	64.2	6.30				
	10000	293.4	10943	45.3	133	12.8				
	100000	293.2	13013	53.9	158	15.5				
	760000	291.7	16548	68.5	201	19.8				
A112 Aerogel Blanket (23, 2, 133)	0.01	293.0	1160	4.800	12.4	1.47	*x*	*n*	A_e	*ρ*
	1	293.2	1299	5.377	13.9	1.64	mm	-	m²	kg/m³
	10	292.7	1626	6.729	17.4	2.06				
Cryogel aerogel composite.	100	292.7	2299	9.514	24.6	2.91	23	2	0.344	133
	1000	293.0	3367	13.934	36.0	4.26				
	10000	293.0	4427	18.318	47.4	5.60				
	100000	292.6	5328	22.047	57.0	6.75				
	760000	293.4	8894	36.803	95.2	11.24				
A151 Fiberglass (49, 2, 16)	0.003	293.0	802	3.32	8.59	1.90	*x*	*n*	A_e	*ρ*
	0.2	294.1	809	3.35	8.67	1.95	mm	-	m²	kg/m³
	1	294.0	903	3.74	9.68	2.13				
Micro-fiberglass batt.	10	293.0	1219	5.05	13.1	2.89	48.6	2	0.386	16
	20	294.2	1487	6.16	15.9	3.59				
	100	293.0	3099	12.8	33.2	7.26				
	400	295.8	5227	21.6	56.0	12.51				
	1000	296.6	6080	25.2	65.2	14.49				

(Continued)

TABLE A.1 (Continued)

Test Specimen	CVP	WBT	Flow	Q	q	k_e	Specifications			
	millitorr	K	sccm	W	W/m²	mW/m-K				
	10000	297.4	7129	29.5	76.4	16.90				
	20000	293.6	7413	30.7	79.5	17.92				
	760000	293.0	10724	44.4	115	25.99				
A102 Glass Bubbles (25, 1, 65)	0.003	292.6	494	2.043	5.9	0.69	x	n	A_e	ρ
	0.1	293.0	495	2.049	5.9	0.70	mm	-	m²	kg/m³
	1	292.9	506	2.096	6.0	0.71				
Type K1 hollow microspheres.	10	293.1	585	2.419	6.9	0.82	25.4	1	0.349	65
Black sleeve.	25	293.3	691	2.861	8.2	0.97				
	50	293.6	875	3.620	10.4	1.22				
	100	293.8	1220	5.048	14.5	1.70				
	350	293.5	2696	11.158	32.0	3.77				
	1000	293.0	5547	22.953	65.9	7.78				
	3000	292.6	9795	40.535	116.3	13.76				
	10000	293.3	14161	58.602	168.2	19.84				
	30000	293.5	16294	67.427	193.5	22.80				
	100000	292.7	17861	73.913	212.1	25.09				
	760000	293.6	18308	75.763	217.4	25.61				
A103 Perlite Powder (25, 1, 132)	0.002	292.6	666	2.756	7.9	0.94	x	n	A_e	ρ
	0.1	292.7	679	2.808	8.1	0.95	mm	-	m²	kg/m³
	1	292.9	712	2.945	8.5	1.00				
High density.	10	293.5	935	3.867	11.1	1.31	25.4	1	0.349	132
Black sleeve	25	293.0	1342	5.555	15.9	1.88				
	100	293.2	2721	11.261	32.3	3.81				
	1000	292.7	9961	41.220	118.3	13.99				
	10000	292.6	19792	81.903	235.0	27.82				
	100000	292.7	23978	99.227	284.7	33.68				
	760000	293.3	24954	103.265	296.3	34.95				
A108 Aerogel Particles (25, 1, 80)	0.002	293.0	1204	4.981	12.6	1.69	x	n	A_e	ρ
	0.1	293.2	1232	5.100	12.9	1.73	mm	-	m²	kg/m³
	1	293.0	1303	5.392	13.6	1.83				
Nominal 1-mm diameter beads.	10	293.0	1746	7.226	18.2	2.45	25.4	1	0.349	80
Black sleeve.	25	293.3	2176	9.004	22.7	3.05				
	100	293.6	3092	12.796	32.2	4.33				
	1000	292.7	5292	21.901	55.2	7.44				
	10000	292.9	6332	26.203	66.0	8.89				
	100000	292.9	7334	30.350	76.5	10.29				
	200000	292.4	7986	33.046	83.3	11.23				
	500000	292.8	9579	39.638	99.9	13.45				
	760000	292.7	10207	42.240	106.5	14.34				
A104 Spray-On Foam BX-265 (25, 1, 42)	0.1	293.0	5527	22.872	65.6	7.75	x	n	A_e	ρ
	1	293.1	5924	24.513	70.3	8.68	mm	-	m²	kg/m³
	10	293.0	9861	40.805	117.1	14.46				
Machined; no rind.	100	293.0	12560	51.974	149.1	18.41	25.4	1	0.349	42
	1000	293.2	13443	55.628	159.6	19.69				
	10000	293.0	13534	56.004	160.7	19.85				
	100000	293.0	13621	56.364	161.7	19.97				
	200000	293.2	13792	57.074	163.8	20.20				
	500000	293.2	13957	57.755	165.7	20.44				
	760000	292.8	14424	59.690	171.3	21.17				

(Continued)

TABLE A.1 (Continued)

Test Specimen	CVP	WBT	Flow	Q	q	k_e	Specifications			
	millitorr	K	sccm	W	W/m²	mW/m-K				
C130 Layered Composite (22, 30, 50)	0.01	297	73	0.300	0.87	0.091	*x*	*n*	*A_e*	*ρ*
	0.02	299	74	0.306	0.89	0.091	**mm**	**-**	**m²**	**kg/m³**
	0.20	299	88	0.363	1.1	0.109				
	0.85	300	105	0.436	1.3	0.130	22	30	0.345	50
Layered Composite Insulation (LCI)	2.00	298	140	0.580	1.7	0.176				
30 layers double-aluminized Mylar	10	298	242	1.002	2.9	0.300				
microfiberglass paper Cryotherm 243	96	300	514	2.128	6.2	0.634				
fumed silica Cabosil 530	998	286	1143	4.728	13.7	1.54				
	9977	286	3449	14.273	41.4	4.91				
	99849	287	7268	30.078	87.2	10.16				
	100005	287	7397	30.610	88.7	10.15	Note: CBT = 86 K (average)			
	525000	296	9923	41.064	119	13.36				

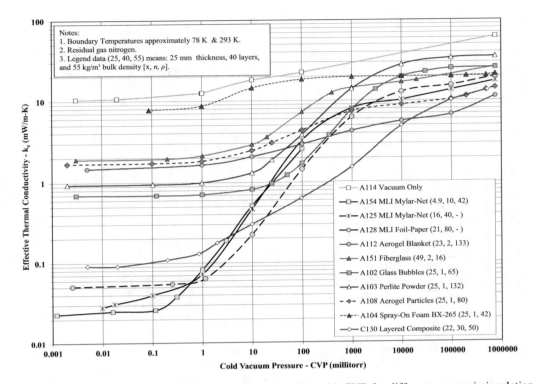

FIGURE A1. Variation of effective thermal conductivity (k_e) with CVP for different cryogenic insulation systems and materials. Boundary temperatures: 78 K and 293 K. Residual gas is nitrogen.

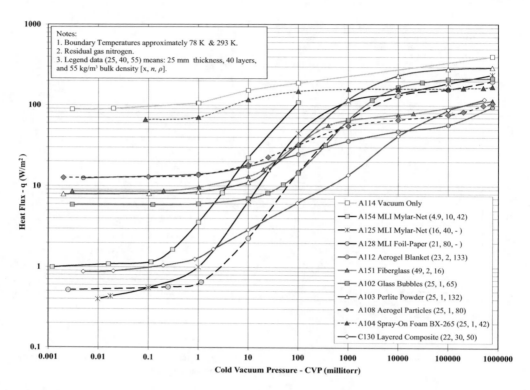

FIGURE A2. Variation of heat flux (q) with CVP for different cryogenic insulation systems and materials. Boundary temperatures: 78 K and 293 K. Residual gas is nitrogen.

Appendix B

Cryostat Test Data for Select MLI Systems

Tables and Plots.

TABLE B.1

MLI Systems Thermal Performance Data Summary: Mylar/paper (MP) and foil/paper (FP) types; each name designation includes total thickness (x), number of layers (n), and layer density (z) as (x, n, z)

Cryostat-100 Test Series (MLI System)	CVP μ	WBT K	Flow sccm	Q W	q W/m²	k_e mW/m-K	x mm	n	z lyrs/mm
MP1-A118 Mylar/Paper Blankets (19, 30, 2.1)	0.005	293	31	0.127	0.375	0.033	19.0	40	2.11
30 layers fold over seam + 10 layers overlap seam	0.1	294	41	0.171	0.506	0.044			
Nov-07	1	293	99	0.410	1.21	0.107			
	10	293	431	1.788	5.30	0.467			
	100	293	2434	10.094	29.9	2.63			
	1022	293	9317	38.635	114	10.1			
	10073	293	13691	56.771	168	14.8			
	100100	294	14112	58.516	173	15.2			
	768986	293	15162	62.870	186	16.4			
MP2-A148 Mylar/Paper 4 layers (5, 4, 0.7)	0.009	293	145	0.602	1.92	0.048	5.4	4	0.74
Layer-by-layer	0.1	293	338	1.402	4.47	0.112			
Staggered seams with 1" overlap typical; Jun-10	1	293	403	1.670	5.32	0.133			
MP3-A149 Mylar/Paper 2 layers (2.4, 2, 0.8)	0.003	293	253	1.049	3.40	0.038	2.4	2	0.82
Layer-by-layer	0.1	293	347	1.439	4.66	0.053			
Staggered seams with 1" overlap typical	0.3	293	786	3.258	10.6	0.119			
Jul-10	1	293	1169	4.847	15.7	0.178			
MP4-A150 Mylar/Paper 1 layer (1.5, 1, 0.7)	0.001	297	604	2.504	8.16	0.057	1.5	1	0.65
One layer of Mylar/paper on cold mass	0.01	296	615	2.550	8.30	0.058			
Jul-10	0.04	297	1179	4.889	15.9	0.111			
	0.1	300	1385	5.743	18.7	0.129			
	0.2	298	1532	6.352	20.7	0.144			
	0.5	301	1859	7.708	25.1	0.172			
	1	302	2456	10.184	33.2	0.226			
FP1-A126 Foil/Paper Rolled (11, 40, 3.6)	0.001	294	46	0.191	0.589	0.030	11.2	40	3.58
Continuous roll-wrapped installation	0.004	294	50	0.209	0.643	0.033			
Vertical roll wrapper device	0.05	294	57	0.237	0.731	0.038			

(*Continued*)

TABLE B.1 (Continued)

Cryostat-100 Test Series (ML1 System)	CVP μ	WBT K	Flow sccm	Q W	q W/m²	k_e mW/m-K	x mm	n	z lyrs/mm
Mar-09	0.2								
	1	294	61	0.253	0.782	0.040			
	3	293	84	0.347	1.07	0.056			
	3	294	136	0.565	1.74	0.090			
	10	293	341	1.415	4.36	0.227			
	30	292	736	3.052	9.41	0.492			
	956	294	8573	35.548	110	5.67			
	10094	294	15243	63.207	195	10.1			
	100091	293	19996	82.912	256	13.3			
	739718	294	23857	98.925	305	15.8			
FP2-A127 Foil/Paper Rolled (16, 60, 3.7)	0.003	293	39	0.163	0.489	0.037	16.1	60	3.73
Continuous roll-wrapped installation	0.08	294	51	0.212	0.638	0.048			
Vertical roll wrapper device	0.3	294	v49	0.203	0.612	0.046			
Apr-09	1	293	66	0.274	0.823	0.061			
FP3-A128 Foil/Paper Rolled (21, 80, 3.8)	0.006	294	32	0.133	0.389	0.038	21.1	80	3.79
Continuous roll-wrapped installation	1	293	54	0.222	0.650	0.064			
Vertical roll wrapper device	10	294	188	0.781	2.29	0.224			
May-09	100	293	1214	5.035	14.8	1.45			
	1055	293	5293	21.947	64.3	6.32			
	10011	293	10943	45.376	133	13.0			
	99320	293	13013	53.960	158	15.5			
	764309	292	16548	68.617	201	19.9			
FP4-A132 Foil/Paper Spiral (17, 40, 2.3)	0.005	294	73	0.301	0.899	0.073	17.5	40	2.29
Spiral wrapped installation	0.008	293	78	0.324	0.967	0.078			
CRS wrap 3.5" wide roll	0.3	293	96	0.397	1.18	0.096			
	1	294	150	0.622	1.85	0.150			
Sep-09	10	294	433	1.797	5.36	0.434			
	100	293	2127	8.820	26.3	2.14			

TABLE B.2

MLI Systems Thermal Performance Data Summary: Mylar/net (MN) and blanket (NB) types; each name designation includes total thickness (x), number of layers (n), and layer density (z) as (x, n, z)

Cryostat-100 Test Series (MLI System)	CVP	WBT K	Flow sccm	Q W	q W/m²	k_e mW/m-K	x mm	n	z lyrs/mm
MN1-A125 Mylar/Net (15, 40, 2.6)	0.01	294	32	0.132	0.399	0.029	15.5	40	2.59
Layer-by-layer	0.02	293	35	0.143	0.431	0.031			
Staggered seams with 1.5" overlap typical	0.1	293	44	0.183	0.551	0.040			
Dec-08	1	293	81	0.334	1.01	0.072			
	10	293	521	2.162	6.52	0.469			
	100	293	3604	14.942	45.1	3.24			
	1040	293	8983	37.248	112	8.08			
	10036	292	11341	47.025	142	10.2			
	99102	295	16447	68.199	206	14.7			
	768585	292	19376	80.341	242	17.5			
MN2-A152 Mylar/Net (8, 20, 2.5)	0.001	293	52	0.214	0.672	0.025	8.0	20	2.52
Layer-by-layer	0.002	293	54	0.222	0.70	0.026			
Staggered seams with 1" overlap typical	0.07	293	100	0.415	1.30	0.048			
Bulk density = 53 kg/m³	0.1	293	103	0.426	1.34	0.049			
Feb-11	0.3	293	114	0.474	1.49	0.055			
	1	293	200	0.829	2.60	0.096			
	3	292	362	1.501	4.71	0.175			
	10	293	1008	4.180	13.1	0.486			
	100	293	6114	25.352	79.6	2.95			
MN3-A153 Mylar/Net (7, 15, 2.3)	0.003	293	66	0.275	0.871	0.027	6.6	15	2.28
Layer-by-layer	0.1	293	138	0.571	1.80	0.055			
Staggered seams with 1" overlap typical	0.3	294	179	0.741	2.34	0.071			
Bulk density = 47 kg/m³	1	293	267	1.108	3.51	0.107			
Mar-11	10	298	1310	5.432	17.2	0.514			
	100	292	7428	30.800	97.4	2.99			
MN4-A154 Mylar/Net (5, 10, 2.0)	0.001	293	75	0.312	0.997	0.023	4.9	10	2.05
Layer-by-layer	0.02	291	83	0.342	1.09	0.025			

(*Continued*)

TABLE B.2 (Continued)

Cryostat-100 Test Series (MLI System)	CVP	WBT K	Flow sccm	Q W	q W/m²	k_e mW/m-K	x mm	n	z lyrs/mm
Staggered seams with 1" overlap typical									
Bulk density = 42 kg/m³	0.1	293	88	0.363	1.16	0.026			
	0.3	293	126	0.520	1.66	0.038			
	1	292	271	1.123	3.59	0.082			
	3	293	623	2.583	8.25	0.187			
	10	290	1688	6.999	22.4	0.514			
	100	285	8220	34.084	109	2.56			
	0.002	325	115	0.477	1.52	0.030			
	1	325	322	1.334	4.26	0.084			
	10	323	1914	7.937	25.3	0.504			
	97	322	9174	38.038	121	2.42			
	0.004	349	143	0.593	1.89	0.034			
NB1-A139 Mylar/Net Blanket-A (43, 40, 0.9)	0.003	305	35	0.146	0.387	0.073	42.7	40	0.94
10 sub-blankets	760000	308	13387	55.509	147	27.4			
Bulk density = 45 kg/m³ Nov-09	0.004	350	51	0.210	0.558	0.088			
NB2-A140 Mylar/Net Blanket-B (64, 60, 0.9)	0.002	306	32.8	0.136	0.332	0.093	63.6	60	0.94
15 sub-blankets	0.002	306	33.1	0.137	0.335	0.094			
Bulk density = 37 kg/m³	0.1	305	68.2	0.283	0.691	0.194			
Dec-09	0.3	305	79.6	0.330	0.806	0.226			
	1.0	305	172	0.713	1.74	0.488			
NB3-A143 Mylar/Net Blanket-B (43, 60, 1.4)	0.002	293	33	0.138	0.366	0.073	42.6	60	1.41
2 sub-blankets	0.1	293	56	0.234	0.621	0.123			
Bulk density = 56 kg/m³	0.3	293	70	0.292	0.775	0.154			
Feb-10	1	293	101	0.417	1.11	0.219			
	10	293	438	1.816	4.82	0.957			
	100	293	2733	11.332	30.1	5.97			
	1074	293	6296	26.106	69.3	13.7			
	10026	293	7274	30.162	80.1	15.8			
	100018	293	7417	30.755	81.7	16.2			
	760000	291	9780	40.553	108	21.5			
NB4-A144 Mylar/Net Blanket-B (23, 60, 2.6)	0.003	305	31	0.127	0.368	0.037	23.0	60	2.61
15 sub-blankets	0.004	305	32	0.131	0.380	0.038			
Bulk density = 97 kg/m³	0.06	304	55	0.226	0.656	0.067			
Mar-10	0.1	304	60	0.250	0.726	0.074			
	0.3	305	72	0.299	0.867	0.088			

TABLE B.3

MLI Systems Thermal Data Summary: Mylar/fabric (MF) and Mylar/silk net (SN) types; each name designation includes total thickness (x), number of layers (n), and layer density (z) as (x, n, z)

Cryostat Test Series (MLI System)	CVP μ	WBT K	Flow sccm	Q W	q W/m^2	k_e mW/m-K	x mm	n	z lyrs/mm
MF1-A145 Mylar/Fabric (6, 10, 1.6)	0.006	294	68	0.282	0.893	0.026	6.4	10	1.57
Layer-by-layer	0.1	293	146	0.604	1.91	0.057			
Apr-10	0.3	293	171	0.707	2.24	0.066			
	1	293	277	1.149	3.64	0.108			
	10	293	1456	6.037	19.1	0.568			
	99	293	7684	31.862	101	3.00			
MF2-A133 Mylar/Fabric G1 Blanket (8, 40, 5.1)	0.005	299	206	0.854	2.68	0.095	7.8	40	5.12
One 40-layer blanket	0.3	300	258	1.070	3.36	0.118			
Bulk density = 146 kg/m^3	1.0	300	310	1.285	4.04	0.142			
Sep-09									
MF3-A134 Mylar/Fabric G2 Blanket (15, 40, 2.8)	0.006	300	56	0.233	0.707	0.046	14.5	40	2.75
Two 20-layer sub-blankets	0.1	300	98	0.406	1.23	0.081			
Bulk density = 82 kg/m^3	0.3	300	115	0.477	1.44	0.095			
Sep-09	1	300	153	0.634	1.92	0.126			
SN1-A177 Mylar/Silk Net Double (24, 20, 0.9)	0.006	293	29	0.120	0.348	0.038	23.5	20	0.85
Layer-by-layer	0.1	293	58	0.242	0.700	0.077			
Staggered seams with 1" overlap typical	1	293	146	0.605	1.75	0.191			
Bulk density = 20 kg/m^3	760000	284	16647	69.027	200	22.8			
Mar-14	0.004	305	34	0.140	0.405	0.042			
	0.004	326	45	0.186	0.538	0.051			
SN2-A178 Mylar/Silk Net Single (15, 20, 1.4)	0.003	293	27	0.114	0.344	0.023	14.7	20	1.36
Layer-by-layer	0.1	293	58	0.240	0.725	0.050			
Staggered seams with 1" overlap typical	1.0	293	138	0.572	1.73	0.118			
Bulk density = 24 kg/m^3 Mar-14									
SN3-A179 Mylar/Silk Net Single (8, 10, 1.3)	0.001	292	41	0.171	0.538	0.019	7.7	10	1.30
Layer-by-layer	0.002	292	42	0.172	0.542	0.019			
Staggered seams with 1" overlap typical	0.1	294	59	0.246	0.774	0.028			
Bulk density = 25 kg/m^3 Apr-14	1.0	293	222	0.921	2.89	0.103			

TABLE B.4

MLI Systems Thermal Data Summary: Mylar/discrete (DX) type and vacuum only (VO); each name designation includes total thickness (x), number of layers (n), and layer density (z) as (x, n, z)

Cryostat Test Series (MLI System)	CVP μ	WBT K	Flow sccm	Q W	q W/m²	k_e mW/m-K	x mm	n	z lyrs/mm
DX1-A142 Mylar/Discrete IMLI (39, 20, 0.5)									
Two 10-layer sub-blankets									
Bulk density = 25 kg/m³									
Jan-10	0.001	292	37	0.152	0.411	0.074	38.7	20	0.52
	0.1	293	107	0.444	1.20	0.216			
	1	292	215	0.893	2.41	0.436			
	10	294	1144	4.745	12.8	2.29			
	100	296	5032	20.867	56.3	10.0			
	1009	297	7107	29.469	79.6	14.1			
	10000	297	9278	38.471	104	18.4			
	0.003	304	51	0.210	0.568	0.097			
	0.1	304	114	0.473	1.28	0.218			
	0.3	305	141	0.586	1.58	0.271			
DX2-A167 Mylar/Discrete LBMLI (38, 19, 0.5)	0.003	293	49	0.202	0.547	0.096	37.7	19	0.50
20 layers LBMLI—first layer doesn't count	0.005	293	52	0.217	0.589	0.103			
Oct-12	0.1	293	135	0.561	1.52	0.266			
	1	293	245	1.015	2.75	0.483			
	10	293	1179	4.888	13.3	2.33			
	100	292	5146	21.336	57.9	10.2			
	994	297	8303	34.428	93.4	16.1			
	9998	297	8477	35.150	95.3	16.4			
	99942	290	8457	35.067	95.1	16.9			
	760000	279	8099	33.582	91.1	17.1			
	0.004	305	62	0.258	0.700	0.116			
	0.006	328	76	0.313	0.850	0.128			
DX3-A164 Mylar/Discrete LBMLI (17, 9, 0.5)	0.005	292	75	0.309	0.926	0.072	16.5	9	0.54
10 layers LBMLI—first layer doesn't count	0.01	292	81	0.337	1.01	0.078			
Bulk density = 30 kg/m³	0.1	290	116	0.481	1.44	0.112			

(*Continued*)

TABLE B.4 (Continued)

Cryostat Test Series (MLI System)	CVP μ	WBT K	Flow sccm	Q W	q W/m²	k_e mW/m-K	x mm	n	z lyrs/mm
Jul-12	1	292	194	0.805	2.41	0.187			
	10	293	1870	7.754	23.2	1.79			
	0.006	324	109	0.454	1.36	0.091			
DX4-A165 Mylar/Discrete LBMLI (8, 4, 0.5)	0.004	292	136	0.564	1.77	0.062	7.5	4	0.53
5 layers LBMLI—first layer doesn't count	0.02	293	135	0.561	1.77	0.062			
Bulk density = 32 kg/m³	0.1	292	231	0.956	3.01	0.106			
Aug-12	1	292	581	2.407	7.57	0.266			
	10	255	3483	14.442	45.4	1.93			
	0.004	304	153	0.634	1.99	0.066			
	0.005	326	201	0.831	2.62	0.079			
VO1-A114 Vacuum w/Black Surface (25, 0, 0)	0.003	293	7447	30.878	88.6	10.5	25.4	0	0.00
Black painted cold mass surface	0.02	293	7620	31.596	90.7	10.7			
25-mm-thick annular space	1	293	8917	36.975	106	12.5			
Apr-07	1	292	8912	36.952	106	12.6			
	10	292	12907	53.518	154	18.2			
	100	293	15961	66.181	190	22.5			
	760000	243	34094	141.369	406	62.5			
VO2-C112 Vacuum w/Stainless Steel (92, 0, 0)	0.6	301	3333	13.820	30.5	12.6	91.8	0	0.00
Stainless steel cold mass surface	1	305	4230	17.540	38.8	15.6			
91-mm-thick annular space	10	301	6545	27.139	60.0	24.7			
Aug-98	86	301	7064	29.291	64.7	26.7			
VO3-C111 Vacuum w/Copper Sleeve (90, 1, 0)	0.25	303	922	3.824	8.49	5.17	90.3	1	0.00
Copper sleeve on cold mass	1	302	2021	8.382	18.6	11.4			
91-mm-thick annular space	10	298	4803	19.917	44.2	21.5			
Aug-98	100	301	5619	23.298	51.7	22.5			

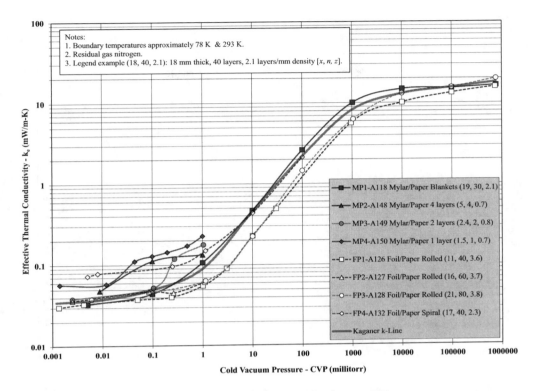

FIGURE B.1 Variation of k_e with CVP: Mylar/paper (MP) and foil/paper (FP) types.

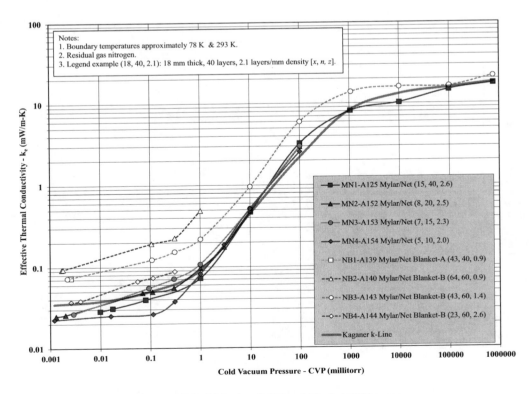

FIGURE B.2 Variation of k_e with CVP: Mylar/net (MN) and blanket (NB) types.

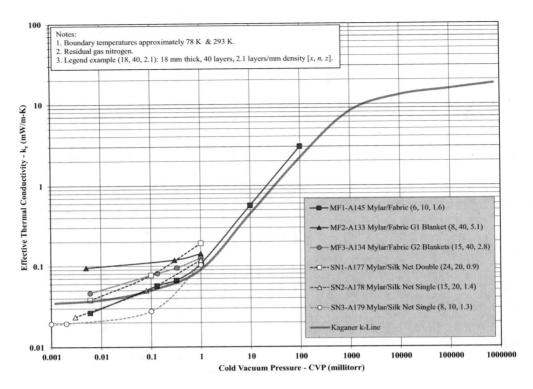

FIGURE B.3   Variation of k_e with CVP: Mylar/fabric (MF) and Mylar/silk net (SN) types.

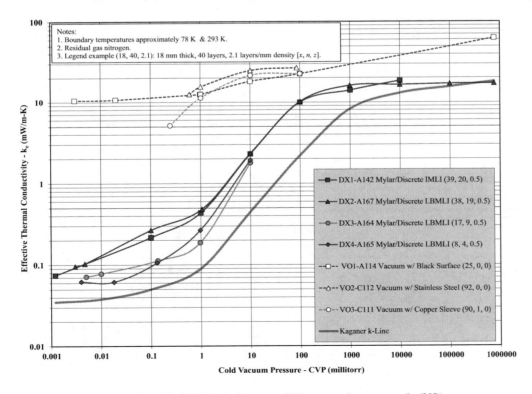

FIGURE B.4   Variation of k_e with CVP: Mylar/discrete (DX) type and vacuum only (VO).

TABLE B.5
Benchmark MLI Thermal Performance Data: Kaganer k-Line

Kaganer k-Line	CVP	k_e
	μ	mW/m-K
Composite of 26 MLI systems	0.001	0.035
From 10 to 80 layers; all types	0.01	0.038
CBT = 78 K	0.1	0.050
WBT = 293 K	1	0.090
Residual gas nitrogen	10	0.450
	100	2.20
	1,000	8.20
	10,000	13.0
	100,000	15.5
	760,000	18.0

TABLE B.6
Benchmark MLI Thermal Performance Data: Augustynowicz q-Band

Augustynowicz q-band	CVP	q_{high}	q_{mid}	q_{low}
		W/m²	W/m²	W/m²
Baseline 13 MLI systems	0.001	1.00	0.665	0.330
From 20 to 60 layers; all types	0.01	1.10	0.735	0.370
CBT = 78 K	0.1	1.52	1.01	0.500
WBT = 293 K	1	3.00	2.00	1.00
Residual gas nitrogen	10	15.5	10.3	5.10
	100	77.0	51.5	26.0
	1,000	184	123	62.0
	10,000	232	155	78.0
	100,000	250	167	84.0
	760,000	257	172	86.0

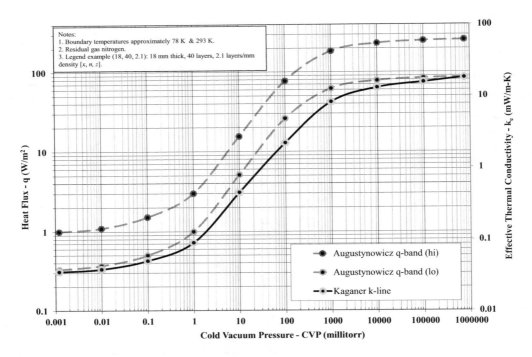

FIGURE B.5 Variation of q and k_e with CVP for benchmark MLI systems: Augustynowicz q-band and Kaganer k-line.

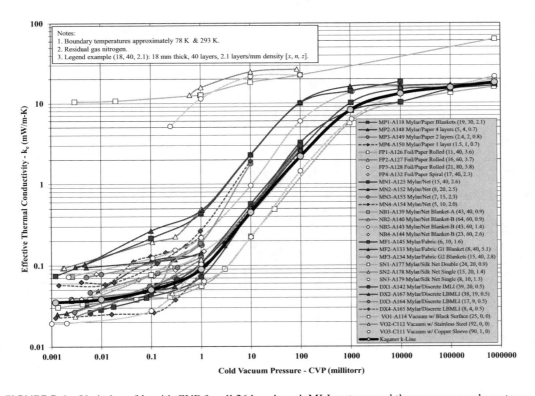

FIGURE B.6 Variation of k_e with CVP for all 26 benchmark MLI systems and three vacuum-only systems.

Appendix C

Thermal Properties of
Solid Materials

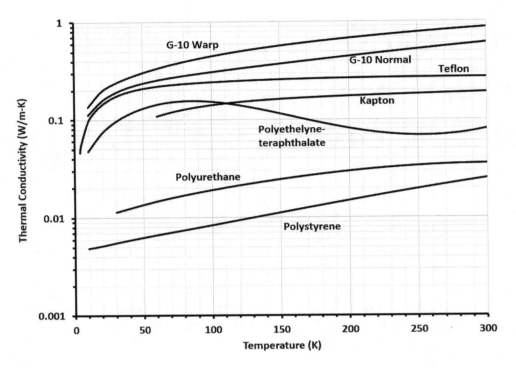

FIGURE C.1   Thermal conductivity of selected non-metal as a function of temperature.

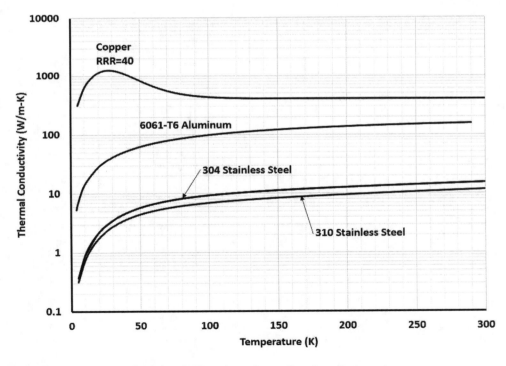

FIGURE C.2   Thermal conductivity of selected metals as a function of temperature.

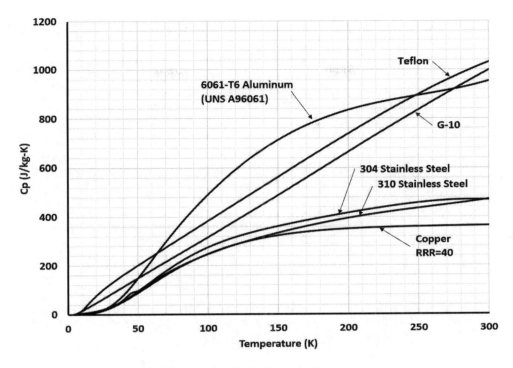

FIGURE C.3   Heat capacity of selected materials as a function of temperature.

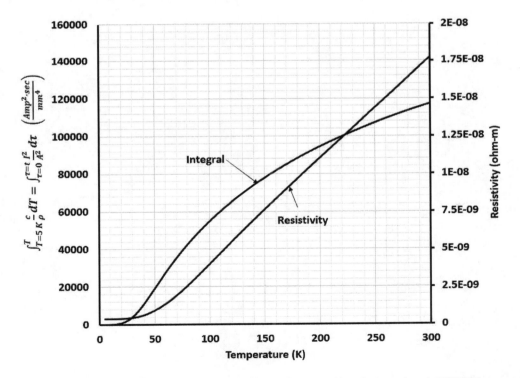

FIGURE C.4   Hot spot integral and electrical resistivity of copper using the equations in NIST Monograph 177, 1992. The adiabatic integral (MIITS) are calculated using the NIST curve fits and numerical quadrature.

Appendix D

Fluid Properties

The fluid properties were generated using REFPROP version 9.1, a product of the National Institute of Standards and Technology (NIST). Developed and maintained by Applied Chemicals and Materials Division Boulder, CO 80305, USA.

LIQUID-VAPOR SATURATION PROPERTIES OF COMMON FLUIDS

TABLE D.1
Argon Liquid Saturation Properties
(Argon Saturated Liquid T = 90.0 to 150.0 K)

Temperature	Pressure	Density	Enthalpy	Entropy	Cp	Therm. Cond.	Viscosity	Prandtl	Surf. Tension
(K)	(MPa)	(kg/m^3)	(kJ/kg)	(kJ/kg-K)	(kJ/kg-K)	(mW/m-K)	(µPa-s)		(mN/m)
87.302	0.10133	1395.4	−117.52	1.375	1.1172	128.46	260.29	2.2638	12.534
90	0.13351	1378.6	−114.49	1.409	1.1212	124.51	240.04	2.1614	11.871
95	0.21305	1346.8	−108.82	1.4696	1.134	117.28	207.87	2.0098	10.661
100	0.32377	1313.7	−103.06	1.5278	1.1537	110.22	181.32	1.898	9.4782
105	0.47224	1279.1	−97.176	1.5841	1.1811	103.26	159.18	1.8207	8.3243
110	0.66526	1242.8	−91.129	1.639	1.2176	96.409	140.45	1.7738	7.2016
115	0.90981	1204.2	−84.874	1.6928	1.2663	89.695	124.35	1.7555	6.1129
120	1.213	1162.8	−78.353	1.7461	1.3324	83.127	110.24	1.7669	5.0617
125	1.5823	1117.9	−71.488	1.7995	1.4253	76.711	97.565	1.8128	4.0527
130	2.0255	1068.1	−64.161	1.8538	1.5638	70.432	85.859	1.9063	3.0919
135	2.5509	1011.5	−56.177	1.9102	1.7903	64.243	74.689	2.0814	2.1879
140	3.1682	943.71	−47.162	1.9712	2.2247	58.057	63.621	2.4379	1.3542
145	3.8896	854.28	−36.192	2.0425	3.3994	52.009	52.06	3.4027	0.61548
150	4.7346	680.43	−17.88	2.1589	23.582	57.631	36.783	15.051	0.043833

TABLE D.2

Argon Vapor Saturation Properties

(Argon Saturated Vapor T = 90.0 to 150.0 K)

Temperature	Pressure	Density	Enthalpy	Heat of Vapor.	Entropy	Cp	Therm. Cond.	Viscosity	Prandtl
(K)	(MPa)	(kg/m^3)	(kJ/kg)	(kJ/kg)	(kJ/kg-K)	(kJ/kg-K)	(mW/m-K)	(μPa-s)	
87.302	0.10133	5.7736	43.618	161.138	3.2208	0.56583	5.6239	7.1686	0.72125
90	0.13351	7.4362	44.572	159.06	3.1763	0.57569	5.8348	7.4127	0.73137
95	0.21305	11.435	46.137	154.96	3.1007	0.59807	6.2452	7.8736	0.75401
100	0.32377	16.859	47.401	150.46	3.0324	0.6269	6.6888	8.3489	0.78249
105	0.47224	24.019	48.32	145.5	2.9698	0.66402	7.1783	8.844	0.8181
110	0.66526	33.287	48.84	139.97	2.9114	0.71223	7.7318	9.3663	0.86279
115	0.90981	45.126	48.898	133.77	2.856	0.77596	8.3764	9.9262	0.91953
120	1.213	60.144	48.413	126.77	2.8025	0.86265	9.1544	10.539	0.99313
125	1.5823	79.194	47.268	118.76	2.7495	0.98571	10.136	11.228	1.0919
130	2.0255	103.56	45.295	109.46	2.6957	1.1717	11.448	12.032	1.2315
135	2.5509	135.39	42.211	98.388	2.639	1.4822	13.337	13.018	1.4467
140	3.1682	178.86	37.472	84.634	2.5757	2.1036	16.391	14.318	1.8377
145	3.8896	244.44	29.757	65.949	2.4973	3.8959	22.53	16.274	2.8141
150	4.7346	394.5	29.402	29.402	2.355	35.468	55.878	21.176	13.441

TABLE D.3
Helium Liquid Saturation Properties.
(Helium Saturated Liquid T = 3.0 to 5.0)

Temperature	Pressure	Density	Enthalpy	Entropy	Cp	Therm. Cond.	Viscosity	Prandtl	Surf. Tension
(K)	(MPa)	(kg/m^3)	(kJ/kg)	(kJ/kg-K)	(kJ/kg-K)	(mW/m-K)	(μPa-s)		(mN/m)
3	0.024062	141.24	-4.6359	-1.1152	2.5656	16.538	3.6856	0.57176	0.21278
3.2	0.032027	139.35	-4.0578	-0.94703	2.7901	17.075	3.6238	0.59213	0.19296
3.4	0.041607	137.22	-3.4237	-0.77592	3.0683	17.542	3.5502	0.62096	0.17271
3.6	0.052956	134.81	-2.7244	-0.59998	3.4049	17.936	3.4679	0.65834	0.15221
3.8	0.066227	132.09	-1.9495	-0.41746	3.8182	18.251	3.3786	0.70682	0.1316
4	0.081581	128.99	-1.0862	-0.22629	4.3431	18.486	3.2829	0.77127	0.11108
4.2	0.099188	125.41	-0.11703	-0.02374	5.0426	18.643	3.1804	0.86026	0.090789
4.2226	0.10133	124.97	0	0	5.137	18.656	3.1684	0.87243	0.088522
4.4	0.11923	121.23	0.98282	0.19418	6.0396	18.732	3.0699	0.98978	0.070915
4.6	0.14191	116.2	2.2514	0.43355	7.6162	18.783	2.9489	1.1957	0.05163
4.8	0.16745	109.9	3.7551	0.70529	10.657	18.836	2.8121	1.5911	0.033118
5	0.1961	101.23	5.6581	1.038	20.341	19.008	2.6439	2.8294	0.015601

TABLE D.4
Helium Vapor Saturation Properties
(Helium Saturated Vapor T = 3.0 to 5.0 K)

Temperature	Pressure	Density	Enthalpy	Heat of Vapor.	Entropy	Cp	Therm. Cond.	Viscosity	Prandtl
(K)	(MPa)	(kg/m^3)	(kJ/kg)	(kJ/kg)	(kJ/kg-K)	(kJ/kg-K)	(mW/m-K)	(µPa-s)	
3	0.024062	4.4554	18.863	23.499	6.7178	6.2367	5.8939	0.79626	0.84258
3.2	0.032027	5.7198	19.408	23.465	6.3859	6.5107	6.3513	0.86182	0.88345
3.4	0.041607	7.2175	19.871	23.295	6.0755	6.8556	6.8196	0.92954	0.93445
3.6	0.052956	8.9852	20.243	22.968	5.7799	7.3003	7.3081	1	0.99894
3.8	0.066227	11.072	20.511	22.46	5.4931	7.8898	7.8185	1.0739	1.0837
4	0.081581	13.544	20.655	21.742	5.2091	8.7001	8.3661	1.1522	1.1982
4.2	0.099188	16.502	20.652	20.769	4.9213	9.8711	8.9752	1.236	1.3594
4.2226	0.10133	16.872	20.641	20.641	4.8882	10.036	9.0495	1.2459	1.3817
4.4	0.11923	20.093	20.464	19.481	4.6216	11.689	9.6919	1.3272	1.6006
4.6	0.14191	24.555	20.032	17.781	4.2989	14.832	10.608	1.4286	1.9975
4.8	0.16745	30.322	19.255	15.5	3.9344	21.376	11.921	1.5453	2.7709
5	0.1961	38.492	17.865	12.207	3.4794	42.082	14.226	1.6905	5.0009

TABLE D.5
Methane Liquid Saturation Properties
(Methane Saturated Liquid T = 100.0 to 190.0 K)

Temperature (K)	Pressure (MPa)	Density (kg/m^3)	Enthalpy (kJ/kg)	Entropy (kJ/kg-K)	Cp (kJ/kg-K)	Therm. Cond. (mW/m-K)	Viscosity (μPa-s)	Prandtl	Surf. Tension (mN/m)
100	0.034376	438.89	-40.269	-0.37933	3.4084	199.56	150.96	2.5784	15.217
105	0.056377	431.92	-23.124	-0.21253	3.4367	192.89	134.63	2.3989	14.229
110	0.08813	424.78	-5.813	-0.05217	3.4692	186.02	120.9	2.2548	13.247
111.67	0.10133	422.36	0	0	3.4811	183.7	116.77	2.2127	12.921
115	0.13221	417.45	11.687	0.10248	3.5064	179.03	109.06	2.136	12.271
120	0.19143	409.9	29.405	0.25207	3.5493	171.95	98.68	2.0368	11.305
125	0.26876	402.11	47.373	0.3972	3.5992	164.83	89.49	1.9541	10.349
130	0.36732	394.04	65.629	0.53846	3.658	157.69	81.298	1.8859	9.4054
135	0.49035	385.64	84.22	0.67639	3.7281	150.55	73.957	1.8314	8.4754
140	0.64118	376.87	103.2	0.81158	3.8129	143.41	67.353	1.7907	7.5608
145	0.82322	367.65	122.65	0.94461	3.917	136.29	61.39	1.7644	6.6634
150	1.04	357.9	142.64	1.0761	4.0474	129.18	55.984	1.7541	5.7852
155	1.295	347.51	163.31	1.2069	4.2145	122.06	51.058	1.7629	4.9285
160	1.5921	336.31	184.8	1.3378	4.4354	114.93	46.541	1.7961	4.0959
165	1.9351	324.1	207.33	1.4701	4.7401	107.74	42.359	1.8636	3.2911
170	2.3283	310.5	231.24	1.6054	5.1872	100.46	38.432	1.9845	2.5186
175	2.7765	294.94	257.09	1.7466	5.9102	93.038	34.659	2.2017	1.7851
180	3.2852	276.23	285.94	1.8991	7.2923	85.482	30.888	2.635	1.1014
185	3.8617	251.36	320.51	2.0765	11.109	78.283	26.784	3.801	0.48857
190	4.5186	200.78	378.27	2.3687	94.012	94.151	20.291	20.261	0.017679

TABLE D.6
Methane Vapor Saturation Properties
Note: Methane Saturated Vapor T = 100.0 to 190.0 K

Temperature	Pressure	Density	Enthalpy	Heat of Vapor.	Entropy	Cp	Therm. Cond.	Viscosity	Prandtl
(K)	(MPa)	(kg/m^3)	(kJ/kg)	(kJ/kg)	(kJ/kg-K)	(kJ/kg-K)	(mW/m-K)	(µPa-s)	
100	0.034376	0.67457	490.21	530.48	4.9255	2.1458	9.8989	3.9139	0.84843
105	0.056377	1.0613	499.31	522.44	4.7631	2.1725	10.528	4.0881	0.8436
110	0.08813	1.5982	508.02	513.84	4.6191	2.2053	11.188	4.2647	0.84064
111.67	0.10133	1.8164	510.83	510.83	4.5746	2.2177	11.415	4.3241	0.84007
115	0.13221	2.3193	516.28	504.59	4.4902	2.245	11.882	4.4436	0.83958
120	0.19143	3.2619	524.02	494.61	4.3738	2.293	12.617	4.6249	0.84055
125	0.26876	4.4669	531.17	483.8	4.2676	2.3508	13.397	4.8092	0.84387
130	0.36732	5.9804	537.67	472.04	4.1695	2.4208	14.232	4.9972	0.85001
135	0.49035	7.8549	543.42	459.2	4.0779	2.506	15.129	5.1903	0.85971
140	0.64118	10.152	548.34	445.14	3.9912	2.6108	16.102	5.3904	0.87402
145	0.82322	12.945	552.32	429.68	3.9079	2.7417	17.166	5.6002	0.89446
150	1.04	16.328	555.23	412.58	3.8267	2.9083	18.344	5.8236	0.92329
155	1.295	20.419	556.89	393.58	3.7461	3.1257	19.67	6.0661	0.96395
160	1.5921	25.382	557.07	372.27	3.6645	3.4189	21.197	6.3361	1.0219
165	1.9351	31.448	555.45	348.12	3.5799	3.8333	23.02	6.6462	1.1067
170	2.3283	38.974	551.54	320.3	3.4895	4.4585	25.314	7.0175	1.236
175	2.7765	48.559	544.52	287.44	3.3891	5.5023	28.463	7.4874	1.4474
180	3.2852	61.375	532.83	246.89	3.2707	7.574	33.484	8.1346	1.84
185	3.8617	80.435	512.49	191.97	3.1142	13.527	44.197	9.1786	2.8092
190	4.5186	125.18	459.03	80.763	2.7937	140.81	120.52	12.236	14.296

TABLE D.7

Nitrogen Liquid Saturation Properties

(Nitrogen Saturated Liquid T = 70.0 to 120.0 K)

Temperature	Pressure	Density	Enthalpy	Entropy	Cp	Therm. Cond.	Viscosity	Prandtl	Surf. Tension
(K)	(MPa)	(kg/m^3)	(kJ/kg)	(kJ/kg-K)	(kJ/kg-K)	(mW/m-K)	(μPa-s)		(mN/m)
70	0.038545	838.51	−136.97	2.6321	2.0145	159.47	220.22	2.782	10.576
75	0.076043	816.67	−126.83	2.7714	2.0311	149.47	176.75	2.4018	9.4163
77.355	0.10133	806.08	−122.02	2.8342	2.0415	144.77	160.66	2.2655	8.8796
80	0.13687	793.94	−116.58	2.9028	2.0555	139.5	145.11	2.1381	8.2844
85	0.22886	770.13	−106.16	3.0277	2.0906	129.66	121.31	1.9559	7.1824
90	0.36046	745.02	−95.517	3.1473	2.1407	119.8	102.82	1.8372	6.1129
95	0.54052	718.26	−84.571	3.263	2.2126	109.95	88.004	1.7709	5.0791
100	0.77827	689.35	−73.209	3.3761	2.318	100.11	75.758	1.7541	4.0856
105	1.0833	657.52	−61.268	3.4882	2.4789	90.272	65.292	1.7929	3.1378
110	1.4658	621.45	−48.486	3.6015	2.7433	80.437	55.993	1.9096	2.2439
115	1.937	578.7	−34.389	3.7198	3.2403	70.626	47.29	2.1696	1.4163
120	2.5106	523.36	−17.87	3.8514	4.5076	61.006	38.425	2.8392	0.67738

TABLE D.8

Nitrogen Vapor Saturation Properties

(Nitrogen Saturated Vapor T = 70.0 to 120.0 K)

Temperature	Pressure	Density	Enthalpy	Heat of Vapor.	Entropy	Cp	Therm. Cond.	Viscosity	Prandtl
(K)	(MPa)	(kg/m^3)	(kJ/kg)	(kJ/kg)	(kJ/kg-K)	(kJ/kg-K)	(mW/m-K)	(μPa-s)	
70	0.038545	1.896	71.098	208.07	5.6045	1.0816	6.3547	4.8835	0.83119
75	0.076043	3.5404	75.316	202.15	5.4667	1.108	6.9138	5.2621	0.84332
77.355	0.10133	4.6121	77.158	199.178	5.409	1.1239	7.1876	5.444	0.85129
80	0.13687	6.0894	79.099	195.68	5.3487	1.1449	7.5057	5.6517	0.86212
85	0.22886	9.8241	82.352	188.51	5.2454	1.1957	8.1483	6.0565	0.88874
90	0.36046	15.079	84.97	180.49	5.1527	1.2655	8.8682	6.4818	0.92497
95	0.54052	22.272	86.828	171.4	5.0672	1.3628	9.706	6.9353	0.97379
100	0.77827	31.961	87.766	160.98	4.9858	1.5026	10.726	7.4285	1.0407
105	1.0833	44.959	87.557	148.83	4.9055	1.7139	12.035	7.9804	1.1365
110	1.4658	62.579	85.835	134.32	4.8226	2.0618	13.834	8.626	1.2856
115	1.937	87.294	81.911	116.3	4.7311	2.749	16.58	9.4395	1.5651
120	2.5106	125.09	74.173	92.043	4.6185	4.6309	21.715	10.624	2.2655

TABLE D.9
Oxygen Liquid Saturation Properties
(Oxygen Saturated Liquid T = 60.0 to 150.0 K)

Temperature	Pressure	Density	Enthalpy	Entropy	Cp	Therm. Cond.	Viscosity	Prandtl	Surf. Tension
(K)	(MPa)	(kg/m^3)	(kJ/kg)	(kJ/kg-K)	(kJ/kg-K)	(mW/m-K)	(μPa–s)		(mN/m)
60	0.000726	1282	−184.19	2.2571	1.6734	193.94	578.07	4.9879	21.053
65	0.002335	1259.7	−175.81	2.3912	1.6772	186.82	457.94	4.1111	19.698
70	0.006262	1237	−167.42	2.5156	1.6781	179.7	371.79	3.4718	18.36
75	0.014547	1213.9	−159.02	2.6313	1.6788	172.58	308.66	3.0025	17.039
80	0.030123	1190.5	−150.61	2.7397	1.6816	165.44	261.22	2.6551	15.737
85	0.056831	1166.6	−142.18	2.8417	1.688	158.27	224.62	2.3956	14.455
90	0.09935	1142.1	−133.69	2.9383	1.6989	151.05	195.64	2.2004	13.193
90.188	0.10133	1141.2	−133.37	2.9419	1.6994	150.78	194.67	2.1941	13.146
95	0.16308	1116.9	−125.12	3.0303	1.7151	143.81	172.12	2.0527	11.953
100	0.254	1090.9	−116.45	3.1185	1.7375	136.55	152.56	1.9411	10.736
105	0.37853	1063.8	−107.64	3.2033	1.7675	129.25	135.93	1.8589	9.5437
110	0.5434	1035.5	−98.641	3.2855	1.8068	121.92	121.52	1.801	8.3784
115	0.75559	1005.6	−89.417	3.3657	1.8585	114.57	108.81	1.7651	7.2423
120	1.0223	973.85	−79.904	3.4444	1.9271	107.23	97.426	1.751	6.138
125	1.3509	939.72	−70.024	3.5222	2.0207	99.912	87.086	1.7613	5.0693
130	1.7491	902.48	−59.662	3.6001	2.1534	92.634	77.571	1.8032	4.0405
135	2.225	860.98	−48.65	3.6791	2.3541	85.404	68.687	1.8933	3.0581
140	2.7878	813.24	−36.695	3.7612	2.6907	78.217	60.223	2.0717	2.131
145	3.4477	755.13	−23.219	3.8498	3.3684	71.056	51.869	2.4589	1.274
150	4.2186	675.48	−6.6709	3.9546	5.4639	64.19	42.9	3.6517	0.51598

TABLE D.10
Oxygen Vapor Saturation Properties
(Oxygen Saturated Vapor T = 60.0 to 150.0 K)

Temperature	Pressure	Density	Enthalpy	Heat of Vapor.	Entropy	Cp	Therm. Cond.	Viscosity	Prandtl
(K)	(MPa)	(kg/m^3)	(kJ/kg)	(kJ/kg)	(kJ/kg-K)	(kJ/kg-K)	(mW/m-K)	(µPa-s)	
60	0.000726	0.046595	54.188	238.38	6.2301	0.94755	4.984	4.5528	0.86557
65	0.002335	0.13853	58.66	234.47	5.9985	0.96671	5.4863	4.9555	0.87318
70	0.006262	0.34573	63.092	230.51	5.8086	0.97798	5.9925	5.3557	0.87405
75	0.014547	0.75227	67.454	226.48	5.651	0.97928	6.5051	5.7533	0.86611
80	0.030123	1.4684	71.694	222.31	5.5185	0.97431	7.0277	6.1486	0.85242
85	0.056831	2.6283	75.749	217.92	5.4055	0.96937	7.5654	6.5423	0.83827
90	0.09935	4.3871	79.551	213.24	5.3076	0.97046	8.1241	6.9355	0.82848
90.188	0.10133	4.4671	79.688	213.058	5.3042	0.97068	8.1456	6.9503	0.82824
95	0.16308	6.9203	83.039	208.16	5.2215	0.98191	8.7113	7.3301	0.82622
100	0.254	10.425	86.155	202.6	5.1445	1.0064	9.3362	7.7281	0.83307
105	0.37853	15.125	88.846	196.48	5.0746	1.0457	10.01	8.1324	0.84953
110	0.5434	21.281	91.054	189.7	5.01	1.1014	10.748	8.5467	0.87583
115	0.75559	29.209	92.717	182.13	4.9495	1.1765	11.571	8.976	0.91264
120	1.0223	39.308	93.754	173.66	4.8915	1.2763	12.509	9.4273	0.96185
125	1.3509	52.109	94.056	164.08	4.8349	1.4109	13.607	9.9112	1.0277
130	1.7491	68.369	93.466	153.13	4.778	1.6002	14.94	10.445	1.1188
135	2.225	89.253	91.744	140.39	4.7191	1.886	16.641	11.061	1.2536
140	2.7878	116.76	88.474	125.17	4.6552	2.3696	18.977	11.823	1.4762
145	3.4477	154.91	82.83	106.05	4.5812	3.3693	22.582	12.881	1.9219
150	4.2186	214.94	72.562	79.233	4.4828	6.6254	29.666	14.721	3.2877

TABLE D.11
Hydrogen Liquid Saturation Properties
(1: hydrogen (normal): V/L sat. T = 20.0 to 30.0 K)

Temperature	Pressure	Density	Enthalpy	Entropy	Cp	Therm. Cond.	Viscosity	Prandtl	Surf. Tension
(K)	(MPa)	(kg/m^3)	(kJ/kg)	(kJ/kg-K)	(kJ/kg-K)	(mW/m-K)	(μPa-s)		(mN/m)
20	0.090717	71.265	−3.6672	−0.17429	9.5697	103.53	13.91	1.2857	1.9754
20.369	0.10132	70.848	0	0	9.7725	103.62	13.49	1.2722	1.9117
21	0.1215	70.115	6.466	0.29876	10.138	103.71	12.815	1.2527	1.8029
22	0.15913	68.893	17.241	0.77472	10.771	103.66	11.843	1.2306	1.6314
23	0.20438	67.592	28.724	1.2556	11.49	103.37	10.971	1.2194	1.4612
24	0.25807	66.199	40.997	1.7437	12.319	102.97	10.181	1.2181	1.2926
25	0.321	64.701	54.161	2.2417	13.298	102.33	9.4577	1.229	1.1258
26	0.39399	63.079	68.346	2.7531	14.487	101.42	8.7872	1.2552	0.96134
27	0.47789	61.305	83.727	3.2825	15.987	100.23	8.1571	1.3011	0.79991
28	0.57359	59.339	100.55	3.8365	17.977	98.731	7.5551	1.3756	0.64237
29	0.68205	57.119	119.2	4.4253	20.807	96.889	6.9677	1.4964	0.49013
30	0.80432	54.538	140.3	5.0661	25.284	94.649	6.3773	1.7036	0.34541

TABLE D.12

Hydrogen Vapor Saturation Properties

(1: hydrogen (normal): V/L sat. T = 20.0 to 30.0 K)

Temperature	Pressure	Density	Enthalpy	Entropy	Cp	Heat of Vapor.	Therm. Cond.	Viscosity	Prandtl
(K)	(MPa)	(kg/m^3)	(kJ/kg)	(kJ/kg-K)	(kJ/kg-K)	(kJ/kg)	(mW/m-K)	(µPa-s)	
20	0.090717	1.2059	446.64	22.341	11.892	450.31	16.996	0.97579	0.68276
20.369	0.10132	1.3322	448.71	22.029	12.037	448.71	17.451	0.99632	0.68718
21	0.1215	1.5701	451.98	21.514	12.312	445.52	18.256	1.0318	0.69582
22	0.15913	2.009	456.43	20.738	12.83	439.19	19.602	1.0891	0.71282
23	0.20438	2.5334	459.88	20.002	13.472	431.16	21.048	1.1482	0.73489
24	0.25807	3.1562	462.24	19.296	14.273	421.24	22.613	1.2097	0.7636
25	0.32100	3.8938	463.37	18.61	15.289	409.21	24.324	1.2746	0.80118
26	0.39399	4.7674	463.10	17.936	16.603	394.75	26.222	1.3441	0.85104
27	0.47789	5.8055	461.18	17.262	18.357	377.46	28.367	1.4196	0.91866
28	0.57359	7.0489	457.3	16.577	20.804	356.75	30.863	1.5036	1.0136
29	0.68205	8.5601	450.92	15.864	24.445	331.72	33.892	1.5999	1.1539
30	0.80432	10.445	441.19	15.096	30.425	300.89	37.817	1.715	1.3798

GASES AT ONE ATMOSPHERE (101.325 KPA)

TABLE D.13
Methane Vapor Properties at 1 Atmosphere (101.325 kPa)
(9: methane: p = 0.10133 MPa)

Temperature	Density	Enthalpy	Entropy	Cp	Therm. Cond.	Viscosity	Prandtl
(K)	(kg/m^3)	(kJ/kg)	(kJ/kg-K)	(kJ/kg-K)	(mW/m-K)	(µPa-s)	
111.67	1.8164	510.83	4.5746	2.2177	11.415	4.3241	0.84007
120	1.6776	529.11	4.7325	2.1756	12.314	4.653	0.82211
130	1.5388	550.71	4.9054	2.147	13.447	5.0441	0.80537
140	1.4223	572.09	5.0638	2.1294	14.605	5.4326	0.79209
150	1.323	593.32	5.2103	2.1182	15.773	5.8185	0.78136
160	1.2372	614.47	5.3468	2.1109	16.946	6.2016	0.77249
170	1.162	635.55	5.4746	2.1065	18.122	6.5813	0.765
180	1.0957	656.6	5.595	2.1044	19.302	6.9574	0.7585
190	1.0367	677.65	5.7087	2.1042	20.503	7.3295	0.75225
200	0.98382	698.7	5.8167	2.1061	21.649	7.6974	0.74881
210	0.93615	719.77	5.9195	2.1098	22.819	8.061	0.74533
220	0.89294	740.9	6.0178	2.1156	24.001	8.4202	0.74219
230	0.85359	762.09	6.112	2.1234	25.199	8.775	0.73943
240	0.8176	783.37	6.2026	2.1333	26.413	9.1253	0.737
250	0.78454	804.76	6.2899	2.1452	27.648	9.4711	0.73488
260	0.75407	826.28	6.3743	2.1593	28.906	9.8125	0.73303
270	0.7259	847.96	6.4561	2.1755	30.189	10.15	0.73141
280	0.69977	869.8	6.5355	2.1937	31.500	10.482	0.73
290	0.67547	891.84	6.6129	2.2139	32.842	10.811	0.72878
300	0.65281	914.09	6.6883	2.2359	34.215	11.136	0.7277
310	0.63162	936.56	6.762	2.2597	35.621	11.456	0.72675
320	0.61178	959.28	6.8341	2.2851	37.06	11.773	0.72592
330	0.59315	982.27	6.9048	2.3121	38.533	12.086	0.72518
340	0.57562	1005.5	6.9743	2.3406	40.041	12.395	0.72452
350	0.55911	1029.1	7.0426	2.3703	41.582	12.7	0.72393
360	0.54351	1052.9	7.1098	2.4011	43.157	13.002	0.72339
370	0.52877	1077.1	7.176	2.4331	44.765	13.3	0.72289
380	0.51481	1101.6	7.2413	2.4659	46.406	13.595	0.72243

TABLE D.14

Nitrogen Vapor Properties at 1 Atmosphere (101.325 kPa)

(8: nitrogen: p = 0.10133 MPa)

Temperature	Density	Enthalpy	Entropy	Cp	Therm. Cond.	Viscosity	Prandtl
(K)	(kg/m^3)	(kJ/kg)	(kJ/kg-K)	(kJ/kg-K)	(mW/m-K)	(μPa-s)	
77.355	4.6123	77.158	5.409	1.1239	7.1876	5.444	0.85129
80	4.4401	80.116	5.4466	1.1131	7.4452	5.6238	0.84078
90	3.8994	91.098	5.576	1.0864	8.4176	6.2971	0.81271
100	3.4833	101.88	5.6896	1.0718	9.3821	6.9588	0.79497
110	3.1509	112.55	5.7913	1.063	10.335	7.6082	0.78255
120	2.8784	123.15	5.8836	1.0573	11.274	8.2449	0.77323
130	2.6503	133.7	5.968	1.0534	12.199	8.869	0.76586
140	2.4564	144.22	6.046	1.0506	13.109	9.4807	0.75981
150	2.2894	154.72	6.1184	1.0486	14.005	10.08	0.75471
160	2.144	165.2	6.186	1.047	14.887	10.668	0.7503
170	2.0161	175.66	6.2494	1.0458	15.755	11.245	0.74643
180	1.9028	186.11	6.3092	1.0449	16.61	11.811	0.74297
190	1.8016	196.56	6.3657	1.0441	17.451	12.366	0.73985
200	1.7107	207	6.4192	1.0435	18.280	12.911	0.73701
210	1.6287	217.43	6.4701	1.043	19.097	13.447	0.7344
220	1.5541	227.86	6.5186	1.0426	19.902	13.973	0.73198
230	1.4862	238.28	6.565	1.0422	20.696	14.49	0.72973
240	1.4239	248.7	6.6093	1.042	21.479	14.999	0.72763
250	1.3667	259.12	6.6518	1.0417	22.251	15.5	0.72566
260	1.3139	269.54	6.6927	1.0416	23.013	15.992	0.7238
270	1.2651	279.95	6.732	1.0414	23.766	16.477	0.72206
280	1.2198	290.37	6.7699	1.0414	24.509	16.955	0.72041
290	1.1776	300.78	6.8064	1.0413	25.243	17.426	0.71886
300	1.1382	311.19	6.8417	1.0414	25.969	17.89	0.7174

TABLE D.15
Helium Vapor Properties at 1 Atmosphere (101.325 kPa)
(7: helium: p = 0.10133 MPa)

Temperature	Density	Enthalpy	Entropy	Cp	Therm. Cond.	Viscosity	Prandtl
(K)	(kg/m^3)	(kJ/kg)	(kJ/kg-K)	(kJ/kg-K)	(mW/m-K)	(µPa-s)	
4.2226	16.872	20.641	20.641	10.036	9.0495	1.2459	1.3817
10	5.0204	55.437	10.242	5.4123	16.894	2.2592	0.7238
20	2.4431	108.42	13.921	5.245	26.201	3.5822	0.7171
30	1.6235	160.7	16.041	5.2163	33.718	4.634	0.71689
40	1.2169	212.81	17.54	5.2062	40.446	5.5422	0.71339
50	0.9735	264.84	18.701	5.2014	46.679	6.3605	0.70875
60	0.81135	316.84	19.649	5.1988	52.554	7.1167	0.70401
70	0.69555	368.82	20.45	5.1972	58.150	7.8275	0.69959
80	0.6087	420.79	21.144	5.1962	63.519	8.5034	0.69562
90	0.54114	472.75	21.756	5.1955	68.698	9.1519	0.69213
100	0.48709	524.7	22.304	5.195	73.714	9.778	0.68911
110	0.44285	576.65	22.799	5.1946	78.587	10.196	0.67398
120	0.40599	628.59	23.251	5.1944	83.333	10.794	0.67279
130	0.37479	680.54	23.666	5.1941	87.967	11.376	0.6717
140	0.34805	732.48	24.051	5.194	92.499	11.945	0.67071
150	0.32487	784.42	24.41	5.1938	96.938	12.501	0.66981
160	0.30458	836.35	24.745	5.1937	101.29	13.048	0.66901
170	0.28668	888.29	25.06	5.1936	105.57	13.584	0.66828
180	0.27077	940.23	25.357	5.1936	109.77	14.111	0.66764
190	0.25653	992.16	25.637	5.1935	113.91	14.631	0.66706
200	0.24371	1044.1	25.904	5.1934	117.98	15.142	0.66654
210	0.23211	1096	26.157	5.1934	122.00	15.647	0.66608
220	0.22157	1148	26.399	5.1934	125.96	16.145	0.66567
230	0.21194	1199.9	26.63	5.1933	129.87	16.636	0.66529
240	0.20312	1251.8	26.851	5.1933	133.72	17.122	0.66496
250	0.195	1303.8	27.063	5.1933	137.54	17.603	0.66466
260	0.1875	1355.7	27.266	5.1933	141.30	18.077	0.66439
270	0.18056	1407.6	27.462	5.1932	145.03	18.547	0.66415
280	0.17412	1459.6	27.651	5.1932	148.72	19.013	0.66393
290	0.16812	1511.5	27.833	5.1932	152.36	19.473	0.66374
300	0.16252	1563.4	28.009	5.1932	155.97	19.93	0.66356

TABLE D.16

Argon Vapor Properties at 1 Atmosphere (101.325 kPa)

(6: argon: p = 0.10133 MPa)

Temperature	Density	Enthalpy	Entropy	Cp	Therm. Cond.	Viscosity	Prandtl
(K)	(kg/m^3)	(kJ/kg)	(kJ/kg-K)	(kJ/kg-K)	(mW/m-K)	(µPa-s)	
87.302	5.7736	43.618	3.2208	0.56583	5.6239	7.1686	0.72125
90	5.5831	45.138	3.2379	0.5606	5.8002	7.3965	0.71489
100	4.9819	50.671	3.2962	0.5474	6.4516	8.2347	0.6987
110	4.5035	56.104	3.348	0.53975	7.0973	9.0621	0.68918
120	4.1122	61.475	3.3947	0.53492	7.7358	9.8782	0.68307
130	3.7853	66.807	3.4374	0.53168	8.3662	10.683	0.6789
140	3.5076	72.112	3.4767	0.5294	8.9883	11.476	0.67592
150	3.2685	77.397	3.5132	0.52774	9.6017	12.258	0.67371
160	3.0605	82.668	3.5472	0.52649	10.206	13.028	0.67203
170	2.8777	87.928	3.5791	0.52553	10.802	13.787	0.67071
180	2.7157	93.18	3.6091	0.52478	11.390	14.535	0.66966
190	2.5711	98.424	3.6375	0.52417	11.969	15.272	0.6688
200	2.4413	103.66	3.6643	0.52369	12.54	15.998	0.6681
210	2.324	108.9	3.6899	0.52328	13.103	16.714	0.66751
220	2.2176	114.13	3.7142	0.52295	13.658	17.421	0.66702
230	2.1205	119.36	3.7375	0.52267	14.205	18.117	0.6666
240	2.0316	124.58	3.7597	0.52243	14.745	18.804	0.66625
250	1.9499	129.81	3.781	0.52222	15.277	19.482	0.66594
260	1.8746	135.03	3.8015	0.52205	15.803	20.15	0.66567
270	1.8049	140.25	3.8212	0.52189	16.321	20.811	0.66544
280	1.7401	145.47	3.8402	0.52176	16.833	21.462	0.66524
290	1.6799	150.68	3.8585	0.52164	17.338	22.106	0.66506
300	1.6238	155.9	3.8762	0.52154	17.837	22.741	0.66491

TABLE D.17

Oxygen Vapor Properties at 1 Atmosphere (101.325 kPa)

(5: oxygen: p = 0.10133 MPa)

Temperature	Density	Enthalpy	Entropy	Cp	Therm. Cond.	Viscosity	Prandtl
(K)	(kg/m^3)	(kJ/kg)	(kJ/kg-K)	(kJ/kg-K)	(mW/m-K)	(µPa-s)	
90.188	4.4671	79.688	5.3042	0.97068	8.1456	6.9503	0.82824
100	3.9946	88.963	5.4019	0.93561	9.0872	7.7122	0.79404
110	3.6106	98.298	5.4908	0.93186	10.043	8.4732	0.78624
120	3.2962	107.6	5.5718	0.92824	10.991	9.2195	0.77863
130	3.0336	116.86	5.6459	0.92481	11.931	9.9515	0.77138
140	2.8107	126.1	5.7144	0.92198	12.861	10.67	0.76488
150	2.619	135.31	5.7779	0.91977	13.782	11.375	0.75911
160	2.4521	144.49	5.8372	0.91808	14.693	12.067	0.75399
170	2.3055	153.67	5.8928	0.91679	15.594	12.747	0.74941
180	2.1757	162.83	5.9452	0.91583	16.485	13.415	0.74526
190	2.0598	171.99	5.9947	0.91512	17.367	14.072	0.7415
200	1.9557	181.13	6.0416	0.91464	18.239	14.718	0.73806
210	1.8618	190.28	6.0862	0.91436	19.102	15.353	0.73491
220	1.7765	199.42	6.1288	0.91426	19.955	15.978	0.73203
230	1.6987	208.57	6.1694	0.91433	20.800	16.593	0.7294
240	1.6275	217.71	6.2083	0.91459	21.636	17.199	0.72701
250	1.562	226.86	6.2457	0.91502	22.464	17.795	0.72485
260	1.5016	236.01	6.2816	0.91563	23.284	18.383	0.72291
270	1.4458	245.17	6.3161	0.91642	24.096	18.963	0.7212
280	1.3939	254.34	6.3495	0.9174	24.900	19.534	0.71969
290	1.3457	263.52	6.3817	0.91855	25.696	20.097	0.71839
300	1.3007	272.71	6.4129	0.91989	26.486	20.652	0.71728

TABLE D.18
Normal Hydrogen Vapor Properties at 1 Atmosphere (101.325 kPa)
(3: hydrogen (normal): p = 0.10133 MPa)

Temperature	Pressure	Density	Enthalpy	Entropy	Therm. Cond.	Viscosity	Prandtl
(K)	(MPa)	(kg/m^3)	(kJ/kg)	(kJ/kg-K)	(mW/m-K)	(µPa-s)	
20.369	0.10133	1.3322	448.71	22.029	17.451	0.99632	0.68718
30	0.10133	0.84743	556.72	26.385	24.348	1.5247	0.67795
40	0.10133	0.62378	663.51	29.459	31.257	1.9886	0.67273
50	0.10133	0.49531	768.71	31.807	37.774	2.406	0.66781
60	0.10133	0.41128	873.5	33.717	43.992	2.7905	0.66519
70	0.10133	0.35184	978.73	35.339	50.045	3.1498	0.66545
80	0.10133	0.3075	1085.2	36.761	56.064	3.4889	0.66818
90	0.10133	0.27314	1193.7	38.038	62.151	3.8115	0.6723
100	0.10133	0.24572	1304.6	39.207	68.369	4.1203	0.6767
110	0.10133	0.22331	1418.3	40.29	74.739	4.4173	0.68057
120	0.10133	0.20466	1534.9	41.305	81.245	4.7042	0.68355
130	0.10133	0.18889	1654.4	42.261	87.848	4.9821	0.68557
140	0.10133	0.17538	1776.6	43.167	94.497	5.2523	0.68678
150	0.10133	0.16368	1901.5	44.028	101.14	5.5154	0.68738
160	0.10133	0.15345	2028.7	44.849	107.73	5.7723	0.68757
170	0.10133	0.14442	2158	45.633	114.24	6.0235	0.68752
180	0.10133	0.13639	2289.4	46.384	120.63	6.2696	0.68734
190	0.10133	0.12921	2422.5	47.103	126.89	6.5109	0.68711
200	0.10133	0.12275	2557.1	47.794	133	6.748	0.68687
210	0.10133	0.1169	2693.2	48.458	138.98	6.981	0.68664
220	0.10133	0.11159	2830.5	49.096	144.8	7.2104	0.68644
230	0.10133	0.10674	2968.8	49.711	150.48	7.4363	0.68627
240	0.10133	0.10229	3108.2	50.304	156.02	7.6591	0.68611
250	0.10133	0.098203	3248.3	50.876	161.42	7.8788	0.68596
260	0.10133	0.094427	3389.2	51.429	166.7	8.0957	0.68582
270	0.10133	0.090931	3530.7	51.963	171.86	8.31	0.68569
280	0.10133	0.087684	3672.8	52.48	176.91	8.5218	0.68555
290	0.10133	0.084662	3815.3	52.98	181.85	8.7313	0.6854
300	0.10133	0.08184	3958.3	53.465	186.7	8.9385	0.68525

Index

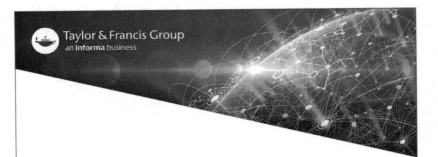